Developments in Atmospheric Science, 5

Radiative Processes in Meteorology and Climatology

Further titles in this series

1. F. VERNIANI (Editor)
Structure and Dynamics of the Upper Atmosphere

2. E.E. GOSSARD and W.H. HOOKE
Waves in the Atmosphere

3. L.P. SMITH
Methods in Agricultural Meteorology

4. O. ESSENWANGER
Applied Statistics in Atmospheric Science

Developments in Atmospheric Science, 5

RADIATIVE PROCESSES IN METEOROLOGY AND CLIMATOLOGY

by

G.W. PALTRIDGE and C.M.R. PLATT

Division of Atmospheric Physics,
Commonwealth Scientific and Industrial Research Organization,
Aspendale, Victoria, Australia

ELSEVIER SCIENTIFIC PUBLISHING COMPANY
Amsterdam - Oxford - New York 1976

ELSEVIER SCIENTIFIC PUBLISHING COMPANY
335 Jan van Galenstraat
P.O. Box 211, Amsterdam, The Netherlands

AMERICAN ELSEVIER PUBLISHING COMPANY, INC.
52 Vanderbilt Avenue
New York, New York 10017

ISBN: ISBN: 0-444-41444-4

Printed in The Netherlands

Preface

Atmospheric radiation is a fairly old discipline which forms the basis of any general investigation of climate and which enters also into the study of individual weather processes. Unfortunately, the modern literature on the subject has developed a certain mystique of its own and is not readily understood by the non-specialist. Apart from the excellent books by Kondratyev, there is a sizeable gap between the single chapter on radiation to be found in a number of text-books on general meteorology and the elegant theoretical treatises on radiative transfer by Goody and by Chandrasekhar. The gap occurs at just the level required by experimental atmospheric physicists or by meteorologists seeking to take account of radiative processes in their models. The present book is pitched at this level. It is aimed at the physicist rather than the mathematician.

Acknowledgements

We would like to thank the following persons and organizations for permission to use various diagrams and tables.

The Royal Society of London for Table 10.1.

Professor H. J. Bolle for Figs. 7.2, 7.3, 7.4 and 7.9.

Weapons Research Establishment, Salisbury, S.A. for Fig. 9.9.

The Clarendon Press, Oxford for Table 2.2. and Fig. 10.2.

The American Meteorological Society, Boston, Mass., for Figs. 1.2, 1.5, 1.11, 1.12, 1.13, 1.17, 4.4, 5.1, 5.2, 5.3, 5.4, 5.5, 5.6, 5.7, 5.8, 5.10, 5.12, 6.11, 7.8, 7.10, 7.18, 7.19, 7.20, 7.23, 8.1, 8.2, 8.3, 8.9, 8.11, 10.3, 10.4, 10.9, and 10.12; Tables 1.2, 1.4, 3.2 and 7.6; Figs. A.4 and A.6.

The Royal Meteorological Society, Bracknell, for Figs. 7.13, 7.14, 7.15, 7.16, 7.17, 7.21, 7.22, 7.27, 8.4, 10.5 and 10.6; Tables 7.2, 7.3 and 7.4.

The American Geophysical Union, Washington, D.C., for Figs. 7.1, 9.1, 9.3, 10.10, A.3, A.13, A.16; Table 6.2.

D. Reidel Publishing Company, Dordrecht, for Figs. 1.6, 4.2 and 4.3.

Pergamon Press, Oxford, for Figs. 5.11, 5.13, 6.1, A.1, A.2, A.5 and A.7; Table 5.1.

Friedr. Vieweg and Sohn, Braunschweig, for Figs. 7.12, 7.25, 7.26, 9.4, 9.6 and 9.7; Table 7.7.

Elsevier Scientific Publishing Company, Amsterdam, for Figs. 1.3 and 6.2; Table 6.4.

Nature (*London*), for Table 3.1.

McGraw-Hill Book Company, New York, N.Y., for Figs. 1.7, 2.2, 2.3, 6.6, 6.7 and 10.1.

The World Meteorological Organization, Geneva, for Fig. 7.28.

The Amerind Publishing Company, New Delhi, for Fig. 6.10.

Tellus for Figs. 2.5, 8.6, 9.10 and A.12.

Icarus for Fig. 8.5 and Table A.5.

Mr. G. Stephens, Department Meteorology, University of Melbourne, for Figs. 8.13 and 8.14.

Academic Press, New York, N.Y., for Table A.2, Fig. 10.8.

Dr. M. P. Thekaekara, NASA, for Fig. 3.1.

Harvard University Press, Cambridge, Mass., for Fig. 3.2.

University of Chicago Press, Chicago, Ill., for Table 1.1.

American Institute of Physics (Optical Society of America), New York, N.Y., for Figs. 9.2., 2.1, A.14, A.9, A.10, A.11; Table A.4.

We would like to thank also Dr. F.A. Berson for certain valuable contributions to Chapter 1.

Contents

Various Conversions and Constants

1 watt (W)	= 1 joule per second (J sec^{-1})
1 calorie (cal)	= 4.1868 joules (J)
	= 2.390×10^{-8} ergs (erg)
	= 1.163 milliwatt hours (mW hr)
1 calorie per minute (cal min^{-1})	= 69.8 milliwatts (mW)
1 calorie per second (cal sec^{-1})	= 4.18×10^{3} milliwatts (mW)
1 calorie per square centimetre (cal cm^{-2})	= 1 langley (ly)
1 langley per minute (ly min^{-1})	= 69.8 milliwatts per square centimetre (mW cm^{-2})
1 joule per second (J sec^{-1})	= 1 watt (W)
	= 10^{3} milliwatts (mW)
1 megajoule (MJ)	= 2.777×10^{5} milliwatt hours (mW hr)
1 calorie per square centimetre per second (cal cm^{-2} sec^{-1})	= 4180 milliwatts per square centimetre (mW cm^{-2})
1 millimetre per hour (evaporation) (mm hr^{-1})	≈ 68.6 milliwatts per square centimetre (mW cm^{-2})
1 millimetre per day (evaporation) (mm day^{-1})	≈ 2.86 milliwatts per square centimetre (mW cm^{-2})
1 lumen (at 0.556 μm)	= 1.47 milliwatts (mW)
1 erg per second (erg sec^{-1})	= 10^{-4} milliwatt (mW)
1 watt per square metre (W m^{-2})	= 0.1 milliwatt per square centimetre (mW cm^{-2})
1 millibar (mb)	= 10^{-3} dyne per square centimetre (dyn cm^{-2})
	= 10^{2} watts second per cubic metre (W sec m^{-3})
1 dyne (dyn)	= 10^{-5} newton (N; 1 N = J m^{-1})

$\hat{h}$ (Planck's constant)	= $(6.62517 \pm 0.00023) \times 10^{-34}$ J sec
$\underset{\sim}{c}$ (velocity light)	= $(2.997930 \pm 0.000003) \times 10^{10}$ cm sec^{-1}
$\hat{k}$ (Boltzmann's constant)	= $(1.38044 \pm 0.00007) \times 10^{-23}$ J K^{-1}
$\underset{\sim}{\sigma}$ (Stefan-Boltzmann constant)	= $(5.6696 \pm 0.0025) \times 10^{-12}$ W cm^{-2} K^{-4}
$\tilde{L}$ (latent heat vaporization)	≈ 590 cal g^{-1} (for water)

List of Principal Symbols

Note: (1) Unless otherwise specified, subscript "g" refers to ground or surface value, subscript "s" refers to standard pressure and temperature conditions.
(2) Average values indicated by bar over symbol, e.g. $\bar{A}$.
(3) ↑ or ↓ refer to vertical components of a parameter.

A	area
A_e	turbulent exchange coefficient
A_m	reciprocal of mechanical equivalent of heat
A_{ij}	two dimensional matrix
B	Planck black-body function — e.g. $B(T)$
$\tilde{B}$	Planck black-body function in flux density units
C	concentration
CC	fractional cloud cover
CR	cooling rate
D	solar diffuse flux density
D_R, D_D	solar diffuse flux density due to Rayleigh (R) or dust (D) scatter
E	global rate of entropy production or exchange
E_0	rate of entropy exchange for an individual latitude zone
$\tilde{E}$	upward vertical flux density of water vapour — evaporation rate
Eq	equation of time
F	radiant flux density
F_0	solar radiant flux density into top of the atmosphere
F_N	net radiant flux density
F_S, F_L	short-wave (solar) or long-wave (terrestrial) flux density
F_{NS}, F_{NL}	net short-wave (solar) or long-wave (terrestrial) flux density
$\tilde{G}$	ground heat flux (vertical)
$\hat{G}$	proportionality constant between atmospheric long-wave emission to space and surface black-body emission
G'	global flux density on a tilted plane
H	irradiance
$\tilde{H}$	vertical sensible heat flux density
$\hat{H}$	scale height of atmosphere
I_0	solar constant (a specific intensity or radiance)
J	radiant intensity
$J(r)$	scattering coefficient of single particle of radius r
JJ	Bessel function
$\tilde{J}$	source function
$\tilde{J}_{sc}$	source function for scatter
J_R	Rayleigh scatter molecular cross-section
$K(r)$	extinction coefficient of single particle of radius r
L	length dimension (also mean free path in Chapter 5)

$\tilde{L}$	latent heat of vaporization of water
M	mass dimension
$\tilde{M}$	magnification factor
$\hat{N}$	number density of molecules
N	radiance
N^+, N^-	upward or downward directed radiances
$N\uparrow, N\downarrow$	vertical stream radiances
P	radiant flux
$\tilde{P}$	phase function
P^0	azimuth independent phase function
P'_n	Associated Legendre Polynomial
Q_a, Q_{sc}, Q_e	extinction efficiencies referred to absorption, scatter and extinction
Q_m	energy per molecule
Q	energy per unit mass
$\tilde{Q}$	weighted transmittance
$\hat{Q}$	time integral of irradiance
$\tilde{R}$	weighted absorptance
R	mean earth radius
R^*	sun-earth distance
Re	reflectance
S	line strength
$\tilde{S}$	$= (\gamma_e + \gamma_r)/2$
T_d	dewpoint temperature
T	temperature
T_p, T_0	planetary temperature
T_e	effective radiant temperature
T_{st}	temperature of lower stratosphere
Tr	transmittance (also Tr_R, Tr_{oz}, Tr_{wv}, Tr_D — cf. τ)
$\tilde{Tr}$	flux transmittance
U	quantity of energy transferred
V	volume dimension; also Junge exponent in aerosol distribution
W	total liquid water in vertical column of unit cross-section (also radiant emittance in Table 2.1)
$\tilde{W}$	photon weight
Z	Rayleigh scatter function (see Chapter 5)
a	absorptance (also a_{oz}, a_{wv}, — cf. τ)
bf	backward to forward scatter ratio
c	velocity of light
c_p	specific heat at constant pressure
d_n	number of days from January 1
d	normalized radius of the sun
$\hat{d}$	planetary albedo of average cloud
e	water vapour pressure
$\tilde{f}$	radiation frequency
$\hat{f}$	ratio of cloud black-body emission to surface black-body emission
f'	fraction of sky seen by tilted plane
f_1, f_2, f_3	functions of
g	acceleration due to gravity
$\hat{g}$	planetary albedo of clear skies
h	height above ground
h_c	average cloud base height
$\hat{h}$	Planck's constant

i, j	indices
j	mass emission coefficient
$\tilde{j}$	volume emission coefficient
k	gas constant per molecule
$\hat{k}$	Boltzmann's constant
l	generalized absorption coefficient
m	mass of molecule
$\hat{m}$	clear-sky total atmospheric absorptance of solar radiation
m_a	absolute air mass
m_r	relative air mass
n	number density
n_c, n	refractive index ($n_c = n_r - in_i$)
n_T	total number density or total number
p	pressure
p_h	pressure thickness corresponding to actual thickness h
p_g	pressure at ground level
q	specific humidity
r	particle radius
r_e	mean effective radius of particles in a distribution
s	path length
s_v	specific volume
t	time
t_h	hour angle
$\hat{u}$	radiant energy density
u	vertical optical pathlength
u^*	scaled optical path length (i.e. scaled absorber amount)
w	liquid water content per unit volume (defined specifically per unit mass in section 4 of Chapter 10)
x, y, z	co-ordinate specification
u, v, w	velocity components in x, y and z direction
x, y	general variables
x	optical path of ozone
y	optical path of water vapour
z	vertical distance
$\hat{z}$	scaled height
$\hat{\alpha}$	Ångström wavelength dependence of turbidity
α	albedo; also line width in Chapter 7 and specific volume of air in Chapter 10
α_R	atmospheric albedo due Rayleigh scatter
$\bar{\alpha}_R$	atmospheric albedo for diffuse flux due Rayleigh scatter
α_R^*	atmospheric albedo for upwelling diffuse flux due Rayleigh scatter
α_0	planetary albedo
$\beta_a, \beta_{sc}, \beta_e$	volume extinction coefficients referred to absorption a, scatter sc, and extinction e (also β_R, β_{oz}, β_{wv} and β_D — cf. τ)
Γ	diffusion coefficient
γ_{sc}	volume scattering function
γ_r, γ_e	coefficients of Rayleigh scatter (see Chapter 4)
γ_{ob}	observed solar altitude
γ	solar altitude
δ	solar declination; also line spacing in Chapter 7
ϵ	emissivity
ϵ_1	flux emissivity of atmosphere as viewed from ground
ϵ_2	flux emissivity of atmosphere above average cloud level

ϵ_{zen}	zenith emissivity
ζ	tilt angle of a plane to the horizontal
θ'	angle between solar direction and normal to a plane
$\hat{\theta}$	fractional cloud cover
θ	zenith angle
θ_0	$= 2\pi d_n/365$
θ_{ob}	observed solar zenith angle
$\kappa_a, \kappa_{sc}, \kappa_e$	mass extinction coefficients referred to absorption a, scatter sc and extinction e
η	$\sec\theta$
λ	wavelength
μ	$\cos\theta$
μ_0, ϕ_0	solar direction in terms of $\cos\theta$ and azimuth angle
ν	wavenumber
ν_0	wavenumber at centre of absorption line
ξ	size parameter $2\pi r/\lambda$
π	3.1416
Π	half-day length
ρ	density
σ	Stefan-Boltzmann constant
τ	optical depth
$\tau_a, \tau_{sc}, \tau_e$	optical depth referred to absorption a, scatter sc, or extinction e
$\tau_R, \tau_{oz},$ τ_{wv}, τ_D	optical depth referred to Rayleigh scatter, ozone absorption, water vapour absorption and dust extinction
τ^*	optical depth from top to bottom of atmosphere
ϕ	azimuth angle; also empirical function in Chapter 7
Φ	latitude
ψ	scattering angle between incoming and outgoing rays
ω	solid angle
Ω_0	unit solid angle
$\bar{\omega}_0$	single scattering albedo
$\langle\cos\psi\rangle$	asymmetry factor for scatter

CHAPTER 1

The Global Picture

1. INTRODUCTION

We begin by regarding the earth and its atmosphere as a single entity subjected to a mean irradiance $\overline{F}_0$ of radiant energy from the sun. It reflects back to space a certain fraction (the planetary albedo $\bar{\alpha}_0$) of solar radiation so that the remainder $\overline{F}_0(1 - \bar{\alpha}_0)$ is absorbed and acts so as to raise the mean planetary temperature $\overline{T}_0$. If further we imagine the planet to be a perfect black body as far as thermal infrared emission is concerned the mean temperature must eventually adopt a value such that the thermal emission of radiant energy $\sigma\overline{T}_0^4$ exactly balances the solar energy absorption. Thus:

$$\overline{T}_0 = [\overline{F}_0(1 - \bar{\alpha}_0)/\sigma]^{1/4} \qquad [1.1]$$

The "climate" of this simple planet is represented by its temperature and is a direct function of solar output. If $\overline{F}_0$ increases the mean temperature will increase appropriately.

It has been shown experimentally that the earth-atmosphere system is indeed in a state of radiant energy balance or very close to it. It is therefore somewhat surprising to the uninitiated that it is not yet possible to predict *a priori* the mean planetary temperature even when the various mathematical difficulties associated with defining mean quantities on a highly variable world are taken into account. This is true in spite of the fact that we know the answer beforehand. As a consequence, we are certainly in no position to predict the change in climate which might occur as a result of any change in the external parameters of the system such as the energy output of the sun. Nevertheless, it is not uncommon for versions of equation [1.1] in various guises to be used to justify predictions of global catastrophe. The basic problem is that $\bar{\alpha}_0$ in its integral or global mean form is an unknown function of $\overline{T}_0$ – i.e. of "climate" as defined by the model of equation [1.1]. Depending on the nature of this function it is quite possible to envisage an earth where an increase in solar output resulted in a decrease of planetary mean temperature. Such a state of affairs is intuitively unlikely but is nevertheless possible in terms of our present knowledge.

The irradiance of solar energy impinging on the top of the atmosphere has a meridional distribution which in the annual mean decreases toward the poles from a maximum at the equator. Owing to the orbital obliquity of the earth the decrease is not nearly as rapid in polar regions as the cosine of the latitude Φ, and the mean polar irradiance is in fact about 0.4 of that at the equator (see Chapter 3). Nevertheless the amount actually absorbed by (mostly) the surface of the earth is roughly proportional to cos Φ as a result of increased reflectivity at higher latitudes, so that the energy input to the system is dominated by the equatorial regions. Thermal emission of long-wave radiation to space and the corresponding planetary temperature also decreases towards the poles but certainly not to zero. (This is an observation, not a theoretical conclusion.) Radiant energy balance dictates that the total outward thermal flux equals the absorbed solar flux. System equilibrium dictates that the energy absorbed at the equator must be exported towards the poles via the atmosphere and ocean so that it can be re-radiated to space. The various phenomena of weather and climate — the dynamics and thermodynamics of atmosphere and ocean — are in a sense simply manifestations of the process by which the required meridional flow is achieved. Their characteristics are determined by the radiant energy input and output at either end of the chain. Again the difficulty is that the overall source and sink properties of radiation are not predictable *a priori*, since the input of solar energy and the output of thermal energy are themselves determined in an unknown way by the processes which they drive. There are two basic problems yet to be solved apart from the sheer complexity of describing all significant motions of the atmosphere and ocean. They are to determine the factors controlling the amount and distribution of cloud over the world, since cloud has a dominating influence on solar radiation reflection and on thermal emission; and to determine the factors controlling the partition of meridional energy flow between the atmosphere and the ocean.

Nevertheless, a natural course to follow in the investigation of climate and climate distribution is to centre attention on the radiative processes and treat the other factors as complementary. (There is, however, an obvious example in section 7 of this chapter which supports the inverse proposition). The conclusion is reinforced by the fact that changes in the state of a radiating body manifest themselves instantaneously. Thus a lessening of radiation of the surface layers of the atmosphere as a result, say, of the advection of colder air will propagate immediately. The stratosphere for instance will sense the change in radiation from below and begin to respond immediately. The mechanical and thermodynamic modes of heat transport are much slower. In this restricted sense, radiation is the most efficient heat conveying agent in the atmosphere and plays a major role in most processes on the large and medium scales. It appears also in the study of weather in the detailed meaning of the word.

Since radiative processes are primarily a matter of climatic concern, it is appropriate at the outset to present the zonal average mean annual observed parameters of climate as a frame of reference. Table 1.1 from Sellars (1965) describes the distribution of the most important quantities. Table 1.2 gives

TABLE 1.1

Zonal means of various climatic parameters according to Sellars (1965)

Latitude zone	O (%)	h (m)	T_g (K)	r (mm)	E (mm)	W (mm)	PE (%)	$\hat{\theta}$ (%)	α_g (%)
80–90°N	93.4	137	249.6	120	42	4.90	6.7	62	61
70–80	71.3	220	257.3	185	145	6.48	7.8	66	46
60–70	29.4	202	266.0	415	333	8.52	13.3	65	24
50–60	42.8	296	273.7	789	469	11.64	18.6	60	14
40–50	47.5	382	280.7	907	641	15.21	16.3	53	12
30–40	57.2	496	287.2	872	1,002	18.95	12.6	46	10
20–30	62.4	366	293.6	790	1,246	26.37	8.2	43	10
10–20	73.6	146	298.3	1,151	1.389	36.73	8.6	47	9
0–10	77.2	158	298.7	1,934	1,235	41.07	12.9	52	8
0–90°N	60.6	284	286.4	1,009	944	23.85	12.1	52	14
0–10°S	76.4	154	298.0	1,445	1,304	40.90	9.7	52	7
10–20	78.0	121	296.5	1,132	1,541	36.66	8.5	48	8
20–30	76.9	156	292.0	857	1,416	29.86	7.9	48	8
30–40	88.8	106	286.7	932	1,256	23.81	10.7	54	8
40–50	97.0	5	281.9	1,226	895	18.10	18.6	66	7
50–60	99.2	5	274.4	1,046	520	12.61	22.7	72	17
60–70	89.6	388	262.2	418	174	6.84	16.7	76	21
70–80	24.6	1,420	243.7	82	45	2.87	7.8	65	63
80–90	0.0	2,272	225.3	30	0	1.56	5.3	54	84
0–90°S	80.9	216	284.6	1,000	1,064	25.49	12.1	57	13
Globe	70.8	250	285.5	1,004	1,004	24.67	12.1	54	13

O = percentage of zone covered by ocean; h = height above sea level; T_g = surface temperature (see also Fig. 1.1 and relevant sources); r = annual rainfall; E = annual evaporation; W = total precipitable water in the mean vertical column of 1 cm^2 cross-section; PE = precipitation efficiency $[(r/365) \div W]$; $\hat{\theta}$ = average cloud amount (see, however, Fig. 1.2 and relevant sources); α_g = surface albedo.

TABLE 1.2

Distribution of average cloud amounts and cloud heights for northern hemisphere during March (from London, 1952). The apportionment is roughly similar for other seasons and for the southern hemisphere (see Telegadas and London, 1954; and Sasamori et al., 1972)

Latitude belt	Ci[1]		As		Ns			St		Cu			Cb		
	amount (%)	height (km)	amount (%)	height (km)	amount (%)	height (km)		amount (%)	height (km)	amount (%)	height (km)		amount (%)	height (km)	
						top	base				top	base		top	base
0–10°N	12.1	9.8	5.0	4.4	6.3	4.4	1.5	7.6	1.5	13.5	2.3	1.5	4.5	5.0	1.8
10–20	10.3	10.4	4.9	4.8	4.1	4.8	1.8	6.6	1.8	11.2	2.8	1.8	4.9	5.7	2.3
20–30	10.6	10.4	6.2	5.0	4.8	5.0	2.0	8.4	2.0	10.7	3.0	2.0	3.3	6.0	2.5
30–40	12.0	10.0	7.1	4.7	7.7	4.7	1.8	14.1	1.8	10.4	2.8	1.8	2.7	5.4	2.3
40–50	10.9	8.7	7.3	4.0	10.9	4.0	1.3	18.2	1.3	9.7	2.3	1.3	3.0	4.7	2.0
50–60	9.6	7.6	6.3	3.6	11.3	3.6	1.2	16.9	1.2	9.1	2.2	1.2	2.8	4.3	1.7
60–70	7.9	7.2	6.0	3.6	11.2	3.6	1.0	17.0	1.0	8.4	2.0	1.0	2.5	4.0	1.5

[1] The symbols for cloud type are London's definition of cirrus (Ci), altostratus (As), nimbostratus (Ns), stratus (St), cumulus (Cu) and cumulonimbus (Cb).

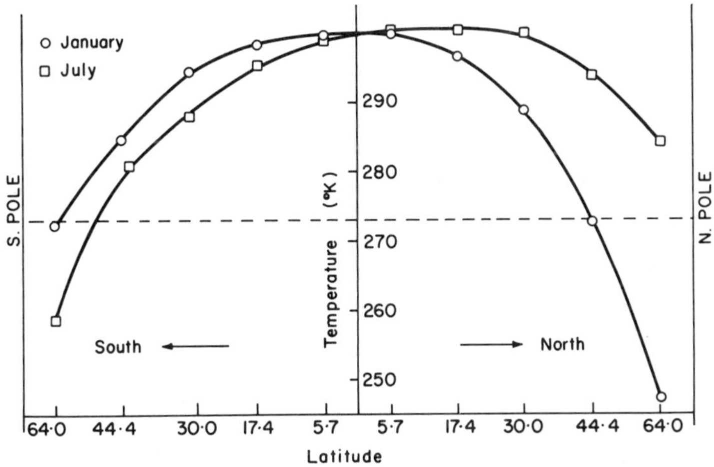

Fig. 1.1. Meridional distribution of surface temperature for January and July from Crutcher and Meserve (1970) for the northern hemisphere and from Taljaard et al. (1969) for the sourthern hemisphere. Latitude Φ on cos Φ scale.

the cloud amount at various levels in the atmosphere as a function of latitude. Figs. 1.1 and 1.2 show the meridional distribution of surface temperature and total cloud cover for the extreme months of January and July. Fig. 1.3 gives latitude-height cross-sections of atmospheric temperature for the same months. The sources for these are given in the captions.

2. BASIC MODELS

We refer now to Fig. 1.4 which is a pictorial presentation in somewhat more detail of the global mean energy balance. A number of such illustrations have been produced since the original version of Dines (1917). (For example, Baur and Philipps, 1935; Albrecht, 1949; Lettau, 1954; London and Sasamori, 1971.) They have improved over the years and have altered slightly as more information has come to hand and a consensus has emerged as to the absolute values of the various energy flows. Nevertheless there is still considerable give and take between the items. The values given here have been tuned slightly so as to be compatible with the later discussions of meridional distributions, but are still well within the known limits of accuracy. The units are in terms of energy flux per unit surface area expressed as a percentage of mean solar input. The mean solar radiation input to the top of the atmosphere $\overline{F}_0$ is one-fourth of the solar constant I_0; i.e. $\overline{F}_0 = 34\,\mathrm{mW\,cm^{-2}}$.

Various authors since Simpson (1928) have produced simple models in an attempt to predict the fundamental climatic variable of surface temperature.

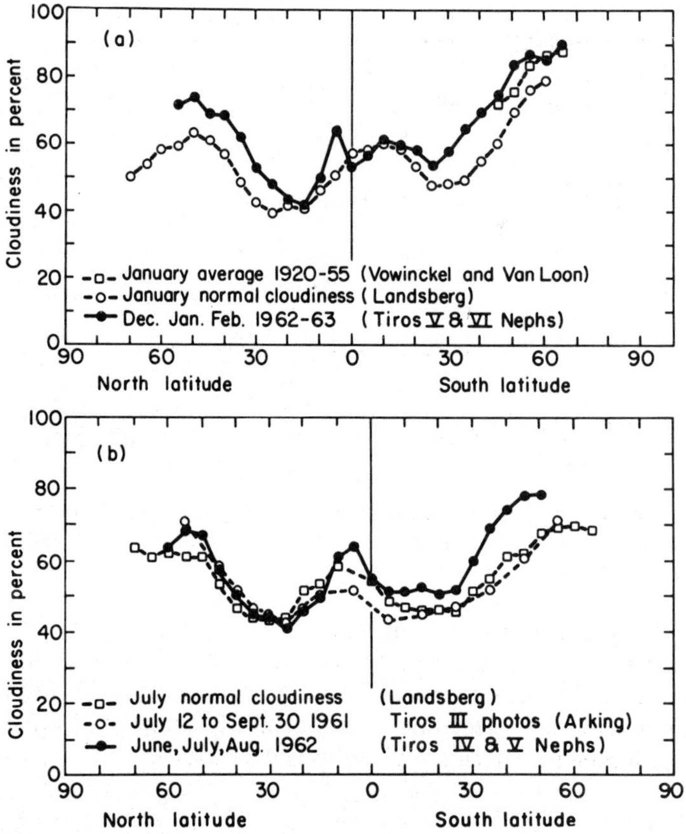

Fig. 1.2. Latitudinal distribution of zonally averaged cloudiness for (a) January and (b) July. The Landsberg and Vowinckel and Van Loon results are from ground observation. Other curves from satellite observations as marked. (After Clapp, 1964.)

Recent examples are those of Adem (1962), Budyko (1969), Sellars (1969) and Saltzmann (1968), which attempt to predict surface temperature as a function of latitude and season. They are based on "the conviction that the steady-state average latitudinal distribution of surface temperature, to a first approximation, should depend entirely on the incoming solar energy, the transparency of the atmosphere to terrestrial radiation, the planetary albedo, the ability of the atmosphere to carry heat and water vapour from source to sink, and the heat storage potential of the oceans and land. The latter factor drops out when average annual temperatures are considered" (Sellars, 1969). As it happens these simple two-dimensional models are not fundamentally different from each other and nor are they fundamentally different from the vastly detailed numerical models now being produced in an attempt to solve the problem. The same basic difficulties arise in all cases. The following example of a simple model by one of us (Paltridge, 1974a, 1975) serves as

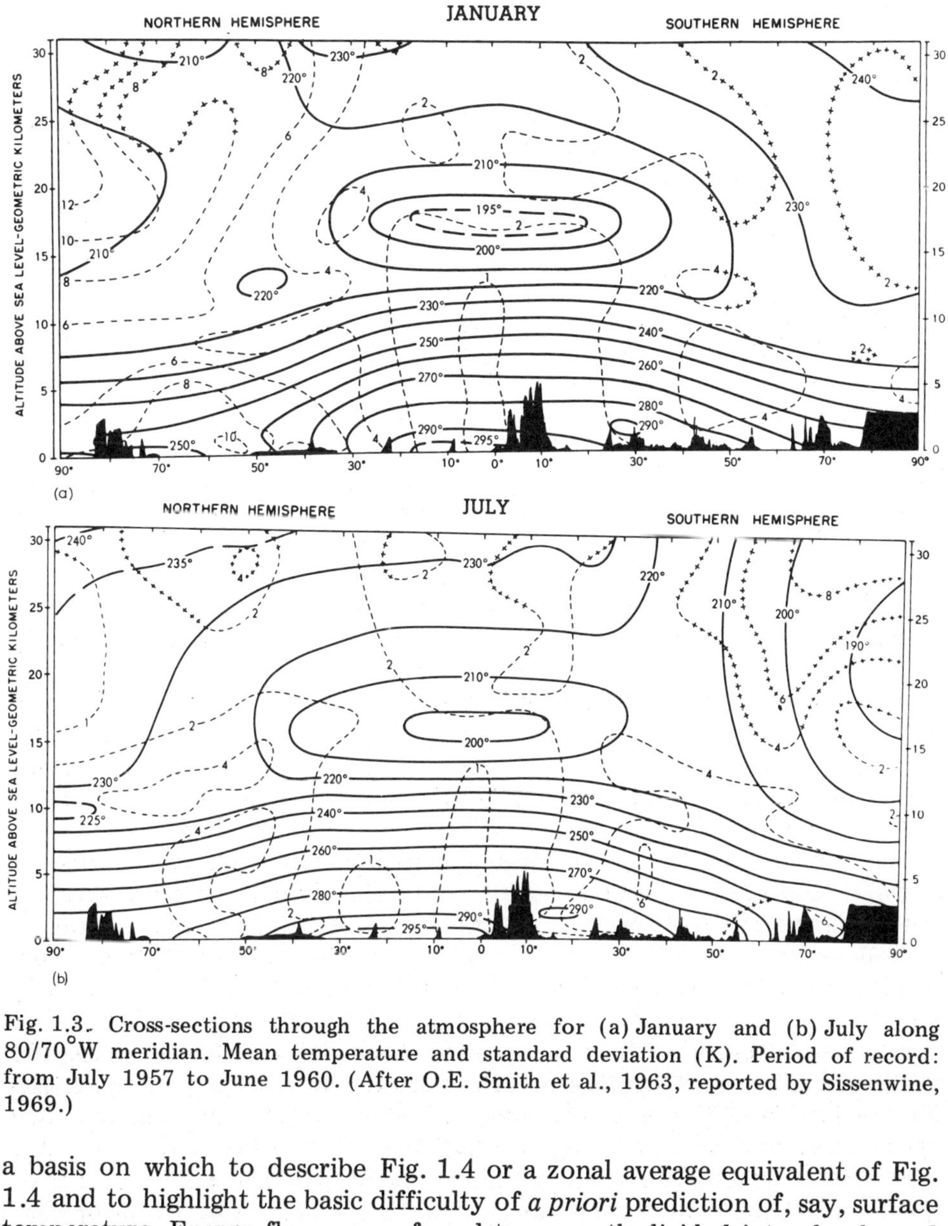

Fig. 1.3. Cross-sections through the atmosphere for (a) January and (b) July along 80/70°W meridian. Mean temperature and standard deviation (K). Period of record: from July 1957 to June 1960. (After O.E. Smith et al., 1963, reported by Sissenwine, 1969.)

a basis on which to describe Fig. 1.4 or a zonal average equivalent of Fig. 1.4 and to highlight the basic difficulty of *a priori* prediction of, say, surface temperature. Energy flows are referred to an earth divided into cloudy and non-cloudy portions of fractional values $\hat{\theta}$ and $1 - \hat{\theta}$ respectively.

Clear skies reflect a certain fraction $\hat{g}(1 - \hat{\theta})$ of F_0 back to space by Rayleigh scatter from molecules, by reflection from the surface and by scatter from dust particles. The apportionment in the global mean is roughly as given in the figure. A fraction $\hat{m}(1 - \hat{\theta})$ is directly absorbed in the

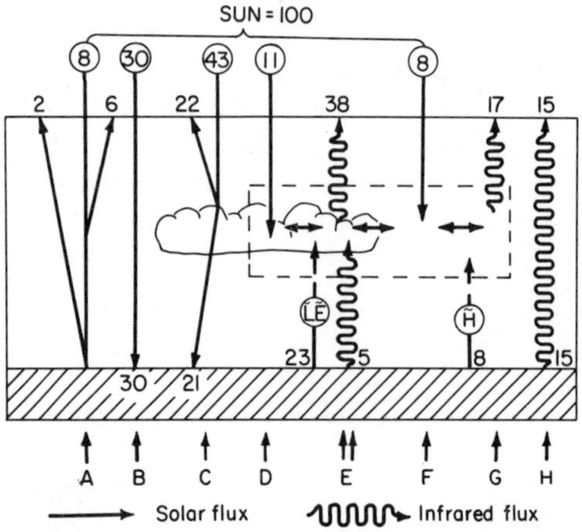

Fig. 1.4. Schematic diagram of planetary, atmospheric and surface energy balance in the global and multi-annual mean. Units in terms of percentage of mean solar input — that is, 100 units = 34 mW cm^{-2}.

atmosphere by water vapour and other trace gases such as ozone. The remainder $F_0(1-\hat{\theta})(1-\hat{g}-\hat{m})$ is absorbed by the earth's surface itself. In cloudy regions there is an additional absorption $\hat{a}\hat{\theta}$ of solar energy within the clouds (in the liquid water drops or ice crystals) and a massive increase in planetary albedo as a result of their high reflecting power. The solar energy absorbed at the surface is $F_0\hat{\theta}(1-\hat{m}-\hat{a}-\hat{d})$ where $\hat{d}$ is the planetary albedo of clouds, of their overlying atmosphere and of the earth and atmosphere beneath them.

For most practical purposes thermal emission from the earth's surface is given by the Stefan–Boltzmann function σT_g^4 of surface temperature T_g. Various gases of the atmosphere (primarily water vapour and carbon dioxide) absorb about 70% of this upwelling radiation. The absorption occurs at wavelengths mostly outside the 8–14-micron (μm) band, which is known for obvious reasons as the *atmospheric window* — the spectral region over which thermal emission from the ground is radiated directly to space. The amount of radiation from the ground transmitted directly to space can therefore be written as $(1-\epsilon_1)\sigma T_g^4$ where ϵ_1 is about 0.7. The amount absorbed by the atmosphere is $\epsilon_1\sigma T_g^4$. By Kirchoff's law the gaseous absorbers of the atmosphere are also emitters over their absorbing wavelengths and at their appropriate temperatures, and furthermore they emit in both the upward and downward directions. The downward emission is returned to and absorbed by the surface. Its exact value is a function of the temperature and water vapour profile of the atmosphere, but to a first approximation the downward flux can be expressed as a function of the surface values of temperature and water vapour mixing ratio — or even of temperature alone. Indeed for the present purpose a value of $\epsilon_1\sigma T_g^4$ is accurate enough. It emphasizes the concept that atmospheric thermal emission returned to the surface derives primarily from the very lowest levels of the atmosphere at wavelengths outside the window region, since the absorption and emission bands are nearly saturated at typical ground level mixing ratios. Thus in clear skies the *net* thermal flux at the surface is mainly a matter of the upward flowing window radiation. In cloudy skies the downward emission of the atmosphere is augmented by radiation from the clouds themselves. They are in most cases close to black bodies at temperatures appropriate to their height. They return radiant energy to the surface primarily via the atmospheric window. They absorb radiation from the surface and from that part of the atmosphere lying below them. In the interests of simplicity, at this stage we assume that there *is* such a thing as average cloud temperature and that this temperature is such that the associated thermal emission is a fixed fraction $\hat{f}$ of surface emission — i.e. $\hat{f}\sigma T_g^4$. The amount of this emission returned to the ground is therefore $(1-\epsilon_1)\hat{\theta}\hat{f}\sigma T_g^4$.

In clear skies the upward thermal flux radiating to space from the atmosphere itself is a most complicated function of the vertical profiles of the absorbing gases and of temperature. Physically, the simplest concept is of the atmospheric window expanding in wavelength with height as the absorbing gases become more tenuous. Thus successively higher layers can radiate directly to space. Bearing in mind that each successive layer radiates at its own temperature, the total emission from the top of the atmosphere has a spectrum completely unlike a black-body spectrum — and is of a total amount $F\uparrow$ corresponding to a much lower radiative temperature T_e if one insists on defining this temperature in terms of the Stefan-Boltzmann law ($\sigma T_e^4 = F\uparrow$). We assume for the moment that the upward atmospheric emission is directly proportional to surface black body emission; e.g. equal to $\hat{G}\sigma T_g^4$ where $\hat{G}$ is of the order of 0.3. Thus in clear skies $F\uparrow = \hat{G}\sigma T_g^4 + (1-\epsilon_1)\sigma T_g^4$.

The upward thermal flux out of the top of the atmosphere above clouds is that from the clouds themselves through the expanded atmospheric window (say $(1-\epsilon_2)\hat{\theta}\hat{f}\sigma T_g^4$ where ϵ_2 is about 0.25) plus the emission from the residual water vapour and carbon dioxide above the clouds. This last will be ignored in the following on the dubious grounds that it is a negligible term in the energy balance.

Thus at the top of the atmosphere where complete radiative energy balance prevails for the world as a whole:

$$F_{NO} = F_0[1-\hat{a}\hat{\theta}-\hat{g}(1-\hat{\theta})]-(1-\hat{\theta})\sigma T_g^4[\hat{G}+(1-\epsilon_1)] - \hat{\theta}(1-\epsilon_2)\sigma T_g^4\hat{f} = 0.0 \qquad [1.2]$$

where F_{NO} is the net radiative input, the first term on the right-hand side is the short-wave component of the net input F_{NSO}, and the last two terms are the long-wave component of the net input F_{NLO}. By various devices and approximations we have produced an equation governing the system of apparently only two unknowns — cloud cover $\hat{\theta}$ and surface temperature T_g.

This is of course not true. All the various "constants" $\hat{g}$, $\hat{d}$, ϵ_1, $\hat{G}$, $\hat{f}$ and ϵ_2 are variables which depend to some extent on the dynamics and thermodynamics of the system. It may be possible to justify assuming them to be constant to a first approximation, but mainly on empirical grounds — not on the grounds of mechanistic calculation. All simple models involve similar assumptions or "parameterization" in order to bypass the necessity for complex description of atmospheric and oceanic dynamics.

This then is the main difference between the "simple" model and the large-scale numerical models currently in vogue. By treating dynamics specifically the large-scale models can in principle calculate the profiles of the atmospheric parameters on which the radiative fluxes depend and hence calculate (implicitly at least) the various "constants" of equation [1.2]. Much of this book is devoted to matters relevant to that problem. It should not be forgotten, however, that even the largest of the numerical models relies heavily on parameterization. Often it is difficult to establish whether a particular parameterization is based purely on empiricism or whether it is solidly founded in physics.

Even the large-scale models cannot yet overcome the basic difficulty for which equation [1.2] is a prime example; namely the problem of having at least one more unknown than there are equations. Using the present example, one can set up a further energy balance equation at the bottom boundary (the surface). The balance is no longer purely radiative and it is necessary to introduce a further variable of the ground-atmosphere energy flux $\widetilde{LE} + \widetilde{H}$ (latent heat plus sensible heat). That is:

$$\widetilde{LE} + \widetilde{H} = F_0[1 - \hat{m} - (\hat{d} + \hat{a})\hat{\theta} - \hat{g}(1 - \hat{\theta})] - (1 - \epsilon_1)\sigma T_g^4 + \hat{f}\hat{\theta}(1 - \epsilon_1)\sigma T_g^4 \qquad [1.3]$$

Selection of a further surface at another atmospheric level in turn requires the splitting of $\widetilde{LE} + \widetilde{H}$ into its separate components and the specification of variables determining the energy flows indicated by the horizontal arrows of Fig. 1.4. Very rapidly one enters the sphere of large-scale three-dimensional numerical modelling whose philosophy is that solution of the problem will be possible when all significant processes can be described in sufficient detail. In principle this is true provided that no such thing as an uncertainty principle is operative. In practice, and in common with the simple models, large-scale numerical models can at present deal only with the partial derivatives obtained by making some assumption equivalent to setting either cloud cover or surface temperature constant.

Fundamentally, the stumbling block is the mechanistic prediction of cloud cover. In turn, as the models enlarge in an attempt to incorporate the dynamics required to predict cloud cover, further problems arise with regard to specifying the meridional energy flows in atmosphere and ocean. The simple models must make some obvious assumption about the appropriate transfer coefficients — such as setting them at a constant value. The complex models must make similar assumptions which are, however, not necessarily so obvious.

3. TEMPERATURES, CLOUD COVER AND SOLAR CONSTANT

Inasmuch as partial derivatives indicate at least the initial tendency resulting from a given change, they are useful for gaining insight into the physics of a problem.

According to Raschke et al. (1973) the mean planetary albedo $\bar{\alpha}_0$ is 0.28. Referring to equation [1.1], differentiation with respect to the solar constant $I_0 (= 4\bar{F}_0 \approx 136\ \text{mW cm}^{-2})$ yields:

$$\partial \bar{T}_0 / \partial I_0 = 0.66\ \text{K/percent} \qquad [1.4]$$

The planetary mean radiative temperature would rise with increasing solar output according to this equation in the unlikely event that such things as cloud cover and the extent of surface ice caps remained unaltered. The corresponding rise in surface temperature $\bar{T}_g$ is larger. Various models produce somewhat different answers which, according to their individual assumptions, lie within the range:

$$\partial \bar{T}_g / \partial I_0 = 0.7\text{–}1.5\ \text{K/percent} \qquad [1.5]$$

Both Sellars' model and the large-scale numerical model of Manabe and Wetherald (1967) put $\partial \bar{T}_g / \partial I_0$ at 1.2 K/percent. The differences between answers are in fact highly significant since the extent of ice cover on the world is presumably dependent on surface temperature and can therefore introduce positive feedback via planetary albedo. In order to make this point, Sellars (1969) incorporated various empirical relations between ice cover and surface temperature. The result in Fig. 1.5 implies that, if cloud cover were to remain constant, a drop of 2% in solar output might be sufficient to switch the world to a completely ice-covered state and to a new equilibrium temperature of 163 K.

The radiative effect of increased cloud is twofold. One the one hand the planetary albedo is increased and the reduction of solar input tends to cool the planet and the surface. On the other hand the emission of long-wave radiation to space is reduced since the "surface" as seen from space through the atmospheric window is of lower temperature. Increased cloud also blocks the atmospheric window and returns additional long-wave radiation to the ground. Thus from the long-wave standpoint both the planet and the surface tend to warm. Whether the *net* result is a cooling or a warming depends on cloud temperature and its geographical position. Furthermore, the effect of cloud on net radiation flux is not the same at the top and bottom of the atmosphere.

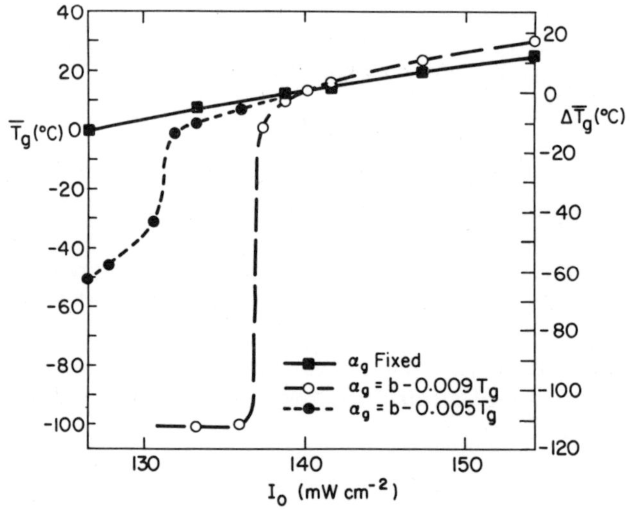

Fig. 1.5. Global mean surface temperature as a function of solar constant I_0 according to the model of Sellars (1965) and various assumed relations between surface albedo α_g and surface temperature T_g (i.e. assumptions about the feedback between ice cover and temperature). The b above varies with latitude but is approximately 2.8.

The net downward radiative flux density F_{N0} at the top of the atmosphere is equal to zero in the mean, but differentiation of equation [1.2] gives:

$$\partial F_{N0}/\partial\hat{\theta} = F_0(\hat{g} - \hat{d}) + \sigma T_g^4[\hat{G} + (1 - \epsilon_1) - (1 - \epsilon_2)\hat{f}] \qquad [1.6]$$

Adopting global mean values of $\hat{g}$, $\hat{d}$, ϵ_1, ϵ_2, $\hat{G}$ and $\hat{f}$ (refer to Table 1.3) a little calculation shows that $\partial F_{N0}/\partial\hat{\theta} \approx -1.0$ mW cm^{-2}/tenth of cloud cover. The fact that it is negative agrees with the simplistic argument as to why the system as a whole is stable — namely, that an increase in cloud cover should reduce global temperature and thereby reduce evaporation. This in turn would feed back so as to lessen cloud cover and restore the original equilibrium. Insertion of zonal average values of the constants of equation [1.6] shows that $\partial F_{N0}/\partial\hat{\theta}$ decreases in absolute magnitude towards the poles and in fact may be positive there as the effect on long-wave emission begins to dominate. Similar calculations on the detailed model of Schneider (1972) show that $\partial F_{N0}/\partial\hat{\theta}$ becomes positive poleward of about 60°.

At the surface, the net downward flux density F_{Ng} equals the upward flux of latent and sensible heat $\tilde{L}\tilde{E} + \tilde{H}$. Differentiation of equation [1.3] leads to:

$$\partial F_{Ng}/\partial\hat{\theta} = F_0(\hat{g} - \hat{d} - \hat{a}) + \hat{f}(1 - \epsilon_1)\sigma T_g^4 \qquad [1.7]$$

whose value in the global mean is still negative (≈ -0.18 mW cm^{-2}/tenth of cloud cover) but only just so. An increase of very high cloud such as cirrus gives a more negative $\partial F_{Ng}/\partial\hat{\theta}$, as can be seen by inserting an appropriately

TABLE 1.3

Zonal mean values of the parameters used in the simple climate model relevant to Fig. 1.19 and to the discussion of equations [1.2] and [1.3]

	Mean latitude									
	64°N	44°N	30°N	17°N	6°N	6°S	17°S	30°S	44°S	64°S
$\hat{d}$	0.55	0.50	0.45	0.43	0.41	0.41	0.43	0.45	0.50	0.55
$\hat{m}$	0.20	0.18	0.17	0.16	0.16	0.16	0.16	0.17	0.18	0.20
$\hat{g}$	0.22	0.18	0.16	0.15	0.14	0.14	0.14	0.14	0.14	0.16
$\hat{a}$	0.04	0.04	0.04	0.04	0.04	0.04	0.04	0.04	0.04	0.04
$\hat{\xi}$	7.17	4.40	3.63	3.29	3.14	3.14	3.29	3.63	4.40	7.17

I_0 (solar constant) = 136 mW cm^{-2}; $F_0 = I_0/\hat{\xi}$; $\epsilon_1 = 0.70$; $\epsilon_2 = 0.25$; $\hat{f} = 0.80$; $\hat{G} = 0.31$.

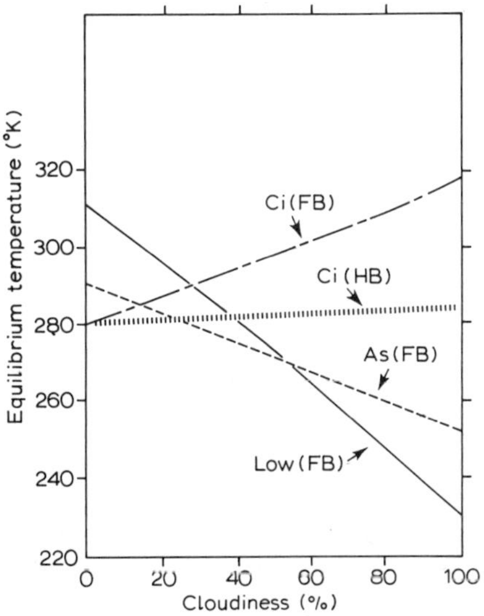

Fig. 1.6. Global mean surface temperature $\overline{T}_g$ as a function of cloudiness of various types (cirrus, altostratus and low cloud) according to Manabe and Wetherald. (After Manabe, 1970.) FB and HB refer to fully black and half black clouds respectively.

lower value of $\hat{f}$. Such a calculation should, however, take account of the lower albedo and lower infrared emissivity of typical high cloud. This has been done by Manabe and Wetherald (1967) using their detailed radiative equilibrium model, and they have produced the curves of $\overline{T}_g$ versus $\hat{\theta}$ shown in Fig. 1.6. The curves imply that $\partial F_{Ng}/\partial\hat{\theta}$ is *positive* for high cloud.

In terms of zonal averages and of zonal average cloud, equation [1.7] shows that $\partial F_{Ng}/\partial\hat{\theta}$ becomes positive poleward of only 40°.

It is of some interest to derive from equation [1.7] the average cloud (base) height h_c at which a change in cloud amount gives no change of the net flux at the surface, even though a change of cloud amount at that height will upset the planetary balance. The parameter $\hat{f}$ is a function of cloud temperature and therefore of height and appropriate values can be found by setting $\partial F_{\mathrm{Ng}}/\partial\hat{\theta}$ equal to zero. Using the mean temperature profiles of the different latitude zones in order to establish h_c, it can be shown that h_c increases from the 800-mb pressure level at the equator to about 500 mb at 65°.

4. TIME CONSTANTS

Equilibrium may exist on a multi-annual mean basis but it is not likely to be the established state on any shorter time scale. Therefore the climatologist is greatly concerned with the time constants of the energy reservoirs of the world. The major reservoir is the ocean. A good summary of the complete system (after Lettau, 1965, 1954) is given in Fig. 1.7 where the major energy transformations are shown and which may be used to determine the relative importance of the energy cycles within the thermodynamic system of the earth-atmosphere. The numerical data are useful for various quantitative estimates. For example, if all energy inputs ceased and the rates of energy expenditure were maintained, the reservoir of mechanical energy (momentum) would be depleted in three days. It is this sort of calculation which is the basis of the fact that purely kinematic weather forecasting models can run quite successfully for a day or so without the necessity to include thermodynamics or radiative inputs. The reservoir of latent heat (precipitable water) would be exhausted in about 12 days; and the reservoir of internal thermal energy in about 100 days. This last is mainly attributable to the ocean and it is worth making the point that for long-term climatic forecasting the attention in numerical modelling should perhaps be directed primarily towards the ocean rather than the atmosphere.

The above time constants can be interpreted as fictitious cycles of turnover of the atmospheric properties. The relative magnitudes of these life cycles show that, in comparison with rainfall and winds, temperature is the most conservative, and will exhibit the smallest, and thus the most regular, temporal and spatial large-scale variation.

5. MEAN MERIDIONAL CROSS-SECTIONS

Figs. 1.8, 1.9 and 1.10 are a set of meridional profiles of the mean annual values of significant radiation fluxes at the top and bottom of the atmosphere. The latitudinal scale is proportional to cos Φ so that equal areas under the curves refer to equal total energy inputs. The curves are theoretical in that they derive from the simple model relevant to equations [1.2] and

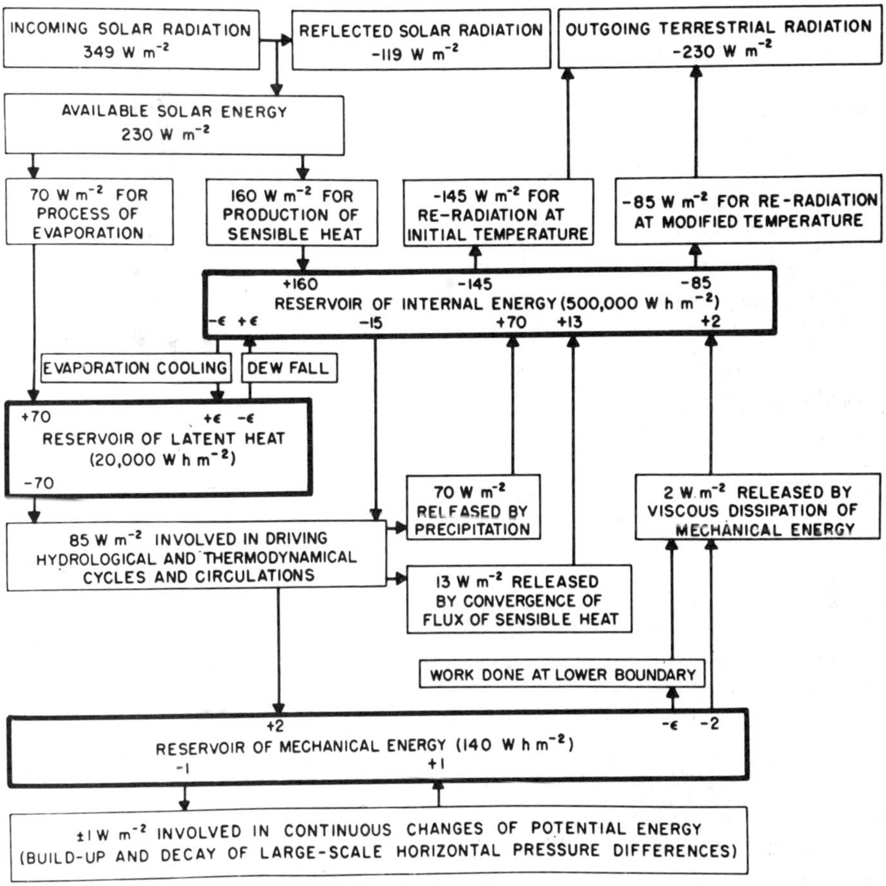

Fig. 1.7. The global mean energy cycle of the atmosphere. A solar constant of $1395\,\mathrm{W\,m^{-2}}$ and a global albedo of 0.34 are assumed. The average total incoming radiation to the globe is 1/4 of the solar constant. ϵ denotes an average rate of less than $0.5\,\mathrm{W\,m^{-2}}$. (After Lettau, 1965.)

[1.3], and are presented here purely as a coherent set to use as a basis of discussion. They are quite consistent with such observational data as is available.

The solid lines of Fig. 1.8 give the downward *net* short-wave flux density at the top and bottom of the atmosphere (F_{NS0} and F_{NSg} respectively), both of which display the dominating influence of latitude which affects the incoming component of the net short-wave flux. On such a scale the meridional variation of albedo is scarcely observable but is nevertheless significant. Note that net flux density is a conservative quantity as a function of height as far as scatter is concerned; and the difference between the two

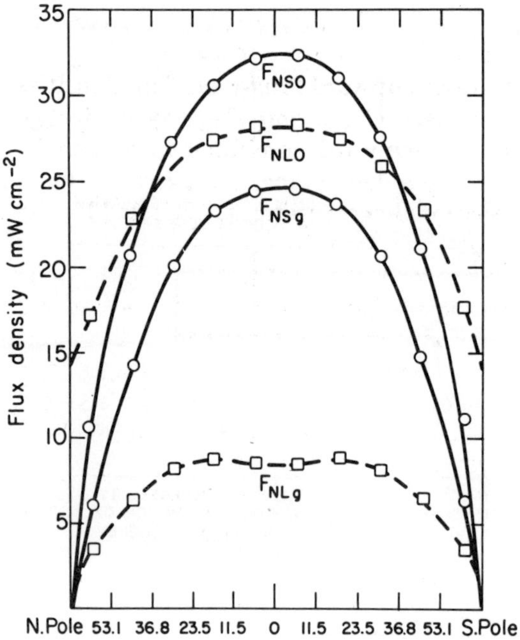

Fig. 1.8. Net short-wave flux densities into planet and surface (F_{NS0} and F_{NSg} respectively); and net upward long-wave flux densities at top and bottom of the atmosphere (F_{NL0} and F_{NLg} respectively). All as function of latitude.

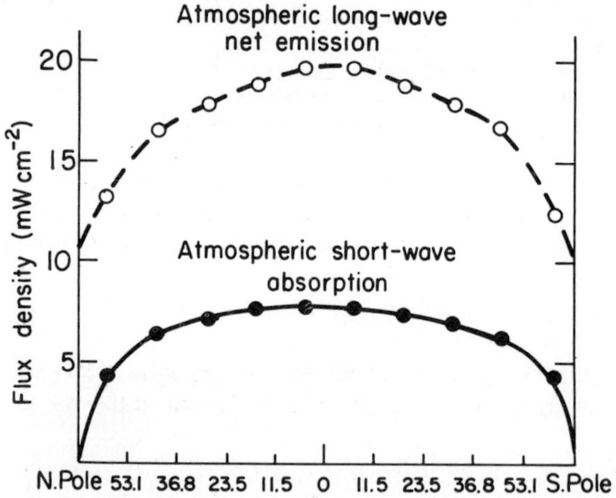

Fig. 1.9. Net long-wave emission (from both top and bottom of the atmosphere) compared with short-wave absorption, as a function of latitude.

curves is entirely a result of the atmospheric absorption of short-wave radiation which is given in Fig. 1.9. The absolute value of this absorption also decreases towards zero at the poles (obviously enough) although the corresponding fractional absorption increases slightly with the increasing air mass through which the solar beam travels.

The dashed curves of Fig. 1.8 are the net upward long-wave flux densities at the top and bottom of the atmosphere (F_{NL0} and F_{NLg} respectively). F_{NL0} is undirectional since there is no downward component and can be used to define the planetary temperature of any zone. F_{NLg} is very much less than that at the top and its bi-directional character is emphasized by the significant dip in equatorial regions — attributable to greater cloud cover there and a consequent increase in the downward component of atmospheric emission. The difference between F_{NSg} and F_{NLg} (i.e. the total downward net flux density into the surface) is positive almost all the way to the poles. This is in direct contrast to the total net flux density of the planet (see Fig. 1.10) which becomes negative poleward of about 35°. The surface net energy balance is made up by the upward fluxes $\widetilde{LE}$ and $\widetilde{H}$, also given in Fig. 1.10.

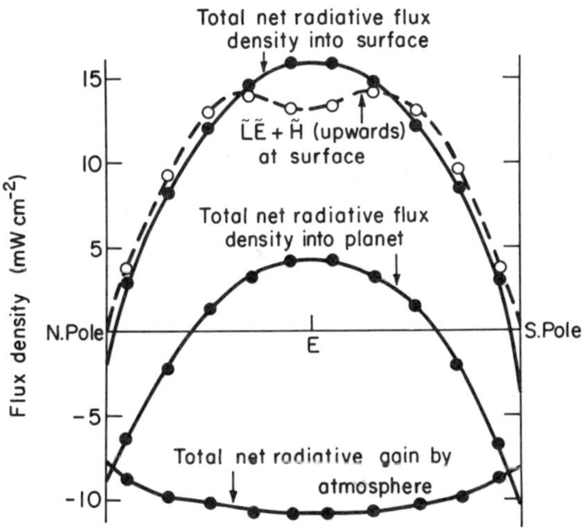

Fig. 1.10. Various net flux densities as a function of latitude.

In summary of Fig. 1.10, the equatorward half of the planet is radiatively warmed and the other half is cooled. The earth's surface is radiatively warmed almost everywhere except the extreme arctic regions. The atmosphere itself is everywhere radiatively cooled by an amount not very dependent on latitude.

There must be meridional energy flows in both the atmosphere and ocean in order to maintain the distribution of vertical flux densities. Such flows, which are consistent with the curves of Figs. 1.8, 1.9 and 1.10, are given later in Fig. 1.19. The separation into atmospheric and oceanic components is discussed in section 8 of the present chapter. The interesting fact which emerges is that the oceanic flux is at a maximum at latitudes where there are no obvious currents to carry the energy. Further, the "baroclinicity" of the ocean is such that a net flow of energy by turbulent eddies is not an obvious mechanism. The fundamental question as to how oceanic energy transport is achieved has yet to be solved.

6. CLOUD AND RADIATION LOSS

Since clouds play a dominating role in determining the radiative properties of the atmosphere, it is instructive at this point to see how clouds affect radiative cooling and warming of the individual layers of atmosphere above and below the middle level of about 500 mb. Such a picture has been given by Cox (1969). In the most general terms, the infrared effect of clouds can be appreciated by ignoring the radiation exchange at wavelengths outside the atmospheric window and imagining them as absorbing layers whose underside is subjected to ground temperature black body radiation only. Thus low-level clouds have little net input from the ground since they are themselves radiating downwards at high temperature not much less than that of the surface. However, their total net loss is large because their upper surface is radiating at high temperature directly to space. The opposite situation occurs for clouds high in the atmosphere, since their low temperature ensures little long-wave loss to space but a large net input from the ground. The following is taken almost verbatim from Cox's paper. "Temperature change" refers to the rate of change of temperature of unit mass of air as a result of radiant energy loss (see also section 7), and, with the atmosphere divided roughly in two, his values are approximately proportional to the total net energy loss from the layer. He deals first with infrared cooling and then with solar heating.

Fig. 1.11 illustrates the effect of cloud on the infrared cooling. Typical temperature and moisture distributions (shown in column *a*) were chosen for the calculation. The different cloud structures were superimposed on those profiles. All clouds were assumed to be black in the sense of being 100% absorbers or emitters of infrared radiation.

Column *b* shows the infrared cooling in a clear atmosphere. The maximum cooling of the lower layer is accomplished by having a cloud low in the layer as in column *c*. The same configuration results in maximum tropospheric mean cooling. The cooling in the lower layer of column *d* represents a mean for the whole layer, thus it contains effects of the radiative warming beneath the cloud and the strong cooling from the cloud top. The net result is moderate cooling of the lower layer. A comparison of columns *b*, *c* and *d* shows that the upper layer is virtually unaffected by the cloud structure in the lower layer. In reality there would be a slight decrease in the cooling of the upper layer as the elevation of the cloud deck beneath is increased.

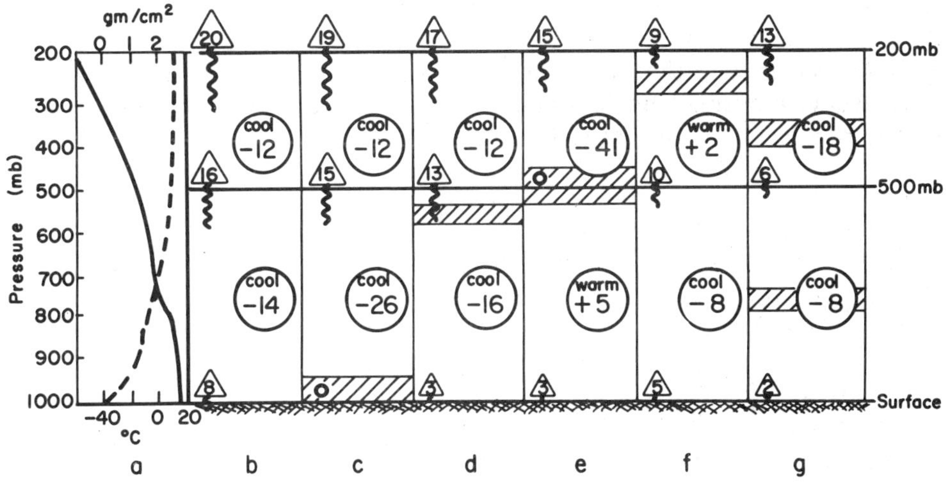

Fig. 1.11. Effect of cloud distribution on infrared cooling. The dashed line in column *a* is cumulative precipitable water ($g\,cm^{-2}$); the solid line is temperature (°C). The arrows represent net radiation ($mW\,cm^{-2}$) and the circles represent the mean infrared radiative temperature change ($°C\,day^{-1} \times 10$) of the layer. Shaded areas depict the presence of clouds which were assumed to be black (emissivity = 1.0). (After Cox, 1969.)

A cloud at 500 mb in column *e* results in the maximum cooling of the upper layer and maximum warming of the lower layer. The extreme cooling of the upper layer is possible because most of the optical mass of water vapour is below 500 mb, and there is only a small downward radiative flux on the cloud from above. The minimum tropospheric cooling is accomplished through the configuration of column *f*. A cloud high in the troposphere acts as a "lid" and does not allow the atmosphere to cool effectively.

The final configuration (column *g*) shows the effect of multilayer clouds. There is relatively small cooling in the lower layer with significant cooling off the top of the upper layer cloud and thus of the upper layer itself.

Water vapour and clouds are the primary variables modulating direct solar absorption. As the optical mass of water vapour increases, so does the solar absorption. High clouds above most of the atmosphere's water vapour tend to decrease solar absorption by reflecting energy back to space before it encounters an appreciable optical mass of water vapour, and low clouds tend to increase the absorption by reflecting solar energy through a longer path length of water vapour and other absorbers.

Both solar and infrared effects are taken into account in Fig. 1.12. The short-wave data are taken from Korb and Möller (1962), and the long-wave data from calculations assuming black clouds. The SW numbers refer to absorbed short-wave energy in the bracketed layer in units of $mW\,cm^{-2}$. The IR numbers refer to net infrared loss of energy in $mW\,cm^{-2}$ from the bracketed layer. The figure shows that despite the decrease of short-wave absorption in the subcloud layer, a marked net warming occurs beneath

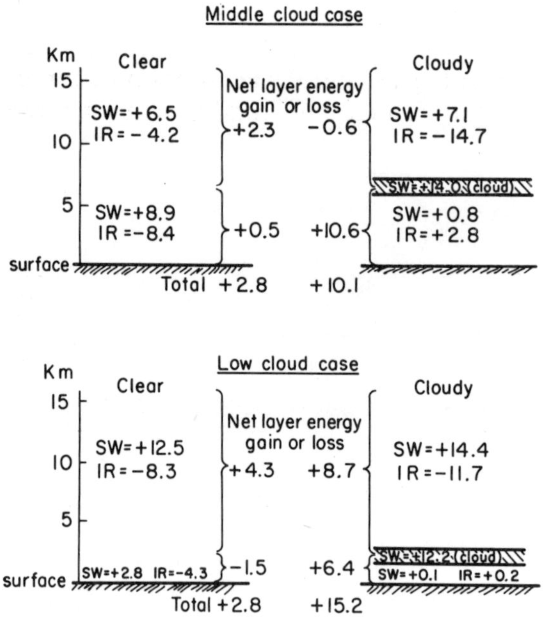

Fig. 1.12. Effects of cloud structure on the vertical distribution of short-wave absorption and total radiative balance (units: $\mathrm{mW\,cm^{-2}}$).

the cloud in both the low and middle cloud cases. This warming is due primarily to the infrared term and may be enhanced by adding the cloud-absorbed short-wave energy to the subcloud layer's net gain.

The most striking feature of Figs. 1.11 and 1.12 is how clouds act to increase the warming of the atmosphere by both short-wave absorption and infrared radiation convergence. In the daytime with high incident solar radiation, low clouds represent the most efficient configuration for tropospheric warming. At night, a high dense cloud represents a more effective warming configuration. Unfortunately there is insufficient data available to say exactly what effects cirrus clouds will have on short-wave warming of the troposphere; however, it is doubtful that the optimum configurations outlined above would change.

7. RADIATIVE COOLING RATES AND GLOBAL DYNAMICS

In the meteorological sense the character of any element of atmosphere is specified completely by its velocity, its temperature T and its pressure p. Temperature and pressure automatically specify density via the equations of state. These five basic parameters (the three velocity components plus T and p) are governed by the five so-called primitive equations — the three kinematic differential equations of motion (one for each velocity component), the equation of mass continuity, and the thermodynamic equation of state. Modelling the motions of the atmosphere involves integration of these equations in time and space and the application of appropriate boundary conditions. Radiation enters the mathematics via the thermodynamic equation:

$$\mathrm{d}Q = c_p \mathrm{d}T - A_\mathrm{m} s_v \mathrm{d}p \qquad [1.8]$$

where $\mathrm{d}Q$ is the total amount of heat received per unit mass of air and is a total differential; s_v is the specific volume and A_m is the reciprocal of the mechanical equivalent of heat. The heat input $\mathrm{d}Q$ is the sum of the inputs resulting from the dissipation of mechanical energy, eddy conduction, phase transition of water, and the radiative heat exchange. Of these, the first two are generally small and the third occurs only at specific times and places within cloud or at the surface. Therefore, referring only to radiative energy input, and putting equation [1.8] in the time derivative form:

$$\mathrm{d}Q/\mathrm{d}t\,|_\mathrm{rad} = c_p \mathrm{d}T/\mathrm{d}t - A_\mathrm{m} s_v \mathrm{d}p/\mathrm{d}t \qquad [1.9]$$

$$= c_p(\partial T/\partial t\,|_\mathrm{rad} + u\partial T/\partial x + v\partial T/\partial y + w\partial T/\partial z) - A_\mathrm{m} s_v \mathrm{d}p/\mathrm{d}t \qquad [1.10]$$

where u, v and w are the three velocity components in the x, y and z directions respectively. Here $\mathrm{d}T/\mathrm{d}t$ is the absolute or Lagrangian derivative appropriate to a (moving?) unit mass of air; $\partial T/\partial t\,|_\mathrm{rad}$ is the local or Eulerian derivative at a given point; and the advective terms $u\partial T/\partial x$, etc., are again locally observed quantities.

Usually therefore, net radiative input is expressed as a heating or cooling rate — for instance, in terms of so many degrees per day. There is, however, room for confusion since the calculated quantity is usually the Lagrangian derivative $\mathrm{d}T/\mathrm{d}t\,|_\mathrm{rad}$ obtained from equation [1.9] on the assumption that the term $A_\mathrm{m} s_v \mathrm{d}p/\mathrm{d}t$ is negligible. Furthermore, it is then often interpreted as the Eulerian parameter $\partial T/\partial t\,|_\mathrm{rad}$.

Since horizontal variations in the atmospheric radiation field are very much smaller than the vertical changes, the net radiant energy input to a unit volume of air can be expressed in terms of the change with height h in the vertical component of net flux F_N. Adopting the "normal" approach, the *cooling rate* CR is thus defined as:

$$CR = (1/c_p\rho)(\partial F_\mathrm{N}/\partial h) \qquad [1.11]$$

$$= (g/c_p\rho)(\partial F_\mathrm{N}/\partial p) \qquad [1.12]$$

which is a convenient means of expressing the magnitude of radiative energy loss but is not physically meaningful for direct insertion in the dynamical equations.

We refer now to Fig. 1.13a and b taken from Dopplick (1972) which give the mean latitude-height contours of total net radiative cooling rate for

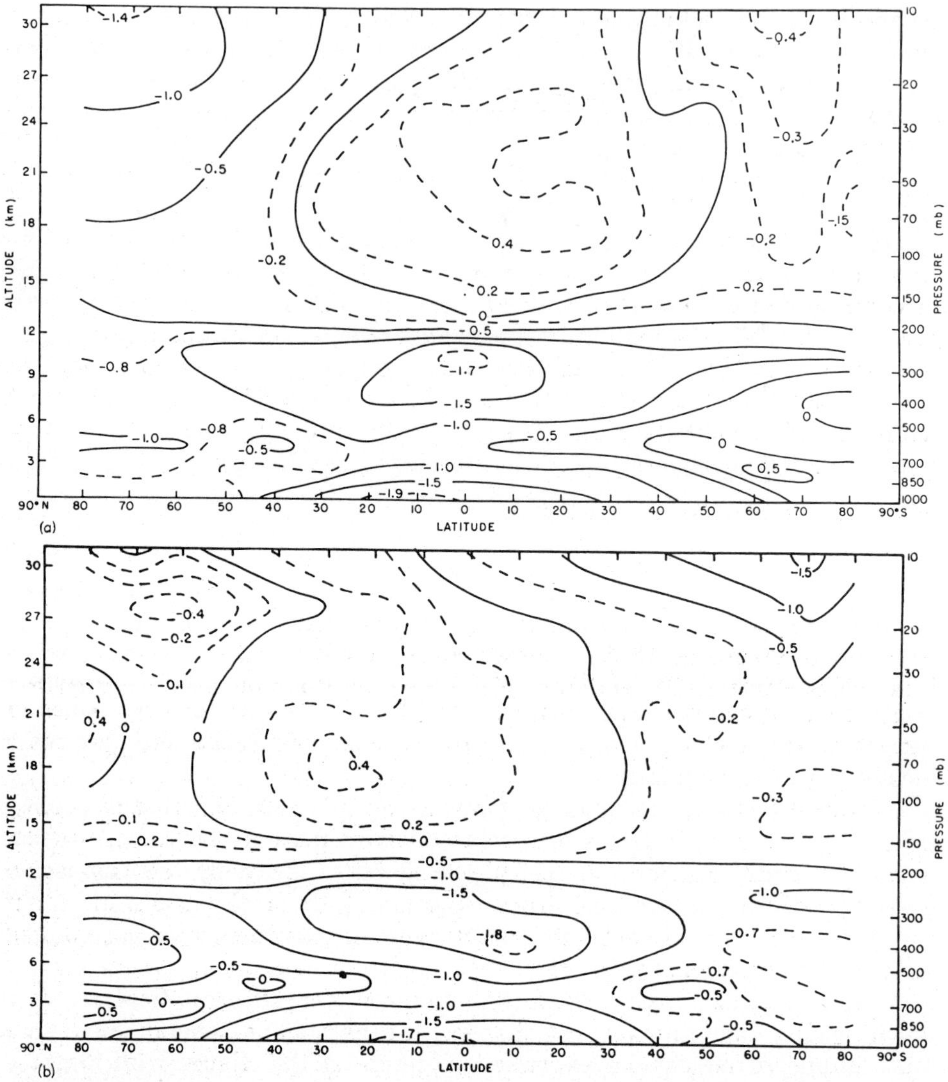

Fig. 1.13. (a) Mean total radiative heating (°C day^{-1}) for December–February. (After Dopplick, 1972.) (b) Mean total radiative heating (°C day^{-1}) for June–August. (After Dopplick, 1972.)

December–February and for June–August respectively. The profiles were calculated theoretically by techniques which will be discussed later, and the information is presented at this time simply to give a broad global picture as a basis for discussion. Fig. 1.13a has been redrawn by us (Fig. 1.14) so as to present the actual energy loss per unit volume. This latter presentation gives a somewhat more obvious picture of the magnitude and position of the energy losses.

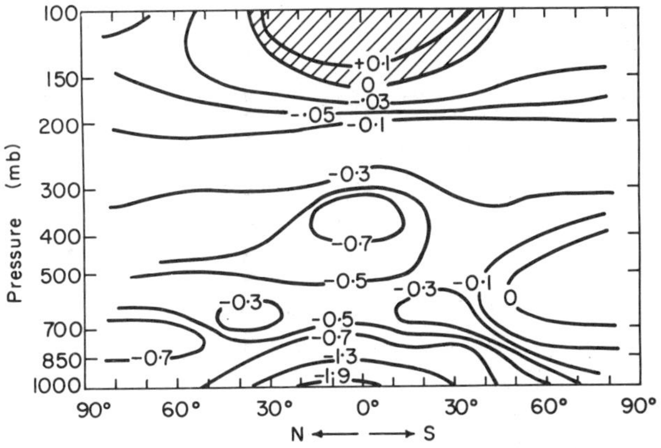

Fig. 1.14. Radiant energy loss per unit *volume* corresponding to Fig. 1.13a (that is, for December–February). Units normalized so that absolute magnitude at the surface equator equals the corresponding cooling rate of Fig. 1.13a.

The most obvious point to make about these diagrams is that the presence of radiation sinks, say, automatically implies that the motions and heat transport mechanisms of the atmosphere must be such as to supply heat to these sinks continually, and that this heat supply is not necessarily related to the mean meridional circulations.

The second obvious point is that although the cooling rates at a point bear some causal relation to the ambient temperature at that point, the reverse proposition is not true. For instance, the tropical troposphere has high cooling rates because of high temperature (and high humidity) and the infrared cooling to space dominates the solar heating. These cooling rates are accentuated to a maximum just below the tropopause by the abundance of middle level cloud which results from the adiabatic expansion occurring in the mean upward motion of the ascending branch of the Hadley cells. Referring to the previous section, such cloud both increases the infrared loss and decreases the solar heating. The tropical stratosphere has high heating rates because of the low temperatures there (relative to the polar stratosphere), the consequently low infrared loss, and the dominance of solar heating. This heating is accentuated with height primarily because of absorption of solar energy by ozone. On the other hand it cannot be said that high temperatures *result* from radiative heating and that low temperatures *result* from radiative cooling because in fact the causation is in the reverse. This fact somewhat belies the earlier suggestion in the introduction that atmospheric dynamics can perhaps be regarded as complementary to the radiation story.

Solar irradiance decreases towards the poles. The associated solar heating decreases and both the stratosphere and troposphere are cooled at all heights. Nevertheless the irradiance in the summer polar regions is sufficient

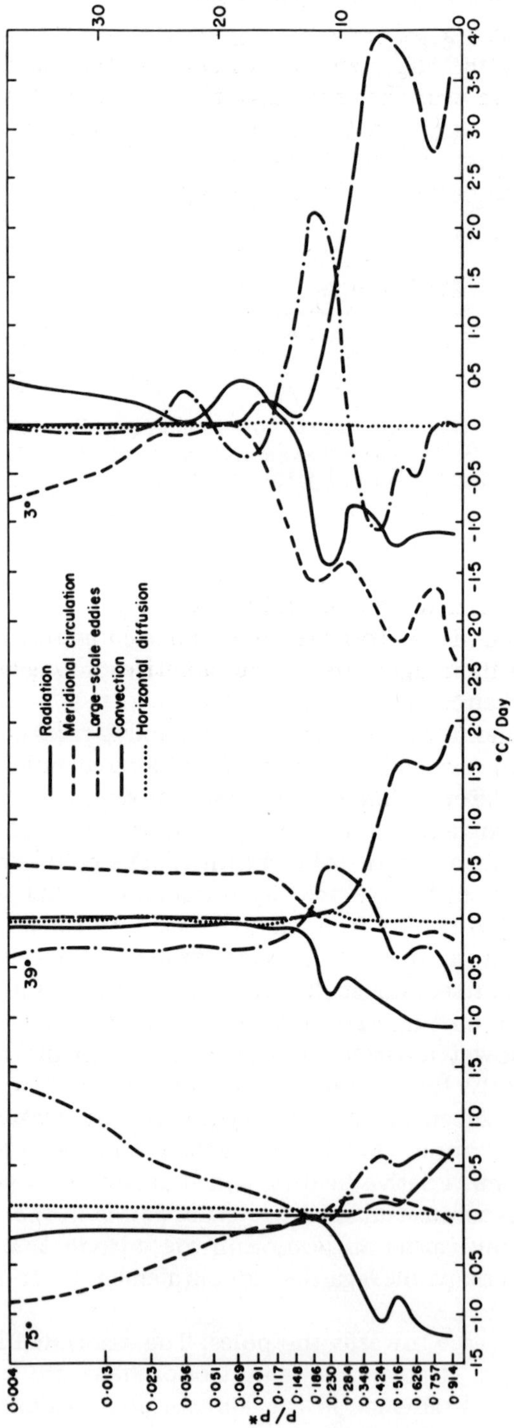

Fig. 1.15. Components contributing to total cooling rate as a function of height at three latitudes (B.G. Hunt, private communication, 1975). The ordinate is pressure normalized with respect to surface pressure.

to reduce the tropospheric cooling to negligible proportions. Correspondingly there is a maximum in infrared cooling towards the winter pole, which again is accentuated by the larger amount of cloud at that place and season.

The high cooling rates near the tropical surface are a function of the high temperature there and are in fact probably underestimated by Dopplick (1972). Recent work (see Chapter 7) has shown that infrared emission by water vapour in the atmospheric window itself is considerably enhanced at high relative humidities and becomes a very significant factor in tropical regions.

The stratosphere is not in radiative energy balance by itself despite the large regions of heating towards the equator. The balance must therefore be made up by heat transport through the tropopause, a process which is not yet well understood. Presumably it is dominated by the upper regions of the Hadley cell which penetrates into the stratosphere. Nevertheless, as will be seen later the presence of a stratosphere can be predicted on purely radiative grounds. In this connection it is interesting that the cooling rate about the 200-mb level is almost constant with latitude — see Fig. 1.12. The actual tropopause height decreases with increasing latitude. This feature and such things as the tropopause gap around 35° must be dynamically determined (Manabe and Hunt, 1968; and see later in Chapter 10).

Fig. 1.15 is from B.G. Hunt (private communication, 1975) and is given here so as to compare the magnitude of the radiative cooling at various latitudes with that of the other balancing dynamical heat transfer processes. The curves are not experimental annual means, but are computed annual means from a large-scale numerical model which incorporates moist convection. They are broadly similar to curves produced earlier by Manabe and Hunt (1968) using a "dry" model. In this diagram, "convection" refers to the heat imported from the ground in the form of latent and sensible heat. It is very large in the tropical troposphere, and the maximum release (primarily of latent heat from the raining clouds of the tropics associated with the Hadley cell) occurs primarily in the upper tropospheric regions to balance the corresponding peak in radiant energy loss. Convection decreases away from the equator as the atmosphere becomes progressively more stable, which in turn can be traced partly to the decrease in radiative cooling in the upper tropospheric levels.

Heat transfer by meridional circulation is significant in the tropical troposphere (again as a result of the Hadley circulation) and is a dominant non-radiative process everywhere in the stratosphere. Meridional heat transfer by large-scale eddies is most significant in the upper troposphere of middle latitudes, and is the major process by which energy is transferred poleward from the Hadley to the Ferrel cells.

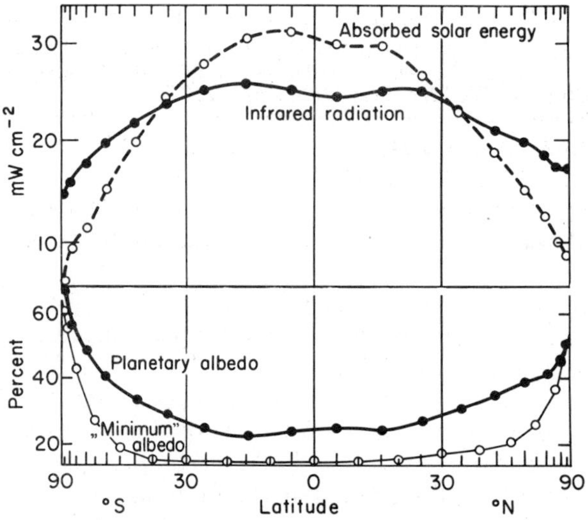

Fig. 1.16. Mean meridional profiles (averages within latitude zones) of components of the planetary radiation budget measured during the period 1962–1966. The abscissa is scaled by the cosine of latitude. (After Vonder Haar and Suomi, 1971.) Minimum albedo from Nimbus 3 data (T.H. Vonder Haar, private communication, 1975).

8. SATELLITE OBSERVATIONS AND LONGITUDINAL ASYMMETRIES

Direct measurements of the components of the planetary energy balance have become available only over the last decade with the advent of satellites. These measurements have confirmed the gross features of earlier theoretical estimates, but have changed the detail in certain significant respects.

Fig. 1.16 gives the experimentally observed mean annual meridional profiles of net short-wave input and of long-wave output from Vonder Haar and Suomi (1971). The associated global and hemispheric means are given in Table 1.4, together with the observed seasonal means.

The most interesting fact to emerge from these measurements is that the net radiation budget of the entire planet, *and of each hemisphere separately*, is in radiative balance, despite the obvious difference in surface albedo of the two hemispheres as a result of the distribution of land masses. The reflectivity of land to solar radiation is much higher than that of ocean. Thus earlier computations suggested that the net input to the southern hemisphere was greater than to the northern, and that as a consequence there must be a significant northward cross-equator flow of atmospheric and oceanic energy to maintain the balance. The satellite observations suggest otherwise, which points to the dominant influence of clouds on the energy exchange with space. Either the amount and distribution of clouds in each hemisphere is different from that observed, or the reflectivity and infrared optical "blackness" or emissivity of clouds is different in each hemisphere.

TABLE 1.4

Mean annual and seasonal radiation budget of the earth-atmosphere system observed from meteorological satellites, 1962–1966 (from Vonder Haar and Suomi, 1971)

	DJF[1]	MAM	JJA	SON	Annual
Global average					
I_0	0.50	0.48	0.47	0.49	0.49
II_a	0.34	0.33	0.34	0.35	0.34
II_r	0.16	0.15	0.13	0.14	0.15
A	31	30	26	28	30
II_L	0.33	0.33	0.34	0.34	0.34
RN_{EA}[2]	0.01	0.00	0.00	0.01	0.007
Northern hemisphere					
I_0	0.34	0.55	0.63	0.42	0.48
II_a	0.24	0.38	0.45	0.31	0.34
II_r	0.10	0.17	0.18	0.11	0.14
A	30	31	28	27	30
II_L	0.32	0.33	0.35	0.34	0.33
RN_{EA}	−0.09	0.05	0.11	0.04	0.006
Southern hemisphere					
I_0	0.67	0.42	0.32	0.56	0.49
II_a	0.46	0.30	0.25	0.40	0.34
II_r	0.21	0.12	0.07	0.16	0.15
A	32	30	23	19	30
II_L	0.34	0.32	0.32	0.34	0.34
RN_{EA}	0.11	−0.03	−0.08	0.06	0.009

I_0 = incident solar radiation (cal cm^{-2} min^{-1}) based on solar constant of 1.95 cal cm^{-2} min^{-1}.
II_a = absorbed solar radiation (cal cm^{-2} min^{-1}).
II_r = reflected solar radiation (cal cm^{-2} min^{-1}).
A = planetary albedo (%).
II_L = emitted infrared radiation (cal cm^{-2} min^{-1}).
RN_{EA} = net radiation budget of the earth-atmosphere system (cal cm^{-2} min^{-1}).

[1] Months of the seasons are indentified by initial letters.
[2] Probable absolute error of ± 0.01 cal cm^{-2} min^{-1}.

Both possibilities are in fact highly probable, bearing in mind that even the simplest parameter of total "cloud amount" for each hemisphere is probably not known to an accuracy of better than ± 10%. The assignment of a fixed mean cloud albedo and emissivity leads to a radiatively defined cloud amount which need bear little relation to observation.

The measurements show that the planet is slightly darker and therefore warmer than expected from earlier computation. Thus the global planetary albedo is apparently just below 30% (Raschke et al. (1973) quote 28%) as compared with one of the latest estimated values of 33% by London and

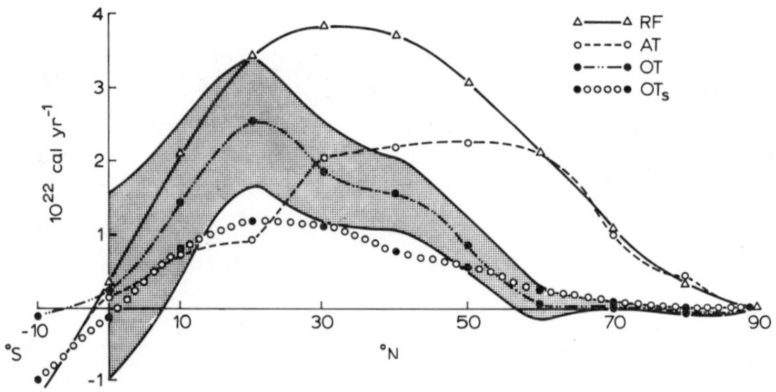

Fig. 1.17. Variation of net energy transport with latitude over the northern hemisphere: RF = total required energy transport inferred from satellite measurements; AT = measured energy transport by the atmosphere; OT = ocean energy transport derived from RF and AT; OT_s = ocean energy transport according to Sellars (1965). Uncertainty in the OT values is denoted by the shading. Minus values indicate net transport to the south. (After Vonder Haar and Oort, 1973.)

Sasamori (1971). The equivalent black-body temperature from equation [1.1] is therefore within about 2° of 254 K. They show also that despite the individual hemispheric net balances, the components of these balances are certainly not symmetrical about the equator. The southern hemisphere subtropics absorb more solar energy than the equivalent northern latitudes, but the southern high latitudes are correspondingly of higher planetary albedo. Cloud amount is again a major factor here, as can be appreciated by comparing in Fig. 1.16 the curve of planetary albedo with that of measured "minimum planetary albedo" (T.H. Vonder Haar, private communication, 1975) which is roughly that of a cloud-free world — i.e. equivalent to $\hat{g}$ in equations [1.2] and [1.3].

The importance of accurate knowledge of the planetary radiative balance lies in the fact that hydrographic data is not yet extensive enough to provide a direct estimate of the meridional energy transport in the oceans. The only method of estimating this flux is as the residual in the subtraction of the measured atmospheric meridional energy flux from the total meridional flux derived from planetary net radiation measurements. Even then the errors are large, as is indicated in Fig. 1.17 by the limits of shading about the ocean transport OT which was derived in this manner by Vonder Haar and Oort (1973). The component errors are indicated in Table 1.5.

Raschke et al. (1973) have provided a global map of the mean annual geographical distribution of surplus and deficit net radiation input to the planet from the zonal average. As a general rule the presence of continents creates a deficit as a result of higher surface albedo. The desert regions of North Africa and Australia are prime examples. Cloudy areas can also be areas of radiation deficit (particularly towards the equator) because of their

TABLE 1.5

Probable error in measurements and estimates of poleward energy transport (units: 10^{22} cal year^{-1}) in the northern hemisphere for the mean annual case (after Vonder Haar and Oort, 1973)

Latitude	E_{RF}	E_{AT}	E_{OT}
90°N	—	—	—
80°N	± 0.02	± 0.12	± 0.12
70°N	± 0.08	± 0.06	± 0.10
60°N	± 0.18	± 0.10	± 0.21
50°N	± 0.32	± 0.16	± 0.34
40°N	± 0.48	± 0.10	± 0.49
30°N	± 0.67	± 0.16	± 0.68
20°N	± 0.88	± 0.08	± 0.88
10°N	± 1.10	± 0.12	± 1.10
Equator	± 1.33	± 0.10	± 1.33
10°S	± 1.56	± 0.24	± 1.58

E_{RF} = error in required transport (i.e. in net radiative input measurements); E_{AT} = error in measured atmospheric transport; E_{OT} = consequent error in oceanic transport (see also shaded area of Fig. 1.17).

higher albedo. However, away from the tropics they can be areas of excess input as a result of their lower radiative temperature.

The point about deserts is not necessarily obvious. From the purely radiative standpoint, a high albedo would suggest low surface temperatures, a reduced long-wave output, and therefore a net radiative input which would not necessarily be in deficit. However, the subtropical deserts are areas of high temperature relative to other places of the same latitude. This comes about because the radiative input to the surface can be balanced by a ground-air sensible heat flux only, since the ground-air latent heat flux is limited by lack of surface water. The surface temperature must be high in order to maintain the sensible heat flow. Further, the low cloud amount of these areas "exposes" the surface so as to allow direct infrared radiation from ground to space. The net planetary radiative input is indeed reduced.

The obvious reason for the existence of deserts is lack of water. In the case of the subtropical deserts the lack of water in turn derives from the low rainfall associated with general atmospheric stability under the influence of the descending branches of the Hadley and Ferrel cells. The feedback via radiation is apparently positive on the grounds that the lesser radiant input to the surface would tend to decrease temperature there and increase the atmospheric stability. On the other hand, the actual higher surface temperatures associated with desert areas should (apparently) introduce an element of negative feedback on the grounds of decreasing the stability, increasing the convective updrafts and thereby increasing the rainfall. The whole affair is extraordinarily complicated, and simple arguments for altering desert climate by artificial manipulation of surface albedo must be examined in great detail and in the light of the circulation patterns of the entire globe.

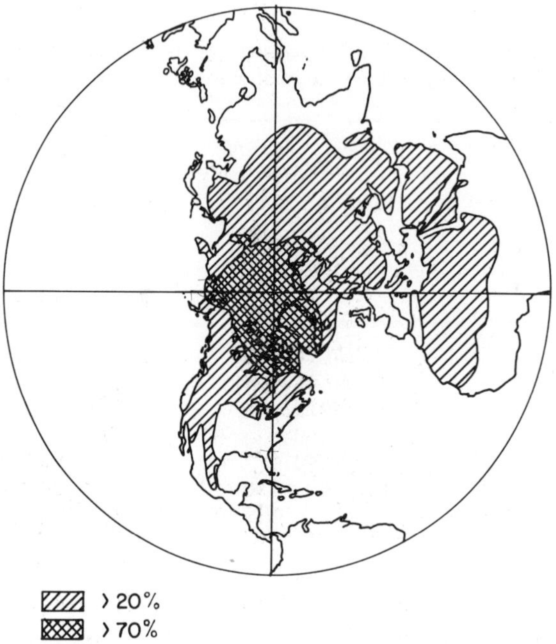

Fig. 1.18. Surface albedo of northern hemisphere in two albedo ranges. (Modified from Posey and Clapp, 1964.)

On the scale of broad meteorology, work on radiation processes has been dominated by consideration of the mean meridional profile. There is a general feeling that radiative processes are usually unimportant on the time scale of individual weather processes or transient synoptic features. This is certainly not true, as various examples in Chapter 10 will show. Further, the advent of the modern three-dimensional models has not altered the general emphasis because such models are still at the stage of incorporating fixed and *zonal-average* cloud amounts. Cloud has a dominant influence on radiation, and the assumption of a fixed zonal-average cloud amount does not allow any definite conclusion about the role of radiation in zonal asymmetries.

There has been much qualitative discussion on the possible feedback between radiation, the geographical distribution of albedo and the persistence of certain weather patterns. The most obvious feature on the global scale is the marked "bi-polar" distribution of surface albedo about the north pole. Fig. 1.18 is a much simplified version of the January surface albedo distribution according to Posey and Clapp (1964). It shows how high surface albedo extends equatorwards in two opposing bulges into Asia and North America. The immediate cause of low temperature in the polar regions is the low solar elevation and solar radiation input, but the belts of high albedo enhance the refrigeration of vast areas in a highly unsymmetrical fashion. This can impress itself on the general circulation in a number of ways. For instance, the east-west discontinuities of albedo indicated in Fig. 1.18 have a significant effect on the maintenance of the Arctic front in winter — especially its persistence into late spring (Sherhag, 1948). The large area of high albedo in Asia is presumably a major factor determining the presence of the Siberian and Alaska-Canadian winter anticyclone. On the time scale of seasons, a large decrease of albedo over extensive areas is brought about in certain subpolar lowlands and on high plateaux and mountains in the middle latitudes. It has been suggested that when one particular sequence of rainstorms in early spring happens to melt most of the winter snow covering a large plateau or mountain range, the effect on the character of the weather in the region during the following summer can be profound. Thus there is evidence (Namias, 1952, 1954; Flohn, 1959) that suddenly reduced surface albedo in the North American high plains and in the Caucasus and Asia Minor mountain complex results in solar energy being consumed in warming the surface and surface layers of the air, enhancing convection and, above all, shifting the major baroclinic zone to the north. This in turn causes the track of subsequent rain-bearing disturbances to be displaced similarly with the result that the region enjoys an early and comparatively settled spring and summer.

There is no counterpart in the southern hemisphere of the bi-polar distribution of surface albedo. However, a feature which has emerged only recently since the advent of satellite cloud observation is the presence of two or three almost permanent cloud bands extending north from the south polar region at fairly specific longitudes. These bands must introduce longitudinal asymmetry in the planetary albedo which is vaguely reminiscent of that in the northern hemisphere. One can legitimately wonder if there is any connection via some requirement for hemispherical symmetry.

If three-dimensional models of the future are to have any hope of simulating synoptic features of relevance to forecasting on a seasonal or climatic time scale, some means will have to be found of parameterizing the synoptic features at a level of detail much less than is presently common. For instance, the time steps will have to be increased to days and weeks rather than the minutes of normal general circulation models. On such a larger time scale the radiation parameterization will become more and more significant. To date the only work in this area has been that of Cox (1969) who examined the radiative heating and cooling patterns associated with certain mid-latitude synoptic features defined in terms of the classic Norwegian model of cyclone development. He examined the stationary front, the nascent cyclone, the warm-sector cyclone and the occluded cyclone. Of these the largest contribution of radiative processes was found in the case of the nascent cyclone where the magnitude of the radiative terms was sufficient to explain the "expected" changes. It should be noted that the vorticity effect was due to horizontal differences in radiative cooling — as opposed to the vertical differences to which most attention is normally paid. Presumably the analysis of data gathered during the GARP Atlantic Tropical Experiment will lead to broad-scale radiation parameterization of major tropical features.

As it happens, it is probably true that much of this sort of parameterization could be done theoretically once the relevant patterns of the normal meteorological variables (especially cloud cover) have been established. Whether it is necessary will depend on whether climate modelling in the future relies on some definite form of pattern recognition and pattern analysis.

9. GLOBAL DYNAMICS AS A PASSIVE VARIABLE

The motions of the atmosphere and ocean obey the basic physical laws such as the conservation of mass, energy and momentum, as embodied in the five primitive equations used as the basis of global modelling. It was pointed out earlier that these laws have not proved sufficient for closure of the problem. It is not yet possible to predict *a priori* either the global mean climate or the global distribution of climate. The modern approach is somewhat analogous to that of predicting the behaviour of a gas by the brute force method of describing the motion of every molecule.

In practice the behaviour of a gas is described by overall equations (e.g. Boyle's law) which take advantage of the statistical properties of a many-body medium and which provide *a priori* prediction of its macroscopic behaviour. The philosophy that such overall governing equations exist is basic to the discipline of physics. It is at least possible that the vastly complicated dynamics of the atmosphere and ocean is governed by some overall "law" which describes the macroscopic behaviour of the system in terms of the external parameters. Since also the significant parameters external to the system are purely radiative energy flows, such a law is likely to be defined in terms of the properties of those flows. The detailed dynamical processes might then be regarded purely as passive variables which adopt values appropriate to satisfying the overall law or boundary condition. Prediction of climate change would not then require direct specification of those dynamics. A number of authors have suggested various overall laws which might be appropriate. Invariably these have been purely qualitative in nature (e.g. Margulis and Lovelock, 1974).

Recently Paltridge (1975) has proposed that the earth-atmosphere system has adopted a configuration such that the entropy exchange between the planet and space is minimized. Entropy E_0 in this context is defined as the macroscopic parameter $(F_{NS} - F_{NL})/T_0$ where $(F_{NS} - F_{NL})$ is the net radiative input to the top of the atmosphere (absorbed short-wave minus the outgoing long-wave) and T_0 is the planetary temperature defined according to the Stefan–Boltzmann law as $T_0 = (F_{NL}/\sigma)^{1/4}$. Thus Paltridge divided the

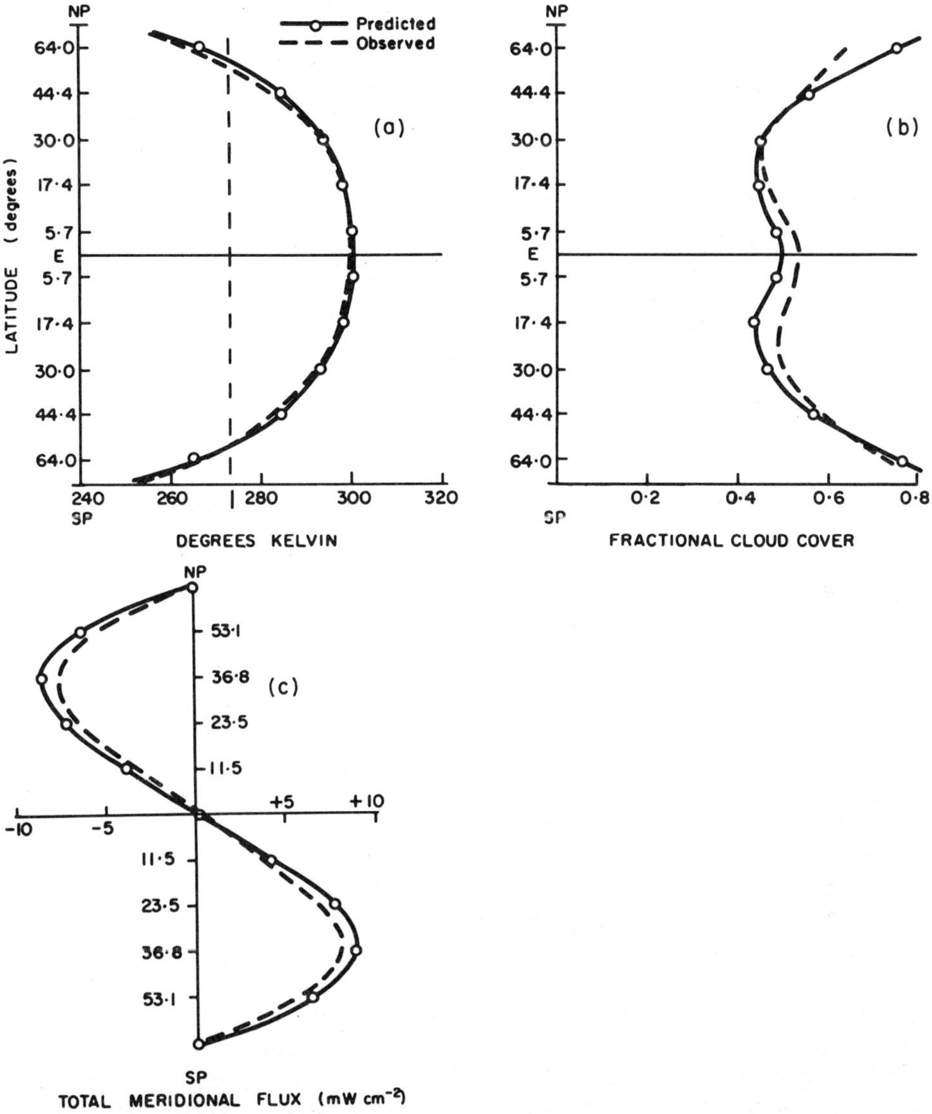

Fig. 1.19. Mean meridional profiles of (a) surface temperature, (b) cloud cover, and (c) meridional energy flux predicted according to a minimum entropy principle, compared with the observed quantities. (After Paltridge, 1975.)

world into ten latitude zones of equal surface area from pole to pole. He devised a model of each zone such that, provided the atmospheric and oceanic meridional flows of energy into a zone were specified, it was capable of calculating the zonal average cloud cover and surface temperature (and hence F_{NS}, F_{NL} and E_0) purely from energy balance considerations. This model was based on the two balance equations at the upper and lower boundaries (equations [1.2] and [1.3] here); and a further equation obtained by splitting the atmospheric balance into two on the basis that, on the broad scale, sensible heat and latent heat are released into and lost by radiation from the lower and upper regions of the atmosphere respectively.

Assembly of an overall model containing the ten individual models of each latitude zone yielded a system which contained nine "unknowns" — the between-zone meridional flows of atmospheric and oceanic energy.

The cloud cover and temperature (and hence E_0) of each zone i is dependent on the meridional flows into and out of that zone. Thus the overall entropy exchange E of the system as a whole ($= \sum_i E_{0i}$) was a function of the nine meridional flows. It was found that there was a unique set of these flows which yielded a minimum in the overall entropy exchange E, and that this unique set had values very close to those observed in the real world. Fig. 1.19 gives the meridional profiles of surface temperature, cloud cover and meridional energy flux corresponding to the minimum in E. They are compared there with the observed values and it can be seen that the correspondence is very good indeed.

This agreement with observation is certainly far from a proof that the overall concept is valid. This would be true even if the model of each individual zone did not require various assumptions which have yet to be justified in detail. There is, however, some further evidence that a minimum principle of this character may be operative. The net radiative input to a zone equals the net meridional energy flux out of that zone, so that entropy exchange with the surroundings (i.e. space) is very close to the net entropy *production* associated with the meridional flow of heat across the boundaries of a given zone. The global total of this entropy production must therefore minimize at the same latitudinal profile of meridional energy flow as that for E. Now it is well known that for ordinary heat conduction in an isotropic medium the entropy production decreases to a minimum at steady state. This is a law which can be proved quite simply. The possible analogy with the present problem is obvious, although the difficulty is that the real atmospheric and oceanic energy flows occur in a medium which is both turbulent and non-isotropic. The problem of partitioning energy input to a turbulent medium between internal and kinetic energy makes it extremely hard to *prove* that the equivalent minimum principle is operative in the real atmosphere.

Thus this particular "law" may or may not prove relevant to the climate problem. Nevertheless it seems intuitively possible that something of the sort will emerge as the final practical answer to the problem of *a priori* prediction of global climate.

CHAPTER 2

Terminology, Basic Laws, and Standards

1. FIELD QUANTITIES

The nomenclature associated with atmospheric radiation has evolved from several disciplines and must be handled with care. Certain terms mean different things to different authors. Even basic quantities such as radiant flux and radiant intensity as used in the general literature are not often the same as the formally endorsed definitions of these terms.

Formal definitions of the various important field quantities are given in Table 2.1. They are abstracted from World Meteorological Organization recommendations (WMO, 1971); and from Kelton et al. (1963) who in turn refer back to the American Standards Association (see also Spiro et al., 1965).

The concepts of radiant emittance and of irradiance are complementary in that they are both measures of the areal density of radiant flux referred to source and receiving surface respectively. The term "flux density" is used in reference to either. They refer to total power emitted or received. Thus the average radiant emittance of a given area of a source is the ratio of total power radiated away from that area, to the area. The *radiant emittance* of a point on the source is the limiting value of the average radiant emittance as the area about the point is reduced in size. Total radiation in this context includes all that which is radiated into the hemisphere (2π sterad) by the area concerned. *Irradiance* is the ratio of the flux of radiation incident on an infinitesimal element of surface containing the point under consideration, to the area of that element. It should be noted, however, that occasions arise when the incident flux arrives from a specific solid angle set by the particular source and in such cases it is obvious (or should be made so) that the concern is not with the full 2π sterad.

Radiant intensity and radiance are measures of the quantity of radiation flowing in a given direction. *Radiant intensity* refers only to a source and furthermore only to a point source. It is the power or flux emitted into an infinitesimal cone containing the given direction divided by the solid angle of the cone. Average radiant intensity is simply the total radiated power divided by 4π sterad. The definition is appropriate only to a point source or to individual points on a source. Thus WMO is careful to refer to the radiant intensity of an extended source as the ratio of radiant flux received at an

TABLE 2.1

Basic field quantities of radiative transfer

Quantity	Definition	Symbol	Unit
Radiation	electromagnetic energy emitted, transferred or received		
Radiant energy	quantity of energy transferred by radiation	U	J
Radiant energy density		$\hat{u}$	$\mathrm{J\,m^{-3}}$
Radiant flux (radiant power)	rate of transfer of radiant energy	$P = \mathrm{d}U/\mathrm{d}t$	W
Radiant emittance (flux density)	radiant flux per unit area emitted from a surface	$F = \{ W = \mathrm{d}P/\mathrm{d}A$	$\mathrm{W\,m^{-2}}$
Irradiance (flux density)	radiant flux per unit area incident upon a surface	$F = \{ H = \mathrm{d}P/\mathrm{d}A$	$\mathrm{W\,m^{-2}}$
Radiant intensity	radiant flux per unit solid angle from a source in a given direction	$J = \mathrm{d}P/\mathrm{d}\omega$	$\mathrm{W\,sr^{-1}}$
Radiance	radiant flux per unit solid angle at a point in a surface per unit projected area of the surface	$N = \mathrm{d}^2P/\mathrm{d}A\cos\theta\,\mathrm{d}\omega$	$\mathrm{W\,m^{-2}\,sr^{-1}}$

elementary surface to the solid angle which this surface subtends at any point of the source, when this ratio is taken to the limit as the distance between the surface and the point is increased.

Radiance refers either to a source or to a point in a receiving surface. Thus the average radiance of an area of a source in a given solid angle is the total flux radiated from that area into that solid angle, divided by the product of the area and solid angle. The limiting value as both the area and solid angle are reduced is the *radiance* at a point of the source in a given direction. Since the actual area of a tortuous surface may be unknown, the area referred to is invariably (although not necessarily) the projected area of the surface on to a plane perpendicular to the direction of measurement. As a received quantity radiance is defined as the ratio of the radiant intensity in the given direction on an infinitesimal element of the surface containing the point under consideration, to the area of the orthogonal projection of this element on a plane perpendicular to the given direction. It should be noted that this definition of radiance breaks down in the case of a parallel beam because of the necessity to divide by a zero solid angle.

It is somewhat confusing that in a very large fraction of the general literature on atmospheric radiation the terms "radiant flux" and "radiant intensity" are used (and sometimes defined) as quantities equivalent to flux density and radiance respectively. The difference is one of areal dimension (L^2) in both cases, with the result that the looseness of terminology is not obvious or generally significant until one has to deal with more detailed problems of radiative transfer involving the various scattering processes.

The important physical relation between irradiance or flux density F and radiance N should be emphasized. Radiation falling on a surface at an angle θ from the normal to the surface gives at the surface an irradiance confined to the solid angle $\mathrm{d}\omega$ equal to $N \cos\theta \, \mathrm{d}\omega$. The total irradiance at the surface is therefore the integral over the half-sphere of which the surface is the diametral plane:

$$F = \int N \cos\theta \, \mathrm{d}\omega \qquad [2.1]$$

In terms of polar co-ordinates θ and ϕ:

$$F = \int_0^{2\pi} \mathrm{d}\phi \int_0^{\pi/2} N \cos\theta \sin\theta \, \mathrm{d}\theta \qquad [2.2]$$

an integral which can be performed only when N is specified for all θ and ϕ. Thus in the special and common case where N is independent of direction — that is, the case of *isotropic radiation*:

$$F = \pi N \qquad [2.3]$$

Finally in this section since it is not immediately obvious, it is worth quoting Lambert's law which states that the radiance emitted by a body is independent of direction provided that the body is a perfect absorber or emitter. Thus to a first approximation the sun appears as a uniformly illuminated disk. The slight darkening towards its edges can be attributed to the imperfect absorbing power of the outer gases in its atmosphere.

1.1. *Monochromatic terminology*

All the above quantities have their counterpart when referred to monochromatic radiation — that is, to radiation confined to a specific and infinitesimal interval of wavelength λ, wavenumber ν or frequency $\tilde{f}$. Such quantities are *spectral* quantities (spectral emittance, spectral radiance, etc.) and are denoted symbolically by an appropriate subscript. For instance the spectral radiance referred to a wavelength is N_λ ($= \mathrm{d}N/\mathrm{d}\lambda$), to a wavenumber is N_ν ($= \mathrm{d}N/\mathrm{d}\nu$) and to a frequency is $N_{\tilde{f}}$ ($= \mathrm{d}N/\mathrm{d}\tilde{f}$). Conversion between the various scales involves the velocity of light c and the relations:

$$\nu = \tilde{f}/c; \qquad \mathrm{d}\nu/\mathrm{d}\tilde{f} = 1/c$$

$$\nu = 1/\lambda; \qquad \mathrm{d}\nu/\mathrm{d}\lambda = -1/\lambda^2 \qquad [2.4]$$

and

$$\tilde{f} = c/\lambda; \qquad \mathrm{d}\tilde{f}/\mathrm{d}\lambda = -c/\lambda^2$$

so that for instance:

$$N_\lambda = (\mathrm{d}N/\mathrm{d}\tilde{f})|\mathrm{d}\tilde{f}/\mathrm{d}\lambda| = N_{\tilde{f}}|\mathrm{d}f/\mathrm{d}\lambda| = cN_{\tilde{f}}/\lambda^2$$

2. ABSORPTION, REFLECTION AND TRANSMISSION

The concept of radiant absorption is unambiguous and generally well understood. It is the process whereby some or all of the radiant energy is transferred to the substance or medium on which it is incident or through which it traverses. Thus the *absorptance* a of a system is simply the fraction of the incident radiation which is absorbed by the system. It cannot be emphasized sufficiently that absorption, unlike the partly analagous processes of reflection or scatter, is always associated with a resultant physical change in the medium. For the atmosphere the significant change is usually of temperature or internal energy. Reflection and scatter have no such effect and the use of "absorption" as an all-embracing term covering the general depletion of radiation in a medium should be discouraged. The word in this case is *attenuation* or *extinction*.

Because it is unambiguous, absorption is used here as the primary illustration of further physical concepts associated with radiative transfer.

When a beam of monochromatic radiation of spectral radiance N_λ is transmitted through an infinitesimal distance ds of an absorbing but non-scattering medium, the spectral absorptance is given by dN_λ/N_λ. If it is assumed that this absorptance is independent of N_λ but is proportional to the density $\rho(s)$ of the absorber and to ds, it follows that:

$$dN_\lambda/N_\lambda = -\kappa_{a\lambda}\rho(s)ds \qquad [2.5]$$

where $\kappa_{a\lambda}$ is thus defined as the spectral absorption coefficient of the medium. Integration along a finite path of length l leads to:

$$N_\lambda = N_{\lambda_0}\exp\left(-\kappa_{a\lambda}\int_0^l \rho(s)ds\right) \qquad [2.6]$$

which is formally known as Beer's law. N_{λ_0} is the initial spectral radiance and $\int_0^l \rho(s)ds$ is the total mass of the absorber per unit cross-sectional area. It has dimensions ML^{-2} {$g\,cm^{-2}$, say} and is defined as the *optical path*. Thus $\kappa_{a\lambda}$ is in fact the *mass absorption coefficient* {$cm^2 g^{-1}$} and should not be confused with the *volume absorption coefficient* $\beta_{a\lambda}$ which is given by $\kappa_{a\lambda}\rho(s)$ and has the dimension L^{-1}. In atmospheric problems it is often the case that both $\kappa_{a\lambda}$ and ρ are functions of position so that equation [2.6] must be modified to:

$$N_\lambda = N_{\lambda_0}\exp\left(-\int_0^l \kappa_{a\lambda}\rho ds\right) \qquad [2.7]$$

The dimensionless quantity $\kappa_{a\lambda}\int_0^l \rho(s)ds$ or $\int_0^l \kappa_{a\lambda}\rho ds$ is called the *optical depth* τ_λ at wavelength λ (in this case referred to pure absorption and its symbol should therefore be $\tau_{a\lambda}$). Optical depth and optical path are often confused in the literature.

The complete exponential of equation [2.6] is the spectral *transmittance* Tr_λ of this particular medium since it was assumed at the outset that no scattering was involved (see later). In summary therefore:

$$N_\lambda/N_{\lambda_0} = Tr_\lambda = 1 - a_\lambda$$

$$= \exp(-\tau_{a\lambda}) = \exp\left(-\int_0^l \beta_{a\lambda}ds\right) = \exp\left(-\kappa_{a\lambda}\int_0^l \rho ds\right) \qquad [2.8]$$

In the case of the atmosphere, because of the very rapid variation of transmittance or of absorption coefficient with wavelength, it is customary for computational reasons to use transmittances which are averaged over some finite band of wavelengths. However, the simple exponential of Beer's law then no longer applies and much of the art and difficulty of manipulating radiative transfer theory centres about resolving this problem.

The reflectance *Re* of a single, specularly reflecting surface and the transmittance *Tr* of a non-scattering medium are well understood as the ratios of reflected or transmitted power to the incident power. Difficulties arise immediately when one is dealing with multiple surfaces, with surfaces which are non-specular reflectors, or with media in which the internal scattering is not easily specified. The problem is to separate in general definition the transmitted, scattered and reflected portions of the radiation; and so in individual cases it is essential to specify the geometry of both the incident power and the receptor. However, it is obvious that in any specific case absorptance can be defined or measured in terms of transmittance and reflectance as:

$$a_\lambda = 1 - (Tr_\lambda + Re_\lambda) \qquad [2.9]$$

or in the case of gaseous media such as the atmosphere which are not usually considered to have reflectance:

$$a_\lambda = 1 - Tr_\lambda \qquad [2.10]$$

3. SCATTERING

The process of scattering figures largely in atmospheric radiation, and the associated terminology is discussed by Kasten and Raschke (1974) among others. We follow here these authors' treatment of the scatter from a volume, although they extend their work so as to define analogous reflection and transmission properties of a surface. This treatment is basic to scattering theory and furthermore provides a useful introduction to the application of the various radiant energy field quantities.

Let a volume element dV {m^3} of a scattering medium be irradiated from a direction specified by zenith angle θ_i and azimuth angle ϕ_i from the solid angle element $d\omega_i$ ($= \sin\theta_i \, d\theta_i \, d\phi_i$) centred around the specified direction ω_i. The radiance $N_i(\omega_i)$ {W m^{-2} sr^{-1}} of the incident radiation produces an irradiance $dH_i(\omega_i) = N_i(\omega_i) d\omega_i$ {W m^{-2}} in dV. Then the radiant intensity $d^2 J_s(\omega_i, \omega_s)$ {W sr^{-1}} of the radiation scattered from dV into a direction specified by $\omega_s \equiv (\theta_s, \phi_s)$ can be described by:

$$d^2 J_s(\omega_i, \omega_s) = \gamma_{sc}(\omega_i, \omega_s) dH_i(\omega_i) dV \qquad [2.11]$$

This equation defines the *volume scattering function* γ_{sc} {m^{-1} sr^{-1}} as a measure of the radiation transferred from the direction ω_i into the direction ω_s. That is, and dropping the formal directional parenthesis:

$$\gamma_{sc} = d^2 J_s / dH_i dV \qquad [2.12]$$

$$= d^2 J_s / N_i d\omega_i dV \qquad [2.13]$$

Note that for parallel incident radiation N_i is undefined so that only equation [2.12] is appropriate. The angle between the directions ω_i and ω_s is given by the expression in Fig. 2.1 and is the *scattering angle* ψ. Integration of the scattering function γ_{sc} over all scattering directions ω_s yields the *volume scattering coefficient* β_{sc} {m^{-1}} as a measure of the total radiation scattered from the single direction ω_i to all directions ω_s. That is:

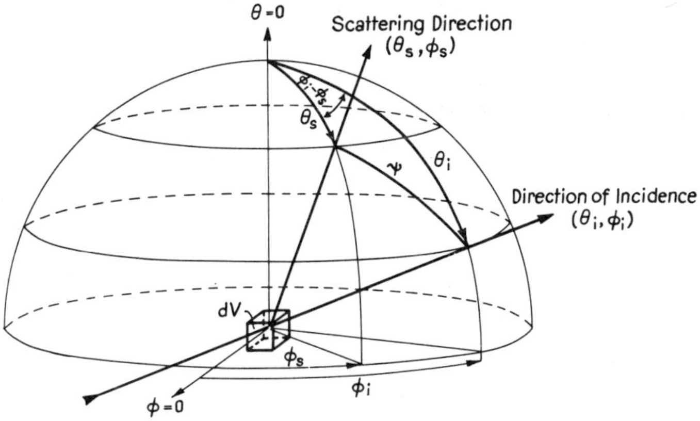

Fig. 2.1. Definition of scattering angle ψ by $\cos\psi = \cos\theta_i \cos\theta_s + \sin\theta_i \sin\theta_s \cos(\phi_i - \phi_s)$. (From Kasten and Raschke, 1974.)

$$\beta_{sc} = \int_{4\pi} \gamma_{sc} \, d\omega_s \qquad [2.14]$$

which, by substitution of equation [2.12] becomes:

$$\beta_{sc} = \frac{\int d^2 J_s \, d\omega_s}{dH_i \, dV} = \frac{d^2 P_s}{dH_i \, dA_i \, dl_i} \qquad [2.15]$$

$$= -\frac{d^2 P_i / dl_i}{dP_i} \qquad [2.16]$$

where dA_i and dl_i are the area element normal to and the path element along the direction of incidence and $d^2 P_s = -d^2 P_i$ is the total radiant flux removed from the incident radiant flux dP_i by scattering in dV along dl_i. This volume scattering coefficient is directly analogous to the volume absorption coefficient discussed earlier in relation to Beer's law. That is, the depletion of radiance along a path in a pure scattering medium is described by:

$$N_\lambda = N_{\lambda_0} \exp\left(-\int_0^l \beta_{sc} \, ds\right) \qquad [2.17]$$

Instead of the scattering function γ_{sc}, the so-called *scattering phase function*:

$$\tilde{P}(\omega_i, \omega_s) = 4\pi\gamma_{sc}(\omega_i, \omega_s)/\beta_{sc} \qquad [2.18]$$

is often used. This is a dimensionless version of the scattering function which is independent of optical path. Referring to equation [2.14] the integral of $\tilde{P}$ over all scattering directions gives the unit sphere 4π, and it can be shown that $\tilde{P}$ is the ratio of the radiance emitted into the direction ω_s to the mean radiance averaged over all directions — i.e. to the radiance of the equivalent isotropic scatter.

Once the scattering function γ_{sc} is defined, it is possible to calculate the radiant intensity per unit volume in the direction ω_s which results from radiation scattered *from* all directions by the volume dV. This intensity is called the *volume emission coefficient in scattering* $\tilde{j}_{sc}(\omega_s)$ {W m^{-3} sr^{-1}} and is given by the integral:

$$\tilde{j}_{sc}(\omega_s) = \int_{4\pi} \gamma_{sc}(\omega_i, \omega_s)\mathrm{d}H_i(\omega_i) \qquad [2.19]$$

$$= \int_{4\pi} \gamma_{sc}(\omega_i, \omega_s)N_i(\omega_i)\mathrm{d}\omega_i \qquad [2.20]$$

This quantity is also equivalent to the radiance per unit path length in the direction ω_s produced by scattering from all directions and is identical with the term "air light" frequently used in atmospheric optics. The actual radiance {W m^{-2} sr^{-1}} in the direction ω_s resulting from radiation scattered from all incident directions is given by the ratio of emission coefficient to scattering coefficient and is called the *scattering source function*:

$$\tilde{J}_{sc}(\omega_s) = \tilde{j}_{sc}/\beta_{sc} \qquad [2.21]$$

$$= \int_{4\pi} (\gamma_{sc}/\beta_{sc})N_i\,\mathrm{d}\omega_i \qquad [2.22]$$

$$= \frac{1}{4\pi}\int_{4\pi} \tilde{P}N_i\,\mathrm{d}\omega_i \qquad [2.23]$$

The relevance of the emission coefficient and source function will become apparent in Chapter 4.

4. THERMAL EMISSION

The derivation and theoretical implications of the Planck radiation distribution function can be found in any of the standard textbooks on thermodynamics or infrared physics and will not be discussed here except to present the basic formulae and the most useful information relevant to direct application. It should be pointed out immediately, however, that the function and its associated laws are derived for situations of thermodynamic equilibrium, which by formal definition concern the state of matter and radiation inside a constant temperature enclosure. Hence the terms *enclosure* or *black-body* radiation. Apart from the fact that the Planck function is therefore derived most fundamentally in terms of energy *density* $\hat{u}$, the implications of thermodynamic equilibrium are far reaching. As will be seen later the condition does not necessarily apply in the upper levels of the atmosphere.

The basic Planck function gives the radiant energy per unit volume per unit frequency interval of unpolarized black-body radiation. As a consequence of thermodynamic equilibrium it is a function of temperature T only:

$$\hat{u}_{\tilde{f}} = 8\pi\hat{h}\tilde{f}^3/\{c^3[\exp(\hat{h}\tilde{f}/\hat{k}T) - 1]\} \qquad [2.24]$$

where $\hat{h}$ and $\hat{k}$ are the Planck and Boltzmann constants respectively. For plane polarized radiation the formula is divided by 2. The equivalent radiance distributions can be derived using [2.4] and are:

$$\begin{aligned} N_{\tilde{f}} &= 2\hat{h}\tilde{f}^3 c^{-2}[\exp(\hat{h}\tilde{f}/\hat{k}T) - 1]^{-1} \\ &= 7.652 \times 10^{-39}\tilde{f}^3[\exp(4.799 \times 10^{-11}\tilde{f}/T) - 1]^{-1} \\ N_\lambda &= 2\hat{h}c^2\lambda^{-5}[\exp(c\hat{h}/\hat{k}\lambda T) - 1]^{-1} \\ &= 2.294 \times 10^{-28}\lambda^{-5}[\exp(1.438/\lambda T) - 1]^{-1} \qquad [2.25] \end{aligned}$$

and:

$$\begin{aligned} N_\nu &= 2\hat{h}c^2\nu^3[\exp(c\hat{h}\nu/\hat{k}T) - 1]^{-1} \\ &= 2.294 \times 10^{-28}\nu^3[\exp(1.438\nu/T) - 1]^{-1} \end{aligned}$$

where the numerical expositions yield radiance in units of $\mathrm{W\,cm^{-2}\,sr^{-1}}$ per unit spectral interval. The distributions are often quoted in terms of the so-called first and second constants of the Planck function, namely, $c_1 = 2\pi c^2\hat{h}$ and $c_2 = \hat{h}c/\hat{k}$ respectively. While these are the endorsed versions of c_1 and c_2 it should be noted that c_1 is occasionally defined otherwise (e.g. $8\pi\hat{h}c$, $\pi\hat{h}c^2$ or $2\hat{h}c^2/\Omega_0$). The equations of [2.25] are presented graphically in Figs. 2.2 and 2.3 for a number of temperatures.

Integration of N_λ over all wavelengths and angles gives the total radiant emittance of or flux density F from a black body as:

$$F = \sigma T^4 \qquad [2.26]$$

where $\sigma\ (= 2\pi^4\hat{k}^4/15c^2\hat{h}^3)$ is the *Stefan-Boltzmann constant.* This constant is of fundamental importance to radiative transfer theory. The latest and most authoritative experimental estimate of its value is $(5.6644 \pm 0.0075) \times 10^{-12}\ \mathrm{W\,cm^{-2}\,K^{-4}}$ (Blevin and Brown, 1971) which is consistent with the theoretical value $(5.6696 \pm 0.0025) \times 10^{-12}\ \mathrm{W\,cm^{-2}\,K^{-4}}$ calculated from the present best values of c, $\hat{h}$ and $\hat{k}$.

Differentiation of N_λ with respect to wavelength allows the maximum of the Planck function to be determined. It occurs at the point defined by:

$$T\lambda_m = 0.28978 \qquad \{\mathrm{cm\,K}\} \qquad [2.27]$$

which is a linear relation between temperature and the wavelength of maximum emission λ_m known as Wein's displacement law (refer again to Figs. 2.2 and 2.3). The spectral radiance at the wavelength of maximum emission is given by:

$$N_{\lambda m} = 4.0947 \times 10^{-16}T^5 \qquad \{\mathrm{W\,cm^{-2}\,\mu m^{-1}\,sr^{-1}}\} \qquad [2.28]$$

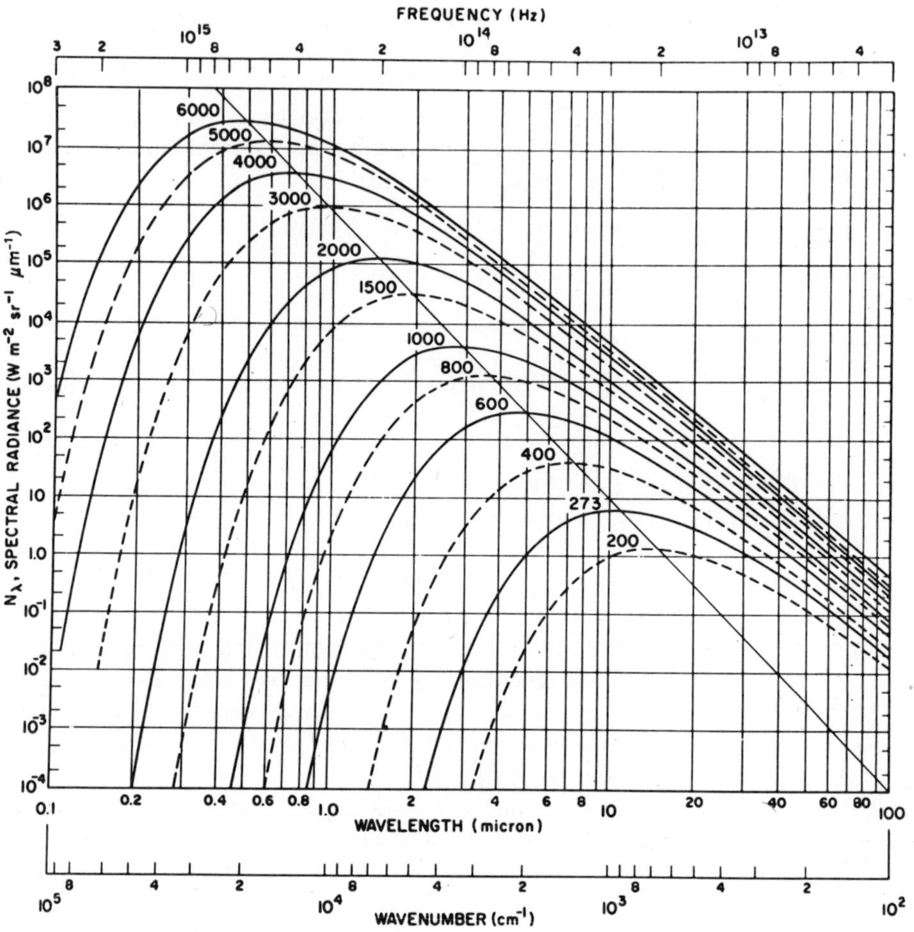

Fig. 2.2. Spectral radiance of a black body, N_λ, at the temperature in K shown on each curve; beyond 100 µm the curves continue to be linear (plotted on logarithmic scales). The diagonal line, intersecting the curves at their maxima, shows Wein's displacement law. (Subdivisions of the ordinate scale are at 2 and 5.) Figure by P.R. Gast. (From AFCRL, 1967, by permission of McGraw-Hill.)

To the uninitiated it is usually somewhat surprising that λ_m and $N_{\lambda m}$ are not the same wavelength and radiance as those corresponding to the maximum in the wavenumber or frequency scale versions of the Planck function. Differentiation in either of these cases leads to:

$$T/\nu_m = 0.5097 \qquad \{\text{cm K}\} \qquad [2.29]$$

with the corresponding radiance at the wavenumber of maximum emission given by:

$$N_{\nu m} = 5.6824 \times 10^{-13} T^3 \qquad \{\text{W cm}^{-1}\,\text{sr}^{-1}\} \qquad [2.30]$$

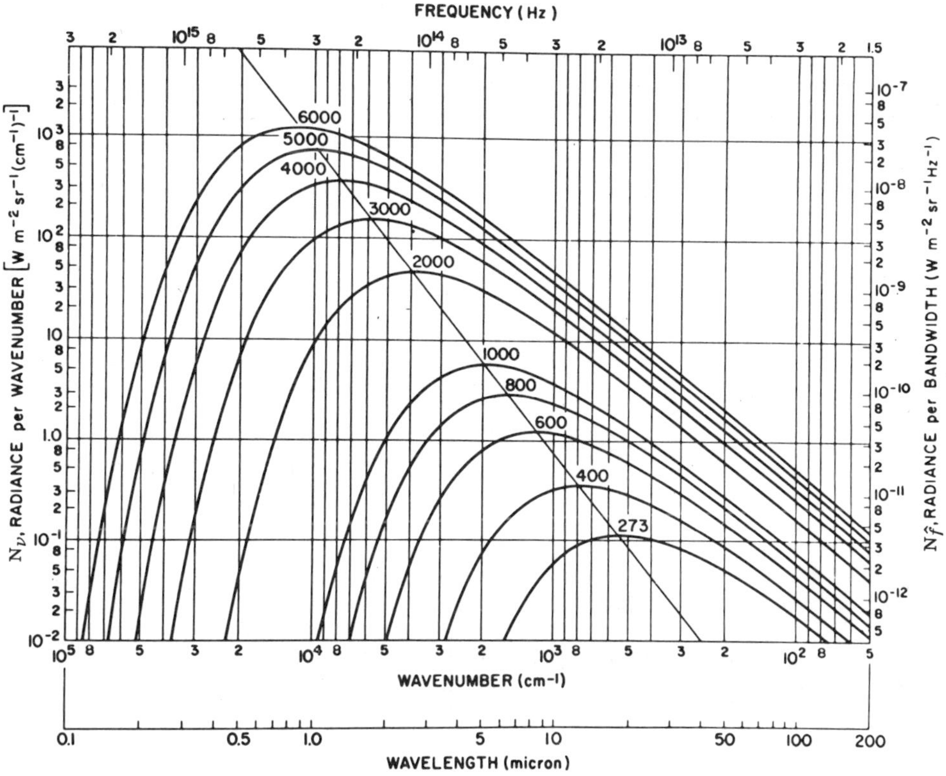

Fig. 2.3. Radiance per reciprocal centimetre, N_ν, and radiance per unit frequency, $N_{\hat{f}}$, for a black body at various temperatures; at frequencies less than 3×10^{12} Hz, wavenumber $< 100\ \mathrm{cm}^{-1}$, the curves continue to be linear (plotted on logarithmic scales). The diagonal line, intersecting the curves at their maxima, shows Wein's displacement law. Figure by P.R. Gast. (From AFCRL, 1967, by permission of McGraw-Hill.)

Comparing the two expressions for the Wein displacement law, $1/\nu_m$ is approximately 1.759 times the corresponding λ_m. At the average terrestrial surface temperature of 300 K, $\lambda_m \approx 9.65\ \mu$m as compared to $1/\nu_m \approx 16\ \mu$m. At solar temperatures (6,000 K), $\lambda_m \approx 0.48\ \mu$m as compared to $1/\nu_m \approx 0.84\ \mu$m.

The difficulties which arise with different scales can be circumvented by various normalization procedures when one is interested in the Planck function *per se* (e.g. Bramson, 1968). In atmospheric physics it is more useful to have to hand the various dimensionless quantities in the last four columns of Table 2.2. They are given there as a function of λT after Goody (1964). The quantity $\int_0^\lambda N_\lambda \mathrm{d}\lambda / \sigma T^4$ — the fraction of black body radiance within the spectral interval 0 to λ — is particularly useful and is plotted graphically in Fig. 2.4. The fraction within any spectral interval can be found by subtraction.

TABLE 2.2

Planck function parameters (from Goody, 1964)
$c_1 = (3.7403 \pm 0.0002) \times 10^{-12}\,\mathrm{W\,cm^{-2}}$; $c_2 = 1.43868 \pm 0.00001\,\mathrm{cm\,K}$

λT (cm K)	$C_2/\lambda T = x$	$\int_0^{\lambda} B_\lambda \mathrm{d}\lambda/B$	$B_\lambda/(B_\lambda)_{max}$	$\lambda B_\lambda/(\lambda B_\lambda)_{max}$	$B_f/(B_\nu)_{max}$
	large x	$\frac{x^3 e^{-x}}{6.4939}$	$\frac{x^5 e^{-x}}{21.201}$	$\frac{x^4 e^{-x}}{4.780}$	$\frac{x^3 e^{-x}}{1.4214}$
0.01	143.85	0.0^{56} 16	0.0^{53} 98	0.0^{54} 30	0.0^{56} 70
0.02	71.925	0.0^{26} 35	0.0^{23} 52	0.0^{24} 33	0.0^{25} 15
0.03	47.950	0.0^{16} 27	0.0^{13} 18	0.0^{14} 17	0.0^{15} 12
0.04	35.962	0.0^{11} 19	0.0^{9} 683	0.0^{10} 84	0.0^{11} 79
0.05	28.770	0.0^{8} 130	0.0^{6} 297	0.0^{7} 458	0.0^{8} 536
0.06	23.975	0.0^{7} 932	0.0^{4} 145	0.0^{5} 267	0.0^{6} 375
0.07	20.550	0.0^{5} 186	0.0^{3} 206	0.0^{4} 444	0.0^{5} 726
0.08	17.981	0.0^{4} 165	0.001 38	0.0^{3} 339	0.0^{4} 635
0.09	15.983	0.0^{4} 872	0.005 63	0.001 56	0.0^{3} 329
0.10	14.385	0.0^{3} 321	0.016 44	0.005 07	0.001 18
0.12	11.988	0.002 14	0.072 64	0.026 87	0.007 54
0.14	10.275	0.007 80	0.186 27	0.080 41	0.026 31
0.16	8.9906	0.019 74	0.345 18	0.170 31	0.063 69
0.18	7.9917	0.039 38	0.520 26	0.288 77	0.121 58
0.20	7.1925	0.066 79	0.683 44	0.421 44	0.196 59
0.25	5.7540	0.161 45	0.946 12	0.729 36	0.426 22
0.30	4.7950	0.273 36	0.997 12	0.922 41	0.646 87
0.35	4.1100	0.383 06	0.922 71	0.995 74	0.814 74
0.40	3.5962	0.481 02	0.800 10	0.986 74	0.922 72
0.45	3.1967	0.564 44	0.671 36	0.931 61	0.979 96
0.50	2.8770	0.633 86	0.554 67	0.855 19	0.999 52
0.60	2.3975	0.737 89	0.373 78	0.691 55	0.969 93
0.70	2.0550	0.808 12	0.253 94	0.548 15	0.896 89
0.80	1.7981	0.856 31	0.175 98	0.434 08	0.811 77
0.90	1.5983	0.890 04	0.124 71	0.346 13	0.728 20
1.00	1.4385	0.914 18	0.090 37	0.278 70	0.651 50
1.20	1.1988	0.945 06	0.050 40	0.186 52	0.523 30
1.40	1.0275	0.962 86	0.030 10	0.129 97	0.425 36
1.60	0.8991	0.973 77	0.019 02	0.093 79	0.350 87
1.80	0.7992	0.980 81	0.012 57	0.089 74	0.293 53
2.0	0.7192	0.985 56	0.008 62	0.053 18	0.248 59
3.0	0.4795	0.995 24	0.001 94	0.017 98	0.126 06
4.0	0.3596	0.997 90	0.0^{3} 655	0.008 08	0.075 60
5.0	0.2877	0.998 90	0.0^{3} 279	0.004 30	0.050 25

TABLE 2.2 (*continued*)

λT (cm K)	$C_2/\lambda T = x$	$\int_0^\lambda B_\lambda d\lambda/B$	$B_\lambda/(B_\lambda)_{max}$	$\lambda B_\lambda/(\lambda B_\lambda)_{max}$	$B_f/(B_\nu)_{max}$
10	0.1438	0.9^3 848	0.0^4 188	0.0^3 579	0.013 54
20	0.0719	0.9^4 801	0.0^5 122	0.0^4 751	0.003 51
50	0.0288	0.9^5 878	0.0^7 318	0.0^5 491	0.0^3 574
100	0.0144	0.9^6 848	0.0^8 200	0.0^6 618	0.0^3 145
	small x	$1-0.05134x^3$	$0.04717x^4$	$0.2092x^3$	$0.7035x^2$

The normalizing factors for use with the table are:
$\pi B = 5.6698 \times 10^{-12} T^4$ W cm^{-2},
$\pi(B_\lambda)_{max} = 1.287 \times 10^{-11} T^5$ W cm^{-3} $= 1.287 \times 10^{-15} T^5$ W cm^{-2} μm^{-1},
$\pi(B_f)_{max} = 5.956 \times 10^{-23} T^3$ W cm^{-2} Hz^{-1},
$\pi(\lambda B_\lambda)_{max} = 4.1736 \times 10^{-12} T^4$ W cm^{-2}.

5. METEOROLOGICAL RADIATION QUANTITIES

Fig. 2.4 illustrates an important practical point in atmospheric physics in that only 0.4% of the total extra-terrestrial solar radiant energy resides in wavelengths longer than 5 μm. On the other hand, only 0.4% of the total infrared radiation from a 250-K black body (cf. the radiative temperature of the earth as seen from outside the atmosphere) resides in wavelengths shorter than 5 μm. Therefore the solar and terrestrial radiation streams can be treated

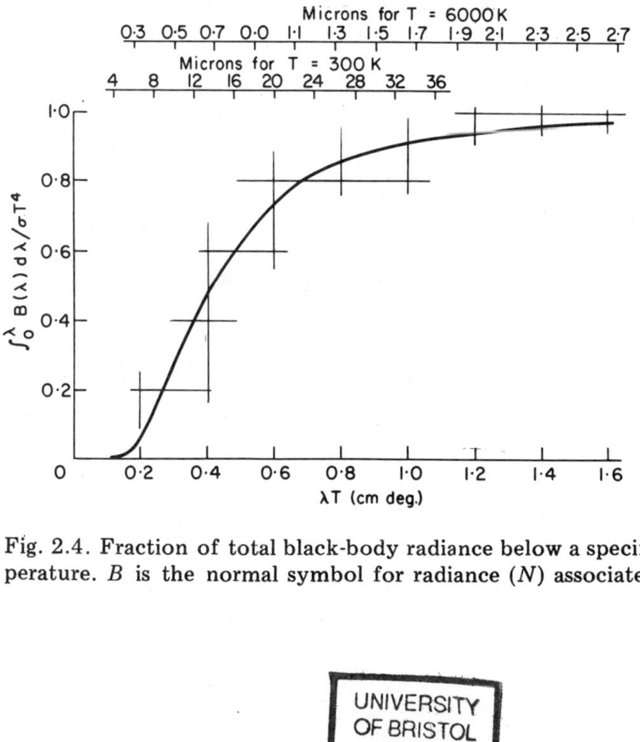

Fig. 2.4. Fraction of total black-body radiance below a specific wavelength for given temperature. *B* is the normal symbol for radiance (*N*) associated with the Planck function.

TABLE 2.3

Fluxes within the atmosphere

Term	Description
Short-wave radiation (solar radiation)	radiation of wavelengths $<$ 2–5 μm
Long-wave radiation (terrestrial radiation)	radiation of wavelengths $>$ 2.5 μm
Direct solar radiation	solar radiation coming from the solid angle of the sun's disc on a surface perpendicular to the axis of the solid angle
Diffuse solar radiation (sky radiation)	downward solar radiation as received on a horizontal surface from a solid angle of 2π with the exception of the solid angle subtended by the sun's disc
Global solar radiation	downward direct and diffuse radiation as received on a horizontal surface from a solid angle of 2π
Vertical component of direct solar radiation	direct radiation as received by a horizontal surface
Atmospheric radiation	radiation emitted by the atmosphere (effectively long-wave radiation) — hence also downward and upward atmospheric radiation; and upward terrestrial radiation which includes both atmospheric and surface radiation
Net radiation	net of the upward and downward components of both short- and long-wave radiation — hence also net solar and net terrestrial radiation

independently for most practical purposes. At the earth's surface the effect of atmospheric absorption is such that the transition point is closer to 2 μm, which is very convenient from the instrumentation viewpoint since this wavelength corresponds almost exactly with the transmission cutoff of glass. Thus glass is a cheap filter for separating short- and long-wave fluxes.

Table 2.3 is a summary of various terms describing the overall fluxes within the atmosphere. It is a much shortened version of the WMO recommendations.

6. STANDARDS

Experimental measurements of radiant flux are relatively inaccurate. Thus the "solar constant" is known to an absolute accuracy only of about ± 2%. It is generally admitted that field measurements of solar radiation at the ground to an accuracy of ± 5% are the result of good and careful work. Laboratory calibration of pyranometers and pyrradiometers for general use is seldom better than ± 2% even in the relative sense. Measurement of spectral irradiance and spectral flux density is even more difficult. This tends to make a mockery of much of the experimental work on atmospheric turbidity, which depends on monitoring the small percentage differences between observed solar spectral irradiance and its theoretical value for a dust-free atmosphere. The problem can be appreciated by referring to the difficulty of defining a basic standard or scale of radiant flux.

In principle an absolute radiation scale can be defined in terms of fundamental physical constants by employing either a standard emitter or a standard detector of radiant flux.

Standard emitters are conventionally high-temperature cavity radiators (black-body sources) and are chiefly employed in calibrating low-level irradiance measurements — that is, of long-wave radiation. In principle the total radiant emittance and spectral distribution of the cavities can be predicted from the Stefan-Boltzmann law and from Planck's law provided their temperature and emissivity are known. The difficulty until very recently has been that even the Stefan-Boltzmann constant has not been "known" to better than 2% in the sense that the various experimental determinations over the years 1921–1968 differed by that amount. Furthermore these values were always higher by up to 2.2% than the theoretical value calculated from best known values of c, $\hat{h}$ and $\hat{k}$. Recent work by Blevin and Brown (1971) has at least achieved agreement between the experimental and theoretical values of σ to within experimental error of about 1%, but the prodigious effort required to achieve this result is indicative of the problems involved in ordinary calibration of standard emitters.

Standard detectors are more suitable for measuring the high irradiance levels associated with solar flux. They are invariably calorimeters in which the heating effect of an unknown flux is compared with that of a heating element (usually electrical) whose energy input is known. The history of radiation standards defined in this way (see, for instance, Latimer, 1973, and Ångström, 1958) is a graphic illustration of the difficulties involved.

At the International Meteorological Conference at Innsbrück in 1905, the pyrheliometer of Ångström (1899) was adopted as a standard instrument for the measurement of solar irradiance. The Ångström scale (ÅS 1905) was defined at that time in terms of a number of these instruments, some of which were (and still are) used for the transference of the scale to others. Some years later the water-stirred pyrheliometer constructed by C. Abbot was used to form the basis of the Smithsonian revised scale of pyrheliometry of 1913 (SS 1913), which is usually transferred via Smithsonian silver disc pyrheliometers.

Comparison of the two scales showed almost immediately that the ÅS 1905 was roughly 3.5% lower than the SS 1913. Subsequent work showed that the SS 1913 as originally defined was in fact about 2.5% high. It was realized also that the instruments which formed the basis of the ÅS 1905 were subject to so-called "marginal error" or "edge effect" which in effect put the ÅS 1905 too low (approximately 2.4% low according to Lindholm, 1958).

Thus at Davos in 1956, the International Radiation Commission recommended a new scale which is a compromise based on the best available knowledge at the time of the difference between the ÅS 1905 and the SS 1913. It is known as the International Pyrheliometric Scale of 1956 (IPS 1956) and is still extant today. In effect, by IPS definition, measurements made on the SS 1913 were decreased by 2.0% (definition A) and those made on the ÅS 1905 were increased by 1.5% (definition B). Theoretically this should have brought the two original scales into agreement. However, subsequent comparisons of instruments calibrated according to the corrected (i.e. IPS definition) Ångström and Smithsonian scales still show a fairly consistent difference of the order of 1%. The situation is displayed graphically on the left of the scale of Fig. 2.5 which is taken from Latimer's (1973) paper. The arrows on the right refer to certain specific instruments and their position on the scale as determined at the time of the two subsequent international comparisons in 1964 and 1970. The spread is some measure of the natural variability

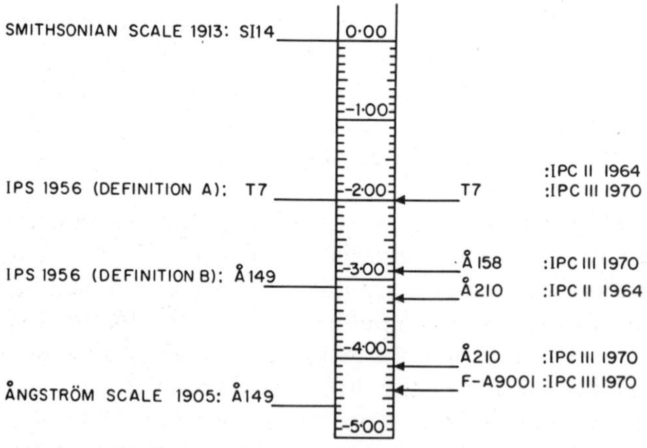

Fig. 2.5. The relationship of the Smithsonian scale of 1913 and the Ångström scale of 1905 to the IPS 1956, derived from comparisons of various pyrheliometers at Davos and Toronto. All positions on the scale are determined with respect to the position of silver disc pyrheliometer SI14. IPC stands for "international pyrheliometric comparison". (After Latimer, 1973.)

between different instruments and between the same instruments at different times.

The recent development of the so-called Absolute Cavity Radiometer (ACR) (e.g. Kendall and Berdahl, 1970; Willson, 1972) allows the definition of an Absolute Radiation Scale (ARS) which, however, has yet to be formally accepted. These radiometers are said to have an absolute accuracy of ± 0.5 mW cm^{-2} at normal solar flux densities of order 100 mW cm^{-2} and the percentage uncertainty decreases with decreasing irradiance. At the last international comparison (Davos, 1970), the two ACR's available gave results which agreed within 0.1% (Latimer, 1972). They gave recorded irradiance levels approximately 1.9% higher than that of the mean of the seven Ångström pyrheliometers chosen at that time to represent the IPS (Marchgraber and Frölich, 1972). It is now becoming highly likely that the IPS is indeed too low by about 2%. It would explain, for instance, why the calibration facility of the CSIRO Division of Atmospheric Physics in Australia has been plagued for some years by an inability to reconcile the short- and long-wave calibrations of commercially made net pyrradiometers.

CHAPTER 3

The Sun

1. RADIANT OUTPUT

The irradiance of solar energy as received on a surface exposed normal to the sun's rays at the mean sun-earth distance (1.496×10^{11} m) and in the absence of the atmosphere is known somewhat hopefully as the *solar constant* I_0. Its quoted value has varied by 7% since 1940. Johnson's (1954) value of $139.5\,\mathrm{mW\,cm^{-2}}$ ($2.00\,\mathrm{cal\,cm^{-2}\,min^{-1}}$) was considered the most authoritative until recently when Thekaekara and Drummond (1971) proposed a standard value based on the weighted mean of data from high-altitude balloons and aircraft, and from spacecraft. Their value of $135.3 \pm 2.0\,\mathrm{mW\,cm^{-2}}$ should be accepted until the present difficulty with the IPS 1956 — to which it is referred — is resolved (see Chapter 2). Direct application of the Stefan–Boltzmann equation yields an average solar radiative temperature of 5762 K.

The associated curve of solar spectral irradiance as a function of wavelength has also "varied" over the years. The most up-to-date version is again that of Thekaekara and Drummond (1971), reproduced here in Table 3.1. They estimate a possible error of ± 5% in the individual values (see NASA, 1971) between 0.3 and 3.0 μm with a greater uncertainty at wavelengths outside this band. Direct application of the Wein displacement law to the somewhat ill-defined wavelength of maximum emission ($\lambda_m \approx 0.47\,\mu$m) leads to a computed solar temperature of 6166 K.

Solar emission occurs over the entire electromagnetic spectrum from X-ray to radiowavelengths. Fig. 3.1 is taken from the above NASA publication and compares the observed or calculated irradiance with that of black bodies at various temperatures radiating according to the Planck distribution. Assignation of a single temperature to the sun must be handled with care.

The intensity of radiation from the sun varies across the disc from a maximum at the centre to a minimum at its edge. The darkening is a consequence of the temperature gradient in the outer layers. The intensity at the edge is less than that at the centre because of the outward decreasing temperature and because radiation from a semi-transparent atmosphere originates, in the mean, at the same optical depth along the line of sight. Over the years solar physicists have amassed a large amount of data on "limb darkening" as a

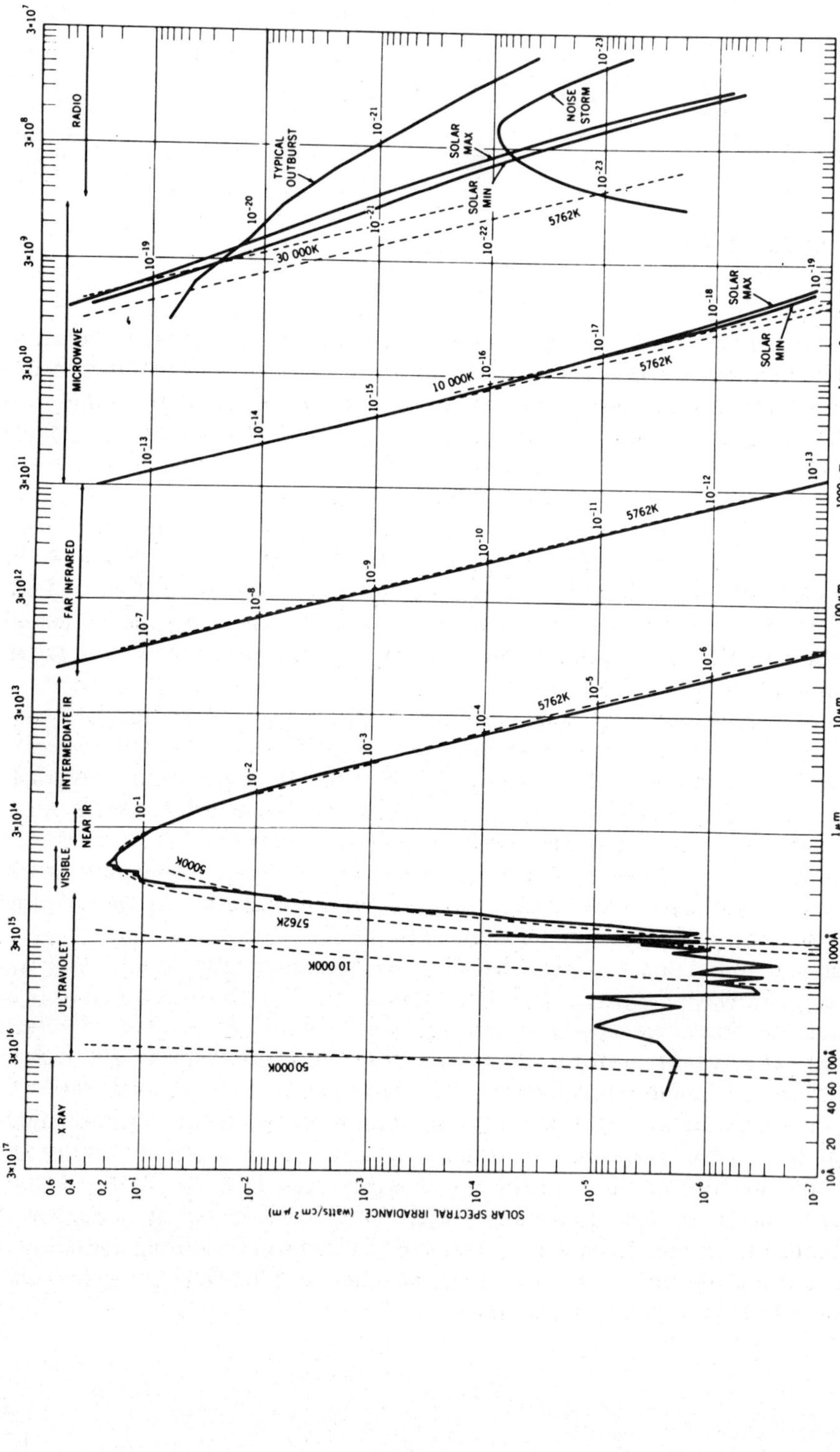

Fig. 3.1. The solar electromagnetic spectrum. (From NASA, 1971.)

TABLE 3.1

Proposed standard of extra-terrestrial solar irradiance (solar constant $I_0 = 135.3$ mW cm^{-2}) (from Thekaekara and Drummond, 1971)

$\lambda(\mu m)$	P_λ	D_λ	$\lambda(\mu m)$	P_λ	D_λ	$\lambda(\mu m)$	P_λ	D_λ
0.12	0.0000	0.00	0.43	0.1639	12.47	0.90	0.0889	63.36
0.14	0.0000	0.00	0.44	0.1810	13.73	1.00	0.0746	69.46
0.16	0.0000	0.00	0.45	0.2006	15.14	0.20	0.0484	78.39
0.18	0.0001	0.00	0.46	0.2066	16.65	0.40	0.0336	84.34
0.20	0.0011	0.01	0.47	0.2033	18.17	0.60	0.0244	88.61
0.22	0.0057	0.05	0.48	0.2074	19.68	0.80	0.0159	91.59
0.23	0.0067	0.10	0.49	0.1950	21.15	2.00	0.0103	93.49
0.24	0.0063	0.14	0.50	0.1942	22.60	0.20	0.0079	94.83
0.25	0.0070	0.19	0.51	0.1882	24.01	0.40	0.0064	95.89
0.26	0.0130	0.27	0.52	0.1833	25.38	0.60	0.0048	96.67
0.27	0.0232	0.41	0.53	0.1842	26.74	0.80	0.0039	97.31
0.28	0.0222	0.56	0.54	0.1783	28.08	3.00	0.0031	97.83
0.29	0.0482	0.81	0.55	0.1725	29.38	0.20	0.0023	98.22
0.30	0.0514	1.21	0.56	0.1695	30.65	0.40	0.0017	98.50
0.31	0.0689	1.65	0.57	0.1712	31.91	0.60	0.0013	98.72
0.32	0.0830	2.22	0.58	0.1715	33.18	0.80	0.0011	98.91
0.33	0.1059	2.93	0.59	0.1700	34.44	4.00	0.0009	99.06
0.34	0.1074	3.72	0.60	0.1666	35.68	4.50	0.0006	99.34
0.35	0.1093	4.52	0.62	0.1602	38.10	5.00	0.0004	99.51
0.36	0.1068	5.32	0.64	0.1544	40.42	6.00	0.0002	99.72
0.37	0.1181	6.15	0.66	0.1486	42.66	7.00	0.0001	99.82
0.38	0.1120	7.00	0.68	0.1427	44.81	8.00	0.0001	99.88
0.39	0.1098	7.82	0.70	0.1369	46.88	10.00	0.0000	99.94
0.40	0.1429	8.73	0.72	0.1314	48.86	15.00	0.0000	99.98
0.41	0.1751	9.92	0.75	0.1235	51.69	20.00	0.0000	99.99
0.42	0.1747	11.22	0.80	0.1107	56.02	50.00	0.0000	100.00

P_λ = solar spectral irradiance averaged over small bandwidth centred at λ, in W cm^{-2} μm^{-1};
D_λ = percentage of the solar constant associated with wavelengths shorter than λ.

function of wavelength. The early work is summarized by Goldberg and Pierce (1959). According to the Ångströms (1971) who analyzed many years of data from the Smithsonian Astrophysical Observatory, the variation of intensity in the meteorologically significant wavelength range (0.3 μm and above) is of the form:

$$J_\lambda = J_{0\lambda}(1 - 0.237d^{2.4}/\lambda) \qquad [3.1]$$

where J_λ is the spectral intensity at the distance d (normalized so that the solar radius = 1) from the centre of the disc, $J_{0\lambda}$ is the spectral intensity at the centre and λ is the wavelength expressed in microns. The form of this function is such that at a normalized radius $d = 0.736$ the spectral distribution of the solar intensity is very similar to that of the total solar radiation. Again according to the Smithsonian measurements, the total intensity integrated over all wavelengths varies as:

$$J = J_0(1 - 0.342d^{2.4}) \quad [3.2]$$

so that the mean solar intensity is only 0.845 of the intensity at the centre of the disc.

Whether the total radiant energy output of the sun varies significantly with time is still an open question. It is well known that the extreme ends of the spectrum are highly variable with sunspot number or other measures of solar activity such as the Baur solar index. The extreme short-wave end of the solar spectrum influences the photochemistry of ozone in the stratosphere among other things, and it is at least possible that solar variation may influence the dynamics of the atmosphere indirectly via this sort of mechanism. Thus there are many reports in the literature of apparent correlations between solar activity and various indices of planetary climate. Recent examples include Wilcox et al. (1974), Stringfellow (1974) and Berson and Kulkarni (1968). Assuming that such correlations exist, it is doubtful whether any of them can be attributed directly to variations in total energy output of the sun. The occasional reference to a direct correlation between measured total solar radiation and the 11-year cycle in sunspot activity must be treated with skepticism in view of the general radiation measurement inaccuracy and to the variability of the atmosphere between sun and instrument. This is true even of the high-altitude balloon measurements reported by Kondratyev and Nikolsky (1970) and Kondratyev et al. (1971) which, taken at face value, suggest a variation of solar output over part of the last sunspot cycle amounting to 2% with the maximum occurring at medium Wolf numbers of about 80 to 100.

Certainly the analysis of *visible* radiation measurements of individual sunspots does not support anything like such a variability. According to Smith and Gottlieb (1974) an extremely large sunspot may have an umbral area of 6×10^{-4} of the solar disc. The intensity may be as low as 0.1 of the normal photospheric intensity. Therefore one such sunspot might reduce the total solar flux by something less than 0.1%. When there are many spots on the surface, the largest total area is of the order of 4×10^{-3}. If one takes the extreme assumption that this whole area is umbra at a tenth of photospheric intensity, one still obtains a diminution of only 0.4% of total solar flux. In fact only about one-sixth of the sunspot area is umbra and the rest is penumbra of relative intensity about 0.7. Again one arrives at a solar flux reduction of at most 0.1%. Further, brightening of the "plage" region near large sunspot groups would tend to make up for part of this deficiency.

The problem as to whether the solar "constant" varies by an amount anywhere up to the present measurement accuracy of a few percent is of fundamental importance to climatology. Solar variations at the extreme ends of the spectrum exist with a variety of periods and have been discussed and summarized by Smith and Gottlieb (1974). Solar flares have lifetimes of the order of minutes. The "slowly varying component" encompasses changes in radio noise over hours or days and is due to the development and disappearance of active regions. Closely related to this is the 27-day period which occurs as a result of the reappearance of active regions as the sun rotates. Finally the 11-year cycle already mentioned reflects the correlation between all solar activity and the sunspot cycle. As it happens the sunspot indices are not necessarily a very good indication of general solar activity, and the attempts to correlate weather phenomena with such indices have arisen mainly because records are available extending back several hundred years.

Two fairly standard astronomical techniques have been used in the search for possible time variations in the solar output. The first makes use of the fact that the sun is a very normal example of the type DG-2 star. Observation of other DG-2 stars and comparing them with standard stellar sources over periods of up to 20 years has failed to reveal any variations in excess of experimental accuracy of about 1%. The second involves monitoring of the visible sunlight reflected from the planets. Of the several fairly short-term monitoring programs over the years (Goldberg and Pierce, 1959), none has yielded variations exceeding the instrumental accuracy of less than 1%.

2. LONG-TERM VARIATIONS

It seems that the one constant factor associated with the motion of the earth about the sun is the period of its revolution. Geological evidence (Zeuner, 1959) indicates that the length of the year has not altered significantly over the last 3 or 4 hundred million years.

The earth revolves around the sun in an ellipse with the sun at one of the foci. The present-day eccentricity of the orbit is small, but significant to the extent that the square of the sun-earth distance R^* (and therefore the solar irradiance at the earth) varies by $\pm$ 3.3% from the mean. To an accuracy better than 10^{-4}:

$$(\bar{R}^*/R^*)^2 = 1.000110 + 0.034221 \cos\theta_0 + 0.001280 \sin\theta_0 + 0.000719 \cos 2\theta_0 + 0.000077 \sin 2\theta_0 \qquad [3.3]$$

where θ_0 is defined in terms of days of the year in equation [3.8]. Thus the point on the orbit closest to the sun (*perihelion*) occurs about January 3 and the complementary point (*aphelion*) occurs about July 5. Other considerations aside, one might therefore expect the southern hemisphere annual surface temperature variation to be greater than the northern, but in fact the reverse is the case (refer to Fig. 1.1 in Chapter 1).

The obliquity of the ecliptic (presently $23\frac{1}{2}°$) is the fundamental cause of the seasons as well as one of the factors modifying the climatic zones. It is easy enough to imagine that, if the obliquity were to increase, the annual mean equator-pole temperature gradient would decrease but the seasonal differences would increase. Indeed, if the obliquity reached as much as 54°, the annual total of solar radiation impinging on the top of the atmosphere would be the same at all latitudes and, other considerations aside, one might therefore expect no meridional variation of annual mean temperature.

Another characteristic of the sun-earth system is the gradual precession of the equinoxes and solstices along the orbit. The cause is to be found in the oblateness of the earth in connection with the obliquity of the ecliptic. The thickened equatorial belt is attracted more intensely than the polar regions which tends to turn the earth's axis upright. Since the earth rotates, its axis in fact precesses in a conical fashion. This has the effect of shifting the cardinal points in the orbit with a period of 26,000 years. However, the orbit as a whole swings round slowly in the opposite direction to the precession and the complete circle of the cardinal points requires only 21,000 years (in the mean).

All three characteristics — orbit eccentricity, orbit obliquity and the precession of the equinoxes — are subject to long-term variability as a result of the attraction by other bodies in the solar system. This variability in turn may affect the value of solar radiation received at a given latitude. The literature on the problem is dominated by the calculations of M. Milankovitch whose orbital calculations have been updated by Van Woerkom (1953). Fig. 3.2 is from the latter's paper, which shows the calculated variations of the three characteristics over the past million years. This time scale is about as far back as one can go using non-relativistic mechanics and in the reasonable certainty that continental drift would not have altered the picture significantly. It can be seen that both eccentricity and obliquity also have a cyclic variation with rough mean periods of 97,000 and 41,000 years respectively.

Broadly, the following points should be appreciated.

(1) Variations in any of the orbital elements do not affect the yearly total of radiation incident on the world as a whole. That is, they do not affect the mean solar constant. It is not immediately obvious that this is true for variations in eccentricity. Nevertheless, it can be shown by integration of the equation for an ellipse that the yearly mean value of $(\bar{R}^*/R^*)^2$, where R^* is the sun-earth distance, is independent of eccentricity provided the period of the system is constant.

(2) Variation in the orbital parameters does not affect the equal partition of incident radiation between the two hemispheres.

(3) The yearly total radiation at any given place is changed only by variation in orbit obliquity. Change poleward of 43° latitude is balanced by an equal and opposite change equatorward of 43°.

(4) Variations in eccentricity and the position of the equinoxes affect only the partition of the yearly insolation between "summer" and "winter". Here these terms refer to the equal half years defined such that each day in summer receives more radiation than any day in winter.

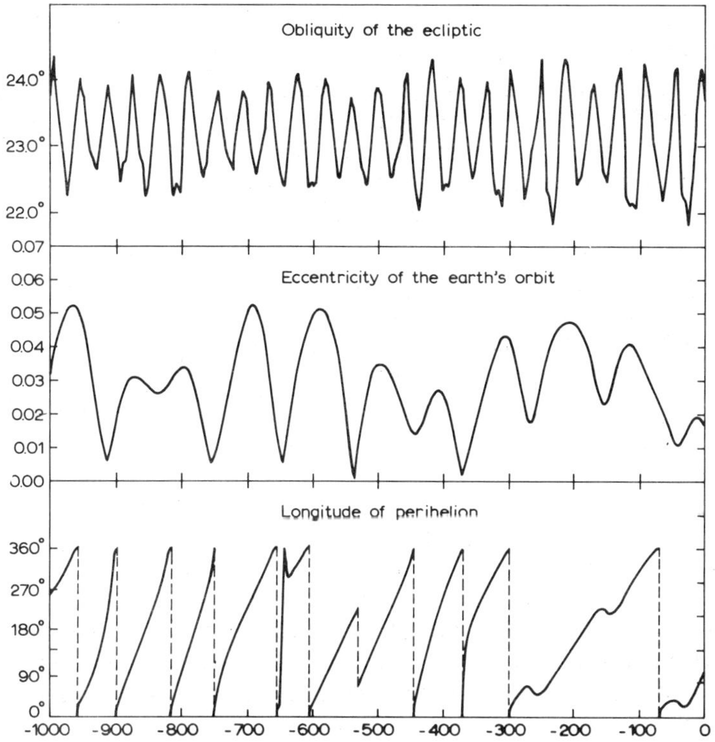

Fig. 3.2. Computed values of the obliquity of the ecliptic, eccentricity and the longitude of the perihelion of the earth's orbit. Abscissa, time before 1950; unit, 1,000 years. (From Van Woerkom, 1953.)

Thus it is quite irrelevant to consider global or hemispheric annual mean "climates" in the context of long-term orbital changes. The effects, if they exist, must be on the *distribution* of climate and must depend heavily on the non-linearities of the system. For instance, if the extent of polar ice is to depend on orbit eccentricity, this extent must be a non-linear function of insolation.

The possibility that orbital changes may produce a significant climate change has been examined by many authors. The work has centred primarily on attempts to explain the advance and retreat of polar ice over the comparatively short period covered by Milankovitch's calculations. However, the arguments are more or less academic since, fundamentally, it is not yet possible to relate a change in insolation to a corresponding change in surface temperature. Thus the curves of Fig. 3.2 can be used to derive the solar insolation variation with time over the past million years at any given latitude. Many such calculations were done by Milankovitch. At 65°N latitude for instance (the approximate present limit of the polar ice cap) the annual mean insolation has oscillated up and down in an irregular manner by about

0.1%. The mean summer radiation has oscillated by about 0.5%. However, Milankovitch's proposition that the variability is sufficient to explain certain changes of the extent of polar ice is questionable to say the least — as was pointed out and discussed in detail by Simpson (1940).

Intuitively, it is more likely that, if the advance and retreat of polar ice is to be ascribed to an extra-terrestrial factor, this factor would be the variability of the solar output itself. On the time scale of the Milankovitch calculations no theroretical reason has as yet appeared to postulate sizeable variations of solar output. However, over time scales of several hundred million years there are grounds for believing that considerable time and periodic changes do occur. Thus Dilke and Gough (1972) have proposed a model which attempts to explain the fact that the solar neutrino flux at the earth is well below that expected from the thermo-nuclear reactions occurring in the sun's core. Their model envisages a periodic mixing within the core which temporarily (for several million years) depresses the neutrino flux and simultaneously depresses the luminosity of the sun's surface by up to 5%. The theory suggests that the sun is presently in the recovery phase of one of these periods and that the minimum in luminosity was located near the beginning of the last major ice age about 3 million years ago.

Finally, Smith and Gottlieb (1974) refer to various stellar models which indicate that the sun has been brightening over the past 5 billion years and will continue to do so over the next 4 billion years at a rate of about 1% every 50 million years.

3. POSITION TERMINOLOGY AND FORMULAE

In order to calculate solar radiation at different times it is necessary to specify the position of the sun in the sky. This is done by defining coordinate systems on the celestial sphere which is an imaginary transparent sphere surrounding the earth and concentric with it (Fig. 3.3). On the celestial sphere are celestial poles which are the points at which the earth's polar axis, when produced, cuts the celestial sphere. They are thus situated directly over the earth's geographical poles. In the same manner, if the plane of the equator is produced outwards, its intersection with the celestial sphere will form the equinoctial or celestial equator. It is convenient to consider that the sun moves round the earth on a path which makes an angle of approximately $23\frac{1}{2}°$ to the plane of the earth's equator. This imaginary path of the sun is called the ecliptic, and the angle between the plane of the ecliptic and the equatorial plane is the obliquity of the ecliptic.

At any given time an observer on the earth's surface has a corresponding position on the celestial sphere called the observer's zenith which is the point of intersection with the celestial sphere of a normal to the earth's surface at the observer's position. The great circle on the celestial sphere, the plane of

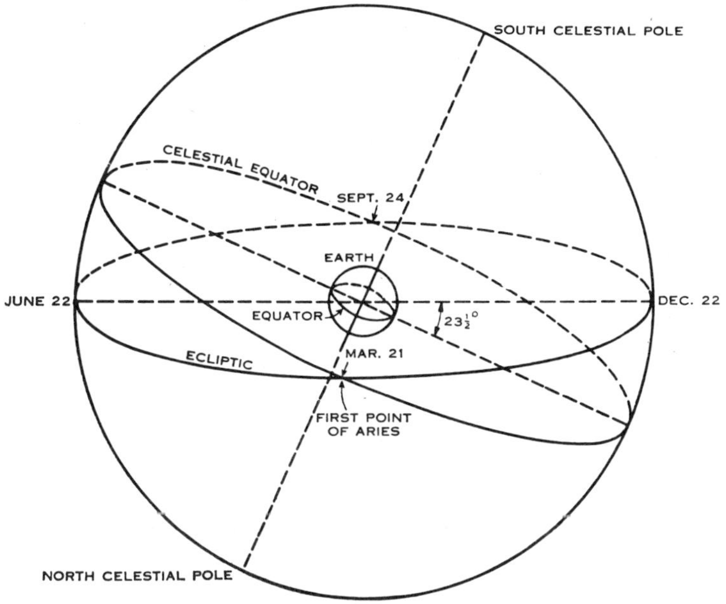

Fig. 3.3. The celestial sphere.

which passes through the centre of the earth normal to the line joining the centre of the earth and the zenith, is known as the observer's celestial horizon. The point at which an imaginary line from a celestial body to the centre of the earth cuts the celestial sphere gives the celestial position of the body.

Parallels of latitude and meridians of longitude on the earth's surface can be produced to form celestial parallels and meridians on the celestial sphere, with reference to which the position of a celestial body can be defined in an analogous way to that used in defining the position of a point on the earth's surface. Terrestrial latitude is measured in degrees north or south from the equator; the celestial equivalent is measured in degrees north or south from the celestial equator and is called the *declination* δ of the body. Latitudes north of the equator are conventionally considered positive and latitudes south of the equator are considered negative; the same convention is used in expression of the declinations. From Fig. 3.3 it can be seen that the solar declination varies from $+23\frac{1}{2}°$ on June 22 to 0° at the equinoxes (March 21 and September 24) and to $-23\frac{1}{2}°$ on December 22. Terrestrial longitude is measured in degrees east or west of the Greenwich meridian, which is the recognized datum line. The analogous celestial measurement is the *right ascension* and the datum line for it is the meridian of the first point of Aries which is the point of intersection of the celestial equator and the ecliptic where the sun's declination changes from south to north. The right ascension of a body is measured eastwards from this meridian and is expressed either as an angle (0–360°) or in time (0–24 hours). Thus the position of a body in space may be defined by its declination and right ascension, this specification being to a good approximation unaffected by the observer's position on the earth.

In solar work and in navigation, it is usually convenient to have another system of defining the position of celestial bodies instead of declination and right ascension. Unlike them, it relates only to the instantaneous position of the observer and specifies altitude, zenith distance, azimuth and hour angle. The altitude γ of a body is its angular height above the observer's celestial horizon and is thus an angle between 0° and 90°. The *zenith distance* θ is the angular depression of the body from the observer's zenith, and is thus the complement of the altitude. The *true bearing* or azimuth ϕ of a celestial body is the angle between the plane of the observer's meridian and the plane of the great circle passing through the zenith and the body measured clockwise from the observer's meridian. It may also be thought of as the angle measured horizontally from north towards east to a point vertically under the celestial body. The azimuth thus assumes values between 0° and 360°. The *local hour angle* or simply *hour angle* t_h of a body is the angle at the celestial pole between the observer's meridian and the meridian of the body measured westwards from the observer's meridian. Thus the sun's hour angle counts from the midday position (since the sun is then on the observer's meridian) and changes 360° per day or 15° per hour.

The most common problem in radiation meteorology is to calculate the zenith distance θ and the azimuth ϕ of the sun for a given geographical position, time of year and time of day. The geometry of the situation leads to the following relations where Φ is the geographical latitude:

$$\sin(90^\circ - \theta) = \sin\gamma = \sin\delta\sin\Phi + \cos\delta\cos\Phi\cos t_h \qquad [3.4]$$

$$\cos\phi = (\sin\delta\cos\Phi - \cos\delta\sin\Phi\cos t_h)/\cos\gamma \qquad [3.5]$$

and:

$$\sin\phi = -\cos\delta\sin t_h/\cos\gamma \qquad [3.6]$$

which are universally applicable provided one adheres to the sign convention.

Since the sun's hour angle t_h counts from midday, solar position for a given place obtained by using equations [3.4] to [3.6] will be symmetrical about solar noon in terms of true solar time — that is, in terms of *local apparent time* which is based on the apparent solar day determined as the elapsed time between two successive transits of the sun across any given meridian. Given the local apparent time at a known geographical position, it is then only necessary to determine the solar declination for the time of year in order to apply the equations.

The maximum change in declination in 24 hours (which occurs at the equinoxes) is less than $\frac{1}{2}^\circ$; so that if accuracy to the nearest degree in calculated solar zenith angle and azimuth is sufficient, a single value of δ can thus be used for each day. Spencer (1971) has derived the following series expression for δ in radians:

$$\delta = 0.006918 - 0.399912 \cos \theta_0 + 0.070257 \sin \theta_0 - 0.006758 \cos 2\theta_0 + 0.000907 \sin 2\theta_0 - 0.002697 \cos 3\theta_0 + 0.001480 \sin 3\theta_0 \quad [3.7]$$

which estimates δ with a maximum error of 0.0006 radians ($< 3'$); or if the final two terms are omitted with a maximum error of 0.0035 radians ($12'$). The angle θ_0 in radians is defined in terms of the day number d_n which ranges from 0 on January 1 to 364 on December 31. That is:

$$\theta_0 = 2\pi d_n/365 \quad [3.8]$$

If greater accuracy is required, reference must be made to the Nautical Almanac published yearly and jointly by the Royal Greenwich Observatory and the U.S. Naval Observatory.

Mean solar time is derived from the motion of an imaginary body called the mean sun, which is considered as moving at uniform angular speed in the celestial equator at a rate equal to the average rate of movement of the true sun. The difference between this fixed time reference and the variable local apparent time is known as the *equation of time Eq* and is also tabulated in the Nautical Almanac. *Eq* may be positive or negative depending on the relative positions of the true and mean suns. Thus:

local apparent time = local mean time + equation of time [3.9]

= clock time + longitude correction + equation of time [3.10]

where the clock time is that referred to the various standard meridians disposed about the world and the longitude correction (4 minutes for every degree) accounts for the difference between the local and standard meridians. This correction is positive if the local meridian is east of the standard and vice versa. Again Spencer (1971) has provided a single series expression for the equation of time in radians:

$$Eq = 0.000075 + 0.001868 \cos \theta_0 - 0.032077 \sin \theta_0 - 0.014615 \cos 2\theta_0 - 0.040849 \sin 2\theta_0 \quad [3.11]$$

which has a maximum error of 0.0025 radians, equivalent to 35 seconds in time.

It should be remembered that the sun subtends a finite angle of $32'$ ($\approx \frac{1}{2}^\circ$) at the earth and the formulae above refer to the centre of the disc. They refer also to the true zenith angle which would be observed through a non-refracting atmosphere. The zenith angle actually observed at the surface is slightly less. The difference ranges from zero at zero zenith angle to about $34'$ at the horizon, is slightly dependent on pressure and temperature, and is tabulated in the Nautical Almanac for θ between 80° and 90° where the effect is most significant. Relevant formulae are discussed in the next section.

4. ZENITH ANGLE AND AIR MASS

When considering absorption and scatter within the atmosphere it is always necessary to know the total mass of the absorbing or scattering substance — the optical path — traversed by the beam of radiation in question. When the reference is to the direct solar beam, the relevant optical path is called the *absolute optical air mass* m_a. That is:

$$m_a(h) = \int \rho \mathrm{d}s \qquad [3.12]$$

where ρ is the density, $\mathrm{d}s$ is the geometrical path element along the solar beam and the integration is taken from the point in question (at a height h above the ground surface) back along the beam to the top of the atmosphere. Further, the mathematics of scatter and absorption usually involve the approximation of a plane-parallel atmosphere where the optical path OP between any two levels along some zenith angle θ is related to the vertical optical path OP_V between the same levels by the multiplying factor $\sec\theta$ independent of azimuth. That is, $OP(\theta) = OP_V \sec\theta$. Again when referring to the direct solar beam, the multiplying factor is called the *relative* optical air mass m_r. It is formally defined:

$$m_r(h) = \int_h^\infty \rho \mathrm{d}s \Big/ \int_h^\infty \rho \mathrm{d}h \qquad [3.13]$$

where $\mathrm{d}h$ is the increment in height from the point of observation at h. To the extent that the atmosphere can be approximated as a non-refracting plane-parallel medium, every $\mathrm{d}s$ in equation [3.13] equals $\mathrm{d}h \sec\theta$ and:

$$m_r(h) = \sec\theta \qquad [3.14]$$

The density in the above formulae is that appropriate to the particular scattering or absorbing process. Thus for Rayleigh scatter where the mass extinction coefficient is defined in terms of unit mass of atmosphere, and for other cases where the reference is to the total mass of the atmosphere, the ρ in equations [3.12] and [3.13] is the total atmospheric density. For water vapour absorption, say, the mass absorption coefficient is defined in terms of unit mass of water vapour and the ρ in the above equations is the density of water vapour. The fact makes no difference to equation [3.14] which is for a plane-parallel atmosphere but becomes significant when earth curvature is taken into account. The error in equation [3.14] due to curvature and refraction of the real atmosphere increases from zero at zero zenith angle. For ground level observations ($h = 0$) and the total atmospheric mass the error is 0.25% at $\theta = 60°$ increasing to 10% at $\theta = 85°$.

The effect of curvature can be appreciated by reference to Fig. 3.4 where R is the radius of the earth (6.370×10^6 m) and $\hat{H}$ is the height of the homogeneous atmosphere of constant density ρ_g equal to the actual surface density. That is, $\hat{H} = p_g/g\rho_g$ where the subscript g refers to the surface observed values of pressure p and density ρ. A little geometry shows that:

$$m_r(0) = \frac{BA}{AC} = \left\{[(R/\hat{H})\cos\theta]^2 + 2(R/\hat{H}) + 1\right\}^{1/2} - (R/\hat{H})\cos\theta \qquad [3.15]$$

where θ is the zenith angle of the sun predicted from the astronomical formulae of the previous section. Refraction effects still limit the accuracy of this equation to 2% at $\theta = 85°$.

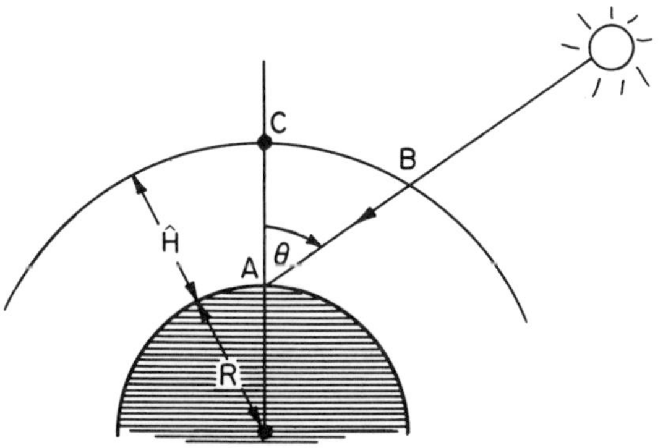

Fig. 3.4. Diagram relevant to air mass in an atmosphere of finite curvature — see text.

The formula applies to the whole atmosphere. When the reference is to restricted layers of the atmosphere it may no longer apply. The important exception involves ozone absorption which occurs mainly in the upper atmosphere above the 20-km level. Again a little geometry (N. Robinson, 1966) leads to:

$$m_r(h) = [1 + (h/R)]/[\cos^2\theta + 2(h/R)]^{1/2} \qquad [3.16]$$

and versions of this formula are used in operational models (see Chapter 5) with the simplification that h is taken as some fixed value appropriate to the mean ozone level at about 22 km. At 22 km, direct application of equation [3.15] yields an m_r which is 1% too high at $\theta = 60°$ and 20% too high at $\theta = 85°$.

Theoretically equation [3.15] should be modified for any process which involves only a particular layer of the atmosphere. For instance solar radiation interaction with water vapour occurs mainly in the lower atmosphere and the appropriate m_r approaches the pure sec θ law. The difference except in the case of ozone is rarely significant mainly because the energies involved at large zenith angles are very small.

In order to account for the refraction which arises from the variation of density with height one must revert to the complete calculation of m_r which, for the surface, involves:

$$m_r = \frac{1}{\rho_g \hat{H}} \int_0^\infty \left[1 - \left(\frac{R}{R+\hat{H}}\right)^2 \left(\frac{n_{rg}}{n_r}\right)^2 \sin^2\theta_{ob}\right]^{-1/2} \rho \mathrm{d}h \qquad [3.17]$$

where n_r is the real part of the refractive index at height h, n_{rg} is the refractive index at ground level and θ_{ob} here is the *observed* solar zenith angle.

Thus relative air mass is obviously a function of the distribution with height of atmospheric density and refractive index and is slightly variable. (It is also a function of wavelength through the wavelength dependence of n_r.) However, the relative air masses (of the whole atmosphere) appropriate to the ARDC standard atmosphere are usually accurate enough and they have been tabulated by various people such as Kasten (1966). Kasten's computations were for a wavelength of 0.7 μm which divides the solar spectrum into two parts of roughly equal energy and he produced also an approximate empirical formula for m_r; namely:

$$m_r = 1/[\sin \gamma_{ob} + a(\gamma_{ob} + b)^{-c}] \qquad [3.18]$$

where γ_{ob} is the observed solar altitude, $a = 0.1500$, $b = 3.885$ and $c = 1.253$. This formula is accurate to better than 0.1% for zenith angles up to 86°. The greatest deviation, 1.25%, occurs at $\theta = 89.5°$.

5. 24-HOUR TOTALS OF RADIATION INPUT

This section provides a classic example of the dangers of dealing with a mean annual world when the processes involved are highly non-linear. Despite the fact that the solar irradiance at the top of the atmosphere decreases to zero at the poles for a world of zero orbit obliquity (cf. the mean real world) the actual mean irradiance at the poles is certainly not zero but is about 0.4 of that at the equator. This comes about because of the finite irradiance at the poles during the summer. Indeed, in terms of 24-hour totals of radiation input, the maximum occurs at the poles in mid-summer where the sun circles around at a constant altitude of 23.5°. Daily (i.e. 24-hour) totals for the world are presented graphically by List (1968). They can be calculated analytically as follows. At latitudes up to $(\pi/2) - \delta$ of the poles, the total 24-hour irradiation $\hat{Q}$ is the time integral of irradiance from sunrise to sunset. That is:

$$\hat{Q} = \int^{(\text{daytime})} I_0 \cos\theta \, \mathrm{d}t \qquad [3.19]$$

which from equation [3.4] becomes:

$$\hat{Q} = I_0 \int_{-\Pi}^{+\Pi} (\sin\Phi \sin\delta + \cos\Phi \cos\delta \cos t_h) \, \mathrm{d}t \qquad [3.20]$$

where Π is the half-day length in radians obtained from equation [3.4] by noting that at sunrise and sunset $\cos\theta$ is zero. That is:

TABLE 3.2

The seasonal and latitudinal distributions of the length of the daytime given in parts of 24 hours and those of the weighted mean value of $\cos \theta$ (from Manabe and Möller, 1961)

Lat.	April	July	October	January
Fractional length of daytime				
5°	0.508	0.517	0.500	0.496
15°	0.521	0.537	0.492	0.471
25°	0.533	0.562	0.483	0.450
35°	0.546	0.596	0.471	0.421
45°	0.562	0.637	0.454	0.361
55°	0.596	0.708	0.437	0.321
65°	0.629	0.837	0.404	0.208
75°	0.750	1.000	0.329	—
85°	1.000	1.000	—	—
$\cos \bar{\theta}$				
5°	0.625	0.587	0.614	0.591
15°	0.618	0.601	0.579	0.549
25°	0.599	0.593	0.524	0.474
35°	0.558	0.567	0.458	0.393
45°	0.501	0.521	0.379	0.317
55°	0.423	0.453	0.282	0.203
65°	0.345	0.369	0.176	0.106
75°	0.241	0.311	0.071	—
85°	0.168	0.318	—	—

$$\cos \Pi = -\tan \Phi \tan \delta \qquad [3.21]$$

Remembering that dt_h/dt is the angular velocity of the earth equal to 2π radians for 24 hours, equation [3.19] can be integrated to give:

$$\hat{Q} = I_0(\Pi \sin \Phi \sin \delta + \cos \Phi \cos \delta \sin \Pi)/\pi \qquad [3.22]$$

provided that the solar constant is expressed in units of energy input per 24 hours. For latitudes poleward of $(\pi/2) - \delta$ the integral of equation [3.19] ranges from $-\pi$ to $+\pi$ during the summer. That is:

$$\hat{Q} = I_0 \sin \Phi \sin \delta \qquad [3.23]$$

For meteorological purposes it is often convenient to make use of the mean zenith angle $\bar{\theta}$ for a day defined as:

$$\cos \bar{\theta} = \int^{\text{(daytime)}} \cos \theta \, dt \Big/ \int^{\text{(daytime)}} dt \qquad [3.24]$$

so that $\hat{Q}$ equals I_0 times the fractional length of the daytime multiplied by $\cos \bar{\theta}$ (Manabe and Möller, 1961). Values of $\cos \bar{\theta}$ and fractional day length for the northern hemisphere are given in Table 3.2 as a function of latitude and season.

CHAPTER 4

Radiative Transfer

1. INTRODUCTION

If it were not for the presence of clouds and aerosols, it is probably true that most practical problems to do with atmospheric radiation and its *meteorological* interaction could be tackled without the elegant formalism which surrounds the general equation of radiative transfer and its solution. However, clouds and aerosols introduce Mie scatter by particles — a process which itself is exceedingly complicated as far as computation is concerned and which depends drastically on particle characteristics which are difficult to measure, let alone predict in an operational model. Thus in order to retain flexibility in approach to individual problems it is essential to have some general framework in which to operate. The theoretical basis is laid in the standard works by Chandrasekhar (1950) and by Goody (1964). The modern literature is saturated with examples of numerical solutions of specific problems and the various techniques are summarized by Hansen and Travis (1974).

This chapter introduces the basic formalism in order that the subsequent chapters can be read in context. It discusses also the processes and theory of Mie scatter. Finally it refers to a few of the common techniques of numerical solution of the general transfer problem. Generally such techniques need to be avoided like the plague in operational models but are necessary both as a basis for the development of more simple parameterization schemes and as test-beds for such schemes when they are proposed.

2. THE TRANSFER EQUATION

Various concepts of radiative transfer were discussed in Chapter 2 by reference to pure absorption of a pencil beam within a medium which itself was not in any sense an emitter of radiation. For such a case the depletion of radiation per unit of distance s is described by the simple differential equation:

$$\mathrm{d}N_\lambda = -\kappa_{\mathrm{a}\lambda}\rho N_\lambda \mathrm{d}s \qquad [4.1]$$

which integrates to Beer's law of exponential absorption. Equation [4.1] is perhaps the simplest case of the general equation of radiative transfer. The solution is generally analytic but can be obtained numerically by an integration starting at $s = 0$ and proceeding directly along the direction of s. Its typical application is to problems concerned purely with extinction of the direct solar beam, where the extinction may in fact arise from pure absorption, pure scatter or a combination of both. The mass absorption and scattering coefficients are simply additive to form the total extinction coefficient $\kappa_{e\lambda}$, and the only real problem apart from specifying the coefficients is the necessity to integrate over wavelength. None of the solar beam extinction processes are wavelength independent so that Beer's law does not apply to total radiance. As it happens, owing to spectral or spatial separation of ozone absorption, water vapour absorption and Rayleigh scatter, these processes can be treated separately in terms of their wavelength-integrated extinctions (see Chapters 5 and 6). Note, however, that even this simplification becomes less useful when clouds and aerosols are present in the atmosphere.

The more common meteorological problem is not so simple. On the one hand the concern is mostly with a field of radiation (i.e. with global flux densities and not a pencil beam of radiation) which necessitates integration over all solid angles. On the other, one must take into account the fact that each volume element may itself be a generator of radiation along the direction of s. This may arise either (a) because the medium is emitting black-body radiation at the wavelength of interest by virtue of its finite temperature, or (b) because each volume element is scattering radiation into the direction of s from sources defined by the scatter from other volume elements. Both (a) and (b) can occur together, but in the *clear-sky* meteorological context it is a fortunate circumstance that process (a) is relevant primarily to terrestrial long-wave fluxes and process (b) is relevant primarily to solar fluxes. Again an important exception occurs in the case of clouds and aerosols. They can effectively negate the various simplifications which are otherwise possible as a result of the process separation.

Considering first the long-wave problem, under the common condition of local thermodynamic equilibrium the differential equation of transfer for a pencil beam becomes:

$$\mathrm{d}N_\lambda = -\kappa_{a\lambda}\rho N_\lambda \mathrm{d}s + \kappa_{a\lambda}\rho B_\lambda(T)\,\mathrm{d}s \qquad [4.2]$$

where $B_\lambda(T)$ is the Planck black-body emission at wavelength λ and temperature T and, by Kirchoff's law, the emission per unit mass is $\kappa_{a\lambda}B_\lambda(T)$. In terms of optical depth $\tau_\lambda(s', s)$, that is of $\int_{s'}^{s} \kappa_{a\lambda}\rho \mathrm{d}s$, the formal solution to this unidirectional problem is:

$$N_\lambda = N_{0\lambda} \exp\left[-\tau_{a\lambda}(0, s)\right] + \int_0^s \kappa_{a\lambda}\rho B_\lambda(T) \exp\left[-\tau_{a\lambda}(s', s)\right] \mathrm{d}s' \qquad [4.3]$$

This is an integral equation where the individual elements of the integration are the contributions to the radiance at s by the emission from each element ds'. The possible simplifications, when dealing with the total radiation field, centre about (1) the assumption of a horizontally plane-parallel atmosphere, and (2) the assumption of isotropy of radiation emission. Taken together they allow an effective "two-stream" approximation whereby the long-wave field can be specified in terms only of the upward and downward components of the flux density $F\uparrow$ and $F\downarrow$ respectively. The associated flux transfer equation can be written down and handled in much the same way as the radiance equation [4.3] with the angular integration already completed. The radiances are replaced by vertical flux densities and the absorption coefficient or emissivity becomes the equivalent *slab* absorption coefficient or emissivity (see Chapter 7). Again the only difficulty in principle is the final integration over wavelength, which in practice is done by dividing the spectrum into the minimum number of bands appropriate to the accuracy required. A further practical problem is the significant variation of $\kappa_{a\lambda}$ with temperature and pressure.

Considering now the pure scattering problem, the pencil beam transfer equation becomes:

$$dN_\lambda = -\kappa_{sc\lambda}\rho N_\lambda ds + \frac{\kappa_{sc\lambda}\rho}{4\pi}\int_0^{2\pi}\int_0^{\pi} \tilde{P}_\lambda(\theta,\phi,\theta',\phi')N_\lambda(\theta',\phi')\sin\theta' d\theta' d\phi' ds \qquad [4.4]$$

where $\kappa_{sc\lambda}$ is the mass scattering coefficient and $N_\lambda(\theta', \phi')$ is the radiance impinging on the volume element at ds from the direction $\theta'\phi'$. The *phase function* $\tilde{P}_\lambda(\theta, \phi, \theta', \phi')$ divided by 4π specifies the fraction of $N_\lambda(\theta', \phi')$ which is scattered into the direction of s specified by θ and ϕ. Semi-analytic solutions of equation [4.4] are possible for the very simplest phase functions such as that of Rayleigh scatter (see Chapter 5) but are generally impossible for cases of Mie scatter. Even numerical solution is difficult since in a sense one needs to integrate the equation in all directions at once. The simplification of isotropy which might enable a two-stream approximation can be applied only to cases of large optical depth where multiple scatter tends to randomize photon direction. The only significant relief derives from the assumption of a plane-parallel atmosphere and from the fact that scattering coefficients are not often dependent on pressure and temperature. The case of swelling aerosols in very humid atmospheres is an obvious exception.

2.1. *The general case*

It is fairly obvious that radiative transfer is described by a general equation of the form:

$$dN_\lambda/ds = -\kappa_{e\lambda}\rho N_\lambda + j_\lambda\rho \qquad [4.5]$$

where j_λ is a parameter formally described as the mass emission coefficient and defines the amount of radiant energy emitted over the wavelength interval $d\lambda$ into the element of solid angle $d\omega$ about the direction of s by unit element of mass. Taking advantage of the concept of optical depth ($d\tau_\lambda = \kappa_{e\lambda}\rho ds$), the equation can be simplified even further to:

$$dN_\lambda/d\tau = -N_\lambda + \tilde{J}_\lambda \qquad [4.6]$$

where $\tilde{J}_\lambda$ is known as the *source function* and is equal to $j_\lambda/\kappa_{e\lambda}$. Referring back to the two specific cases of pure scatter and of local thermodynamic equilibrium, the corresponding source functions are respectively:

$$\tilde{J}_\lambda = \frac{1}{4\pi}\int_0^{2\pi}\int_0^{\pi} \tilde{P}_\lambda(\theta,\phi,\theta',\phi')N_\lambda(\theta',\phi')\sin\theta'\,d\theta' d\phi' \qquad [4.7]$$

and:

$$\tilde{J}_\lambda = B_\lambda(T) \qquad [4.8]$$

Where both processes are simultaneously involved:

$$\tilde{J} = \frac{\bar{\omega}_0}{4\pi}\int_0^{2\pi}\int_0^{\pi} \tilde{P}_\lambda(\theta,\phi,\theta',\phi')N_\lambda(\theta',\phi')\sin\theta'\,d\theta' d\phi' + (1-\bar{\omega}_0)B_\lambda(T) \qquad [4.9]$$

Here the absorption to scatter ratio is specified by the *single scattering albedo* $\bar{\omega}_0$. It is the fraction of the total extinction of a pencil beam at the point in question which is attributable to pure scatter. Thus:

$$\bar{\omega}_0(\tau_\lambda) = (d\tau_\lambda)_{sc}/(d\tau_\lambda)_e \qquad [4.10]$$

$$= \kappa_{sc\lambda}/\kappa_{e\lambda} \qquad [4.11]$$

$$= 1 \text{ for pure scatter}$$

Note that equation [4.9] assumes $\bar{\omega}_0$ to be independent of direction so that it can be taken outside the integral. Further, by definition here the phase function is always normalized so that $\int \tilde{P}d\omega = 4\pi$. Chandrasekhar (1950) makes the direct definition $\int \tilde{P}d\omega/4\pi = \bar{\omega}_0$.

In principle therefore, the information required to calculate the radiation field of any given medium is the distribution of extinction coefficient κ_e, single scattering albedo $\bar{\omega}_0$ and the phase function $\tilde{P}$ all as a function of wavelength, together with the appropriate boundary conditions.

In the most common application to a plane-parallel atmosphere it is convenient to measure distance in terms of the vertical co-ordinate z. Then θ becomes formally identified with zenith angle away from the vertical and ϕ with azimuth. In turn one refers to the *vertical* optical depth τ defined by

$\int \kappa_{e\lambda} \rho dz$. The general transfer equation (equation [4.6]) becomes:

$$\cos\theta \, dN(z,\theta,\phi)/\kappa_e \rho dz = -N(z,\theta,\phi) + \tilde{J}(z,\theta,\phi) \qquad [4.12]$$

or:

$$\mu dN(\tau,\mu,\phi)/d\tau = -N(\tau,\mu,\phi) + \tilde{J}(\tau,\mu,\phi) \qquad [4.13]$$

where $\mu = \cos\theta$ and the parameter of wavelength is dropped for simplicity. Further, it is convenient in solar flux applications to preserve the distinction between the direct solar beam and the resulting diffuse flux density. The practical situation is one where the upper boundary of the medium is illuminated by a parallel beam of irradiance πF from the direction (μ_0, ϕ_0). (For instance πF equals the spectral value of the solar constant $I_{0\lambda}$ when the medium comprises the complete atmosphere.) The source function can then be expressed:

$$\tilde{J}(\tau,\mu,\phi) = \frac{\bar{\omega}_0}{4\pi}\int_0^{2\pi}\int_{-1}^{+1} \tilde{P}(\tau,\mu,\phi,\mu',\phi')N(\tau,\mu',\phi')\,d\mu' d\phi'$$

$$+\frac{F\bar{\omega}_0}{4}\tilde{P}(\tau,\mu,\phi,\mu_0,\phi_0)\exp(-\tau/\mu_0) \qquad [4.14]$$

It is standard notation to refer to upwelling radiances as N^+ and downcoming radiances as N^-. With this notation the typical boundary conditions for the complete atmosphere of total vertical optical depth τ^* would be:

$$\left.\begin{aligned} N^+(\tau^*,\mu,\phi) &= \text{surface reflection specification} \\ \text{and } N^-(0,\mu,\phi) &= 0 \end{aligned}\right| \qquad [4.15]$$

The discussion above refers to equations appropriate to pencil beams. When considering the complete radiation field it is necessary to replace all the quantities by the corresponding matrices.

2.2. *Approximate solutions for operational models*

The situation of atmospheric scatter which is the simplest mathematically is that where the horizontally plane-parallel medium supports a vertical flux of radiation which is azimuth independent. Such a state of affairs might exist deep within fairly dense clouds where the direct beam has lost its identity as a result of intense scatter. The equation of transfer [4.13] becomes:

$$\mu dN(\tau,\mu)/d\tau = -N(\tau,\mu) + \frac{\bar{\omega}_0}{2}\int_{-1}^{+1}\tilde{P}^0(\mu,\mu')N(\tau,\mu')\,d\mu' \qquad [4.16]$$

where:

$$\tilde{P}^0(\mu,\mu') = \frac{1}{2\pi}\int_0^{2\pi} \tilde{P}(\mu,\phi,\mu',\phi')\,\mathrm{d}\phi' \qquad [4.17]$$

and the factor 1/2 in front of the integral of equation [4.16] arises from the azimuth integration of the source function.

Although it is highly unreal physically — especially in cases of Mie scatter by particles — we assume first that the phase function is completely isotropic (i.e. $\tilde{P}^0(\mu,\mu') = 1$ in all directions). Then:

$$\mu \mathrm{d}N(\tau,\mu)/\mathrm{d}\tau = -N(\tau,\mu) + \frac{\bar{\omega}_0}{2}\int_{-1}^{+1} N(\tau,\mu')\,\mathrm{d}\mu' \qquad [4.18]$$

Suppose further that the upward and downward angular radiance distributions have a particular form such that the upward and downward flux densities can be specified by only one radiance parameter for each hemisphere. For instance, we might assume the radiances in the upward and downward hemispheres to be individually isotropic, and then $F\!\uparrow = \pi N^+$ and $F\!\downarrow = \pi N^-$ where N^+ and N^- are the radiances constant with angle in the corresponding hemispheres. Mathematically this is equivalent to a "two-stream" assumption whereby N^+ and N^- are set equal to zero everywhere except in the vertical directions where the radiances are $N\!\uparrow$ and $N\!\downarrow$ and are defined to be equal to $F\!\uparrow/\pi$ and $F\!\downarrow/\pi$ respectively.

Once such a form has been assumed, the description of the complete radiation field can be reduced to two equations involving, say, the upward and downward "stream" radiances $N\!\uparrow$ and $N\!\downarrow$ respectively. That is:

$$+\tfrac{1}{2}\mathrm{d}N\!\uparrow/\mathrm{d}\tau = -N\!\uparrow + \frac{\bar{\omega}_0}{2}(N\!\uparrow + N\!\downarrow) \qquad [4.19]$$

and:

$$-\tfrac{1}{2}\mathrm{d}N\!\downarrow/\mathrm{d}\tau = -N\!\downarrow + \frac{\bar{\omega}_0}{2}(N\!\uparrow + N\!\downarrow) \qquad [4.20]$$

where the factor $\pm\frac{1}{2}$ on the left-hand sides allows (approximately) for the mean obliquity of the rays to the vertical direction and $N\!\uparrow$ and $N\!\downarrow$ are effectively the average radiances in the upward and downward hemispheres. The solution of equations [4.19] and [4.20] is easily written down analytically for the case where the medium is a pure scatterer (i.e. $\bar{\omega}_0 = 1$) in terms of the reflectance *Re* and transmittance *Tr* of the layer in question. That is:

$$Re = \tfrac{1}{2}\tau/(1+\tfrac{1}{2}\tau) \qquad [4.21]$$

$$\text{and } Tr = 1/(1+\tfrac{1}{2}\tau) \qquad [4.22]$$

for a layer where the underlying layers or surface have zero reflectance. It is interesting physically that, as the optical thickness of the layer (cloud?) increases, the reflectance approaches unity while *Tr* tends to zero. A cloud illuminated from above does not tend to a state where equal flux emerges from its upper and lower boundaries. It is notable also that even over the thickest clouds the experimentally measured albedos rarely exceed 0.8 which indicates that cloud absorption is certainly not negligible (Twomey, 1970). A fundamental problem at the present time is whether the solar absorption within clouds can be attributed entirely to the associated water vapour, or whether absorption by the drops themselves must be included.

The solution of equations [4.19] and [4.20] is a little more complicated when absorption is included but is still analytic. They are given here by equations [5.17] and [5.18] of Chapter 5, where the model is brought a little closer to the reality of a Mie scattering medium by separating the phase function into forward and backward hemispheres which may be individually isotropic but which in any event are such that the ratio of backward to forward scatter is specified by some fraction *bf*. This "backward to forward" scatter ratio is related to what is called the *asymmetry factor* $\langle \cos \psi \rangle$ defined as the integral over all solid angles of $\tilde{P} \cos \psi$. That is:

$$\langle \cos \psi \rangle = \int_{-1}^{+1} \tilde{P}(\psi) \cos \psi \, \mathrm{d}\,(\cos \psi) \qquad [4.23]$$

and is the quantity normally quoted for a medium. It is the minimal information required about the phase function which allows some accuracy in approximate solutions of a particular transfer problem. It is equal to zero for isotropic scatter and to $+1$ and -1 for complete scatter into the forward and backward directions respectively. Physically it is the difference between the *flux* densities in the forward direction and the backward direction arising from the scattering by a volume (or particle) when the incident flux is normalized to 1. The fraction *bf* above equals $\frac{1}{2}(1 - \langle \cos \psi \rangle)$.

The above solutions are those of the original classic "two-stream" approximation of Shuster and Schwarzchild (Shuster, 1905) which has been applied extensively in early work on the subject of cloud-radiation interaction (e.g. Hewson, 1943; Nieburger, 1949). In its unmodified form it is incapable of handling, say, the variation of cloud albedo with solar zenith angle. However, the separation of the source function of the transfer equation into its direct and diffuse components (equation [4.14]) goes a very long way towards removing this difficulty. First, it should be realized that the two-stream approximation and the so-called Eddington approximation with which it is occasionally confused are indeed very similar in concept, application and the range of validity in their results. Both simply assume that the zenith angle dependence of the upward and downward radiances N^+ and N^- have a particular form which allows easy integration over angle to give the corresponding upward and downward flux density. They are illustrated in Fig. 4.1 together with the expressions for the flux densities. Each approximation inherently contains its own phase function in terms of the asymmetry factor $\langle \cos \psi \rangle$. Thus for the two-stream case (Fig. 4.1b):

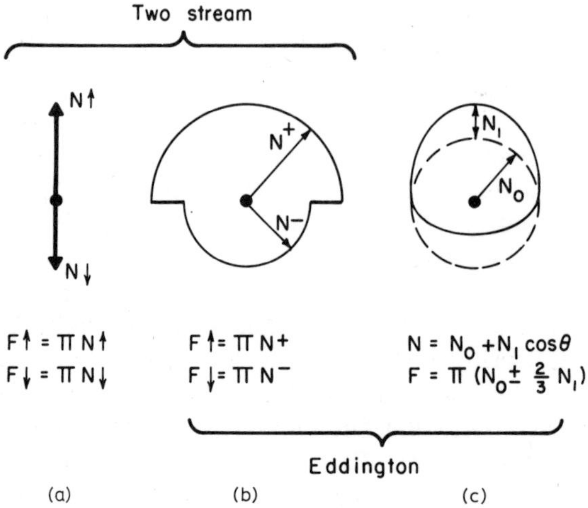

Fig. 4.1. Two-stream and Eddington approximations. (a) and (b) are mathematically equivalent, (b) and (c) are conceptually equivalent.

$$\tilde{P}(\psi) = 1 \pm \langle \cos \psi \rangle \quad [4.24]$$

and for the Eddington approximation (Fig. 4.1c):

$$\tilde{P}(\psi) = 1 + 3 \langle \cos \psi \rangle \cos \psi \quad [4.25]$$

Both approximations lead to simple analytic formulae for the reflection and transmission functions even when the direct beam is explicitly included. These solutions are given for the Eddington approximation by Shettle and Weinman (1970). Here we record those of the two-stream approximation of the relevant transfer equation (cf. equation [4.14]):

$$\mu \mathrm{d}N(\tau, \mu)/\mathrm{d}\tau = -N(\tau, \mu) + \frac{\bar{\omega}_0}{2} \int_{-1}^{+1} \tilde{P}^0(\mu, \mu') N(\tau, \mu')\, \mathrm{d}\mu'$$

$$+ \frac{\bar{\omega}_0}{4} \tilde{P}^0(\mu, \mu_0) F \exp(-\tau/\mu_0) \quad [4.26]$$

which are (Radiation Commission, 1975):

$$Re(\mu_0, \tau) = \frac{G[(1 - \exp(-2s\tau/\mu_0)]}{1 - G^2 \exp(-2s\tau/\mu_0)} \quad \text{for } \bar{\omega}_0 < 1 \quad [4.27]$$

$$= \frac{bf \cdot \tau/\mu_0}{1 + bf \cdot \tau/\mu_0} \quad \text{for } \bar{\omega}_0 = 1 \quad [4.28]$$

and:

$$Tr(\mu_0, \tau) = \frac{\exp(-2s\tau/\mu_0)(1 - G^2)}{1 - G^2 \exp(-2s\tau/\mu_0)} \quad \text{for } \bar{\omega}_0 < 1 \quad [4.29]$$

$$= 1 - \frac{bf \cdot \tau/\mu_0}{1 + bf \cdot \tau/\mu_0} \qquad \text{for } \bar{\omega}_0 = 1 \qquad [4.30]$$

where $bf = \frac{1}{2}(1 - \langle \cos \psi \rangle)$ and here only:

$$\left.\begin{aligned} f &= 1 - bf \\ r &= 1 - \bar{\omega}_0 f + \omega_0 \cdot bf \\ s &= [(1 - \bar{\omega}_0 f)^2 - \bar{\omega}_0^2 bf^2]^{1/2} \\ \text{and } G &= (r - s)(r + s)^{-1} \end{aligned}\right| \qquad [4.31]$$

Despite the fact that the two-stream and Eddington approximations are limited in application to cases where the scatter is not highly anisotropic, they are highly valuable because of the simplicity of their analytic solution. Further, they retain a measure of directness in physical conception which allows easy and unambiguous formulation of the boundary conditions in specific cases. Thus if it can possibly be justified within the accuracy of operational models (and sometimes when it cannot) they are used as a method of parameterizing cloud-radiation interaction on the basis that the prediction or observation of cloud amount and cloud character is inherently inaccurate to much worse than 10%. Generally they give answers to better than about 5% for quite anisotropic scattering ($\langle \cos \psi \rangle$ less than 0.8 say) when absorption is negligible and when the layer optical depth is greater than about 5. This last is usually the case for water clouds in the low atmosphere, but is certainly not true for aerosol layers and thin cirrus.

Note that the formulae above assume a non-reflecting medium or surface below the layer in question. The situation for non-black underlying layers can be handled by a simple adding procedure to be described in the last section.

The stream concept is valuable, and the obvious first move to improve accuracy is to divide the radiation field into more than the two discussed above. For instance, a four-stream model can be devised (Liou, 1974) which has a relatively simple and quite usable analytic solution. With a few important exceptions most of the formal numerical solutions to the radiative transfer problem involve concepts which are equivalent to that of multiple streams of a number $2n$, say. The problem then boils down to solving $2n$ simultaneous equations.

3. MIE SCATTER

In a perfectly clear atmosphere the scattering processes are restricted to scatter by molecules (Rayleigh scatter) for which the phase function is known precisely. Semi-analytic solutions of the transfer equations are relatively easy to derive, and in the meteorological context the grossest parameterizations are usually adequate since the density distribution of molecules in the atmosphere is not significantly variable (see Chapter 5).

Cloud and aerosol particles are a different matter. Their scattering properties depend on their size, shape and the real and imaginary parts of their refractive index. Since the particles occupying a particular volume normally have a size *distribution* (such that the total number of particles $n_T = \int_0^\infty n(r)\,dr$ where $n(r)$ is the number of particles of radius r), the mass (or volume) extinction coefficients and the phase function of a unit volume of the medium must be calculated from appropriate integrals over all the particles. That is:

$$\beta_{sc} = \kappa_{sc}\rho = \int_0^\infty J(r)\,n(r)\,dr \qquad [4.32]$$

$$\beta_e = \kappa_e\rho = \int_0^\infty K(r)\,n(r)\,dr \qquad [4.33]$$

and:

$$\tilde{P}(\psi) = \frac{1}{\kappa_{sc}\rho}\int_0^\infty n(r)\,J(r)\,\tilde{P}(r, \psi)\,dr \qquad [4.34]$$

where $J(r)$, $K(r)$ and $\tilde{P}(r, \psi)$ are the scattering and extinction coefficients and the phase function appropriate to the individual particle of radius r. Thus the initial problem is to evaluate the single particle properties. In operational models the volume coefficients will probably be pre-calculated on the basis of standard particle distributions with the only likely *predictable* variable being n_T. Since, however, the distribution $n(r)$ depends on n_T, it will still be necessary to parameterize the volume coefficients as a function of n_T by a relation which is rather more sophisticated than simply multiplying by n_T.

Atmospheric particles fall broadly into four categories.

(1) Solid aerosols of mean size between 0.1 and 1 μm; they need not be spherical and their refractive index may be highly variable.

(2) Haze water drops of mean size between 0.1 and 1 μm; they would be spherical and have known refractive index.

(3) Cloud water drops of mean size between 1 and 10 μm; again spherical and of known refractive index.

(4) Cloud ice particles of mean size between 1 and 100 μm; irregular shape but known refractive index.

Although it is possible in principle to take account of irregular particle shape in any detailed treatment, normal practice is to assume all particles to be spherical on the basis that random orientation would ensure that they behave in the mean like spheres of equivalent size. This assumption is questionable — especially for large flat particles which on fluid dynamical grounds might be expected to orient themselves horizontally. Non-random orientation might be evident in a peculiar variation of direct solar beam intensity as a function of solar zenith angle (Paltridge, 1969); it is also indicated by the occasional appearance of the "mock-sun" in a field of thin cirrus cloud.

Mie theory provides a "complete" solution to the problem of scatter and absorption by an isolated sphere. It is exact insofar as the model of the sphere is exact. That is, the theory assumes a homogeneous sphere with a sharp discontinuity of refractive index at the surface; and the extent to which actual water drops, say, approach this ideal is difficult to assess. Nevertheless the theory has proved highly successful in application and certainly in the atmospheric context there seems to be no need to modify the basic model. The extinction by a sphere is described in terms of its *size parameter* $\xi = 2\pi r/\lambda$ and of its complex refractive index $n_c = n_r - in_i$.

The single particle scattering and extinction coefficients $J(r)$ and $K(r)$ are normally quoted in terms of an expression for the corresponding efficiency Q_{sc} and Q_e which are simply the ratios of $J(r)$ and $K(r)$ to the geometric cross-section πr^2. Thus:

$$Q_{sc} = \frac{J(r)}{\pi r^2} = \frac{2}{\xi^2} \sum_{n=1}^{\infty} (2n+1)(a_n a_n^* + b_n b_n^*) \qquad [4.35]$$

$$Q_e = \frac{K(r)}{\pi r^2} = \frac{2}{\xi^2} \sum_{n=1}^{\infty} (2n+1) \text{ Real } (a_n + b_n) \qquad [4.36]$$

and the asymmetry factor $\langle \cos \psi \rangle$ is given by:

$$\langle \cos \psi \rangle = \frac{4}{\xi^2 Q_{sc}} \sum_{n=1}^{\infty} \left[\frac{n(n+2)}{n+1} \text{ Real } (a_n a_{n+1}^* + b_n b_{n+1}^*) + \frac{2n+1}{n(n+1)} \text{Real } (a_n b_n^*) \right] \qquad [4.37]$$

The a_n and b_n are complex quantities depending only on the size parameter and refractive index and are described by expressions containing spherical Bessel functions (Van de Hulst, 1957). These expressions can be replaced by computer-compatible recursion relations — see for instance Diermendjian (1969) or Kattawar and Plass (1967) — which are too detailed for reproduction here.

The total number of terms required in the above expressions is just slightly larger than the size parameter ξ because a_n and b_n rapidly approach zero as n becomes larger than ξ. Physically this occurs because the nth term in each of the series corresponds to a light ray passing the centre of the sphere at a distance $n\lambda/2\pi$. Higher terms correspond to rays missing the sphere entirely.

The complete single particle phase function is given by:

$$\tilde{P}(r, \psi) = \frac{\lambda^2}{2\pi J(r)} (S_1 S_1^* + S_2 S_2^*) \qquad [4.38]$$

where S_1 and S_2 are functions of ξ and n_c via the coefficients a_n and b_n and of ψ via the associated Legendre polynomial P_n'. That is:

$$S_1(\psi) = \sum_{n=1}^{\infty} \frac{2n+1}{n(n+1)} \left(a_n \frac{\mathrm{d}P_n'}{\mathrm{d}\psi} + b_n \frac{P_n'}{\sin \psi} \right) \qquad [4.39]$$

and:

$$S_2(\psi) = \sum_{n=1}^{\infty} \frac{2n+1}{n(n+1)} \left(a_n \frac{P_n'}{\sin \psi} + b_n \frac{\mathrm{d}P_n'}{\mathrm{d}\psi} \right) \qquad [4.40]$$

The problem of the meteorologist is to sort out from the complex behaviour of Mie scatter various general characteristics which might serve as a basis for simple parameterization of the scatter from a *volume* containing a dispersion of particles. For a detailed description of the dependence of the various parameters (Q_{sc}, Q_e, $\langle \cos \psi \rangle$, etc.) on the characteristics of an aerosol or hydrosol, the reader is referred to Hansen and Travis (1974). Here, a few major points are made by reference to Figs. 4.2 and 4.3 redrawn from their paper.

Fig. 4.3 is a plot of Q_{sc} for non-absorbing aerosol ($n_i = 0$) as a function of the size parameter $2\pi r_e/\lambda$. Here the Q_{sc} refers to a *volume* scattering efficiency for a given aerosol size distribution. That is:

$$Q_{sc} = \int \pi r^2 Q_{sc}(r)\, n(r)\, \mathrm{d}r \Big/ \int \pi r^2 n(r)\, \mathrm{d}r \qquad [4.41]$$

and the radius in the size parameter is the "effective" radius:

$$r_e = \int r \pi r^2 n(r)\, \mathrm{d}r \Big/ \int \pi r^2 n(r)\, \mathrm{d}r$$

The parameter b referred to in the figure is a measure of the variance of the size distribution about the effective radius and increases as the spread of particle sizes increases; $b = 0$ for an aerosol consisting of particles of one size only (a monodisperse aerosol), and Q_{sc} is then the same as Q_{sc} for a single particle.

With increasing particle size Q_{sc} tends to a value of 2. That is, the effective scattering cross-section is twice the geometrical cross-section as a result (primarily) of diffraction. Further, even a small dispersion of drop sizes about a mean is sufficient to wipe out the oscillatory nature of the scatter associated with a single particle. Thus for problems involving pure scatter of *solar* radiant flux where most of the energy is at wavelengths less than 2 μm, and for *clouds* where the mean radius normally exceeds about 4 μm, it is generally permissible to assume an overall doubling of the geometric

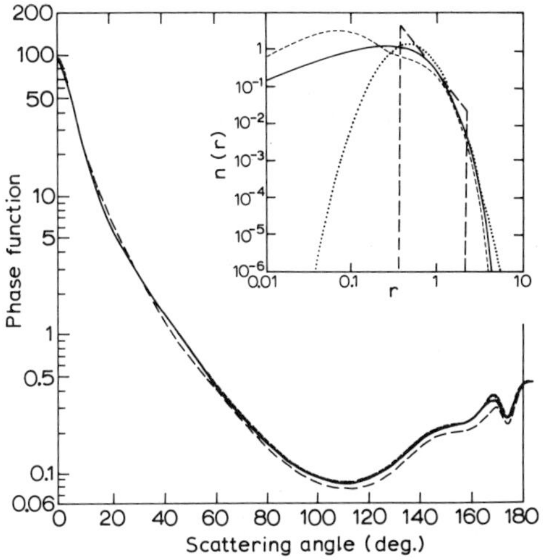

Fig. 4.2. Phase function for four distributions (shown in inset) each of which has an r_e of 1μm and an effective variance of 0.25. (After Hansen and Travis, 1974.)

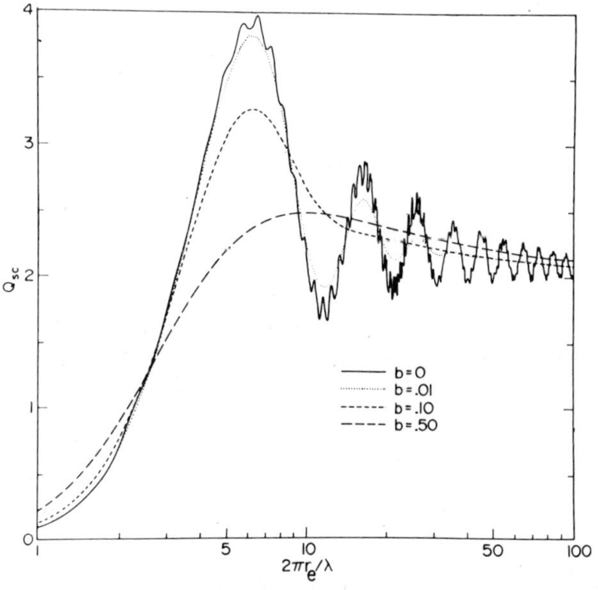

Fig. 4.3. Efficiency factor for scattering, Q_{sc}, as a function of the effective size parameter, $2\pi r_e/\lambda$. A standard size distribution $n(r) = \text{const.}\ r^{(1-3b)/b} \exp(-r/br_e)$ was used with several values of the variance b; $n_r = 1.33$, $n_i = 0$. (After Hansen and Travis, 1974.)

cross-section (the "large drop" approximation). Most parameterizations of solar radiation and cloud interaction make use of this simplification. The asymmetry factor $\langle \cos \psi \rangle$ also approaches a fairly constant value for size parameters in the same range ($> \approx 10$); that is, to the value of about 0.87 which can be predicted from pure geometrical optics applied to large and non-absorbing particles. As it happens, the imaginary part of the refractive index of water is very small at all wavelengths less than $2\,\mu$m. It ranges upwards from about 10^{-7} at $0.2\,\mu$m to 10^{-3} at $2\,\mu$m — see Irvine and Pollack (1968) and the Appendix for detailed tables of the optical constants of water and ice. Thus the effect of absorption on Q_{sc} and $\langle \cos \psi \rangle$ is quite small and can usually be ignored in short-wave interaction with cloud. Note, however, that short-wave absorption is not *necessarily* negligible in clouds and an $\bar{\omega}_0$ of $(1.0–10^{-3})$ has a considerable effect on overall cloud albedo (Twomey, 1970). Cloud drops are not necessarily pure water.

The phase function for scatter by particles is not significantly variable with the *shape* of the size distribution (e.g. Fig. 4.2), but becomes more and more forward peaked as the effective radius or size parameter ξ increases. The half-width of the phase function in the forward direction is proportional to ξ^{-1} and is relatively insensitive to n_r and n_i (Kattawar and Plass, 1967). Note that this tendency is not reflected in $\langle \cos \psi \rangle$ which, as already mentioned, is fairly constant for $\xi > 10$. Note also that although the ratio of forward to backward scattered radiance may be of order 500, the forward to backward flux density ratio is only of order 10. The lesson here is that the distinction between two- and three-dimensional problems needs to be handled with care.

Thus although the presence of clouds in the atmosphere vastly complicates consideration of solar fluxes because the relevant transfer equations are difficult to solve, the scattering coefficients themselves are often susceptible to simple parameterization by using the "large drop" approximation. This is *not* generally true for aerosol particles where the size parameter is much smaller than that of typical cloud drops and where n_i (i.e. absorption) may be significant.

On the other hand, when considering long-wave fluxes and the calculation of the emissivity of clouds from fundamental data on drop size distribution, the relevant coefficients are not generally susceptible to simple approximation because large ξ and small absorption do not apply. The transfer equation itself may be easy to solve but the coefficients are difficult to determine. One possibly useful simplification is that single particle absorption $(1 - \bar{\omega}_0)$ is closely proportional to n_i over most of the size parameter range which is relevant (Kattawar and Plass, 1967; and see Chapter 9).

Tables of volume or mass scattering and absorption coefficients, of $\langle \cos \psi \rangle$, and of the complete phase function are available in the literature for various types and size distributions of particles (see for instance Diermendjian, 1969, and Penndorf, 1956).

4. NUMERICAL SOLUTIONS

We describe here only two of a number of possible methods of potentially accurate numerical solution of the radiative transfer problem. They are the most common, the easiest to understand, and the most generally applicable.

4.1. *Monte Carlo method*

This is a direct simulation of the actual process of radiation propagation. The computer simply fires large numbers of photons at the medium from the direction of the source and traces the path of these photons individually as they are scattered or absorbed. The "bull at a gate" solution to an individual problem would be to count the number of photons which arrive at a point or points in the medium where the flux density is required. In principle it is only necessary to keep firing photons until one has a statistically significant sample at the detector.

At first glance one would assume that such a technique would be extraordinarily wasteful of computer time, and without modification of the above this is true. However, efficient Monte Carlo programs incorporate a number of refinements which reduce calculation (i.e. the required number of photons) to a level quite compatible with other methods of solution. Further the technique is virtually the only method which can handle problems of arbitrary geometry which arise for example in situations of broken cloud (McKee and Cox, 1974). Unlike other methods it has no difficulty in handling arbitrary single scattering albedos or the most anisotropic phase function. Within reason it can produce in a single computer run the fluxes at any position in the medium — that is, not just the fluxes which emerge from the top or bottom of a layer. The only real disadvantage arises because of the statistical nature of the answers. The accuracy in predicted flux at a point is proportional to the square root of the number of photons used, so that extreme accuracy can become expensive. For much the same reason the method cannot easily be reduced in complexity to form the basis of a simple parameterization scheme in operational models. In that context it can be used only as a test-bed for schemes based on other methods.

The general technique was described in detail by Cashwell and Everett (1959) and was used to calculate light scattered by a point source within the atmosphere by Collins and Wells (1965). It has been applied extensively in a series of papers by Kattawar and Plass to the study of atmosphere-cloud-ocean radiation interactions (e.g. Plass and Kattawar, 1969). The basis of a typical program is as follows.

Suppose it is required to calculate the flux density into a detector on the bottom of a layer which is illuminated from above by the direct beam of the sun. The first photon enters the layer and travels a certain distance s before striking a scattering or absorbing centre. The probability PR of a photon travelling a certain distance without interaction is a function of the optical depth of the medium. Since the probability is equal to the fraction of radiation transmitted through the distance:

$$PR = \exp(-\tau_e) = \exp(-\int_0^s \kappa_e \rho ds) \qquad [4.42]$$

Thus for the first photon, the computer selects a random number, sets it equal to PR and calculates the distance s from the relation:

$$\int_0^s \kappa_e \rho ds = \tau_e = -\ln(1-PR) \qquad [4.43]$$

At the point of interaction the photon may be scattered or absorbed. If it was absorbed, the "bull at a gate" approach would simply remove the photon and start again with a new one. However, it is much more efficient to reduce the weight of the photon, initially unity, by an amount proportional to the absorption to extinction ratio of the elementary volume — that is, to a weight $\widetilde{W}$ equal to $\bar{\omega}_0$ — and carry on with the same photon. It will be scattered in some direction α away from its initial direction which is again determined by a probability distribution associated with the local phase function (e.g. Fig. 4.4). That is, the probability PP of a photon being scattered between 0 and α is given by:

$$PP(\alpha) = (1/2) \int_0^\alpha \widetilde{P}(\psi) d(\cos\psi) \qquad [4.44]$$

The computer again chooses a random number, sets it equal to $PP(\alpha)$, and finds α by solving equation [4.44] for the upper limit of the integration. The corresponding "azimuth" angle is independent of phase function and is chosen randomly between 0 and 2π. The photon continues on its way in its new direction and with its new weight. The process is repeated until the photon weight reduces to an insignificant level or until it leaves the layer altogether or until it arrives at the detector. The computer then repeats the whole process with other photons until there are sufficient weighted photons in the "bin" of the detector.

The technique is still inefficient and two further refinements are normally added. First, at each scattering event the probability P_B of scatter directly into the detector is calculated according to:

$$P_B = \widetilde{W}_n \widetilde{P}(\bar{\theta}, \bar{\phi}) \exp(-\tau_{Bn}) d\omega/4\pi \qquad [4.45]$$

where $d\omega$ is the solid angle subtended by the detector at the point of scatter, τ_{Bn} is the optical depth from that point to the detector, $\widetilde{W}_n$ is the current weight of the photon and $\widetilde{P}(\bar{\theta}, \bar{\phi})$ is the value of the phase function at the angle $(\bar{\theta}, \bar{\phi})$ defining the direction of the detector from the direction of travel before collision. The value of P_B is added to the other P_B's already in

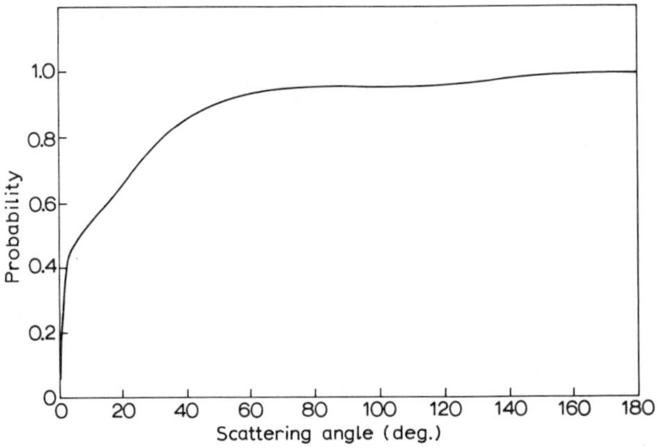

Fig. 4.4. Cumulative probability of photon scatter as a function of scattering angle. (After McKee and Cox, 1974.)

the "bin" of the detector, and the total flux at the end of a run is determined by the sum of all the P_B's arising from all scattering events by all the photons. In this way each photon is being used throughout its life. Second, any photon which escapes from the layer altogether is forced to undertake a collision before it leaves the medium, but has its weight reduced by an amount equal to 1 minus the probability of that event occurring.

Various refinements can be added in specific situations to improve accuracy or save computer time. For example, with spherical atmospheres it may be useful to reverse the photon path and trace backwards from the detector to the source (Collins et al., 1972).

4.2. *The adding method*

This technique has the advantage of being simple in concept and can be reduced in complexity for purposes of operational models. Indeed it is virtually a necessity in models which contain a number of plane-parallel layers where each layer has separately specified reflection and transmission properties. The basis is simple. If the reflectance and transmittance of two adjacent layers are known separately, the reflectance and transmittance of the combination of the two can be calculated. When the technique is used as a test-bed model one can divide the plane-parallel medium or atmosphere into layers so thin that the scattering and absorption of each layer is effectively that of single molecules or particles — that is, such that there is no multiple scatter within an individual layer and the phase function is that appropriate for single particles. It is then a matter of adding one layer at a time and calculating the new combined transmission and reflection at each addition. The layer thickness to ensure single scatter must be very small ($\tau \approx 10^{-15}$ to 10^{-20}) so that the pure adding procedure would take a long time. However, one can divide the medium into layers of significant thickness τ_T but which are sufficiently thin to be approximated as homogeneous layers. Then it is a matter of building up from $\tau = 10^{-15}$, say, to τ_T by a doubling procedure — by adding at each step a layer of the same thickness as the combined thickness from the previous step. One reaches the thickness τ_T very quickly and one can then carry on with pure adding over the required number of layers of thickness τ_T for which the reflection and transmission have been computed by doubling.

The adding procedure can best be illustrated by referring directly to the typical case of a multilayer atmosphere where the reflectance Re_n and the transmittance Tr_n have already been specified or calculated for each of the several coarse layers n of total number n_T. Suppose also that the fluxes are diffuse and have been described by some approximation akin to the classic two-stream concept. The general problem is to determine the flux profile of the atmosphere and in particular (for meteorological problems) the absorption within each layer.

Consider two adjacent layers where the subscripts a and b refer to the top and bottom layers respectively. It is easy to see that for downcoming radiation:

$$Re_{ab} = Re_a + Tr_a Re_b Tr_a/(1 - Re_a^* Re_b) \qquad [4.46]$$

$$Tr_{ab} = Tr_a Tr_b/(1 - Re_a^* Re_b) \qquad [4.47]$$

where the superscript asterisk (*) refers to illumination from below and the denominator in each equation accounts for multiple reflections between two layers. Similar equations apply for upwelling radiation:

$$Re_{ab}^* = Re_b + Tr_b^* Re_a^* Tr_b^*/(1 - Re_a^* Re_b) \qquad [4.48]$$

$$Tr_{ab}^* = Tr_b^* Tr_a^*/(1 - Re_a^* Re_b) \qquad [4.49]$$

Then, in the manner of Lacis and Hansen (1974), the following three steps can be carried out to evaluate the upward and downward flux densities at the interface between each layer.

(1) The layers are added, going down, to obtain $Re_{1,n}$ and $Tr_{1,n}$ for $n = 2$ to $n_T + 1$ and $Re_{1,n}^*$ and $Tr_{1,n}^*$ for $n = 2$ to n_T.

(2) The layers are added, going up, to obtain Re_{n_T+1-n,n_T+1} for $n = 1$ to $n_T - 1$ starting with the bottom or ground layer which may have an albedo α_g. That is $Re_{n_T+1} = \alpha_g$ and $Tr_{n_T+1} = 0$. After each layer is added, step 3 is performed.

(3) As two composite layers (say 1, n and $n+1$, $n_T + 1$) are added, the upward and downward fluxes U_n and D_n at the interface are determined:

$$U_n = Tr_{1,n} Re_{n+1,n_T+1}/(1 - Re_{1,n}^* Re_{n+1,n_T+1}) \qquad [4.50]$$

and:

$$D_n = Tr_{1,n}/(1 - Re_{1,n}^* Re_{n+1,n_T+1}) \qquad [4.51]$$

where U_n and D_n are fractions of the total incident flux.

The absorption over a given layer n is simply the difference between the *net* flux densities at the top and bottom of the layer.

The general adding procedure when the radiant energy field is to be specified precisely (in a manner more sophisticated than the classic two-stream approximation) involves the same principles as above. The difference is that the radiant fields and the reflection and transmission functions are generally matrix quantities and the "recipe" for adding is not so obvious from first principles. The technique is discussed by Hunt and Grant (1969) and the following "recipe" is from Hansen and Travis (1974):

$$
\begin{aligned}
\mathbf{Q}_1 &= \mathbf{Re}_a^* \mathbf{Re}_b \\
\mathbf{Q}_n &= \mathbf{Q}_1 \mathbf{Q}_{n-1} \\
\mathbf{S} &= \sum_{n=1}^{\infty} \mathbf{Q}_n \\
\mathbf{D} &= \mathbf{Tr}_a + \mathbf{S}\exp(-\tau_a/\mu_0) + \mathbf{S}\mathbf{Tr}_a \\
\mathbf{U} &= \mathbf{Re}_b \exp(-\tau_a/\mu_0) + \mathbf{Re}_b \mathbf{D} \\
\mathbf{Re}(\tau_a+\tau_b) &= \mathbf{Re}_a + \exp(-\tau_a/\mu)\,\mathbf{U} + \mathbf{Tr}_a^* \mathbf{U} \\
\mathbf{Tr}(\tau_a+\tau_b) &= \exp(-\tau_b/\mu)\,\mathbf{D} + \mathbf{Tr}_b \exp(-\tau_a/\mu_0) + \mathbf{Tr}_b \mathbf{D}
\end{aligned}
\qquad [4.52]
$$

where again the radiant energy field has been divided specifically into direct and diffuse components. Thus the $\mathbf{Tr}(\tau_a + \tau_b)$ is the transmittance relevant to diffuse flux emerging from the bottom of the two layers; and the total flux density from the bottom is given by the diffusely transmitted flux plus $\pi F \mu_0 \exp[-(\tau_a + \tau_b)/\mu_0]$ where the total incident irradiance in the direct beam is πF. Also, $F\mu_0\mathbf{D}$ and $F\mu_0\mathbf{U}$ are the diffuse downward and upward radiances at the interface between the two layers. The summation for $\mathbf{S}$ is over the number $n-1$ which indicates the number of times reflected radiation crosses the interface between layers going up. The sum can generally be terminated after only a few terms. A similar recipe to equation [4.52] exists for upwelling radiation.

CHAPTER 5

Solar Radiation within the Atmosphere

1. INTRODUCTION

The meteorologist is concerned primarily with the absorptions of solar radiation by ozone, water vapour, clouds and (to some extent) aerosols; all of which can and do occur at any height in the atmosphere and all of which involve absorption processes which must be treated separately.

From the meteorological standpoint it is only the absorbed solar radiation which is of direct interest since it is this energy which is involved in the thermodynamics of the system. However, since the amount absorbed at any level depends on the amount available at any level, almost all calculations of absorption involve as well a consideration of the conservative processes of scatter and reflection. In principle the full radiative transfer equation must be solved for any given situation. This in turn requires a full knowledge of the absorption and scattering coefficients as a function of wavelength for all atmospheric constituents, and a knowledge of how these coefficients vary with temperature and pressure.

In practice, the accuracy of such calculations is limited by the accuracy of the measured or computed values of the absorbing or scattering constituents. The primary limiting factor is the specification of clouds. Thus it is quite possible to solve the transfer equation to almost any degree of precision provided sufficient computer time is available, and in clear sky situations to an *accuracy* quite compatible with the media specification. However, since cloud amount is rarely known to better than 10% (to say nothing of cloud geometry) it is obviously futile in such situations to waste time on highly precise calculations of scatter and absorption. The work of the radiation scientist is therefore something of an art whereby the approximations necessary to solve the transfer equation in reasonable time are matched to the accuracy of the problem specification. He is concerned with parameterizing the problem so that the significant absorption processes are realistically dependent on the variable constituents. In this context the major value of a precise and complex treatment of the transfer equation is simply that it provides a test-bed for the parameterizations.

Figs. 5.1 and 5.2 from Braslau and Dave (1973) are an example of a precise numerical solution of the transfer equation as applied to a standard dust-free mid-latitude clear-sky atmosphere (McClatchey et al., 1970). The

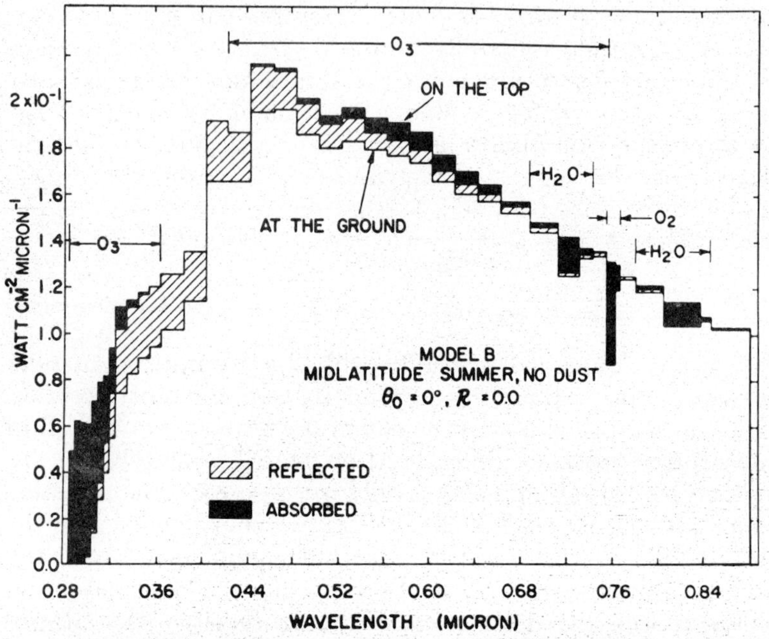

Fig. 5.1. Computed spectral distribution of the 0.285–0.88 μm region of the solar flux absorbed within the atmosphere, reflected back to space, and transmitted to a perfectly absorbing ground by a dust-free standard atmosphere, with total ozone of 0.318 cm atm and total water vapour of 2.925 g cm^{-2} in the vertical column of the atmosphere. Sun at zenith.

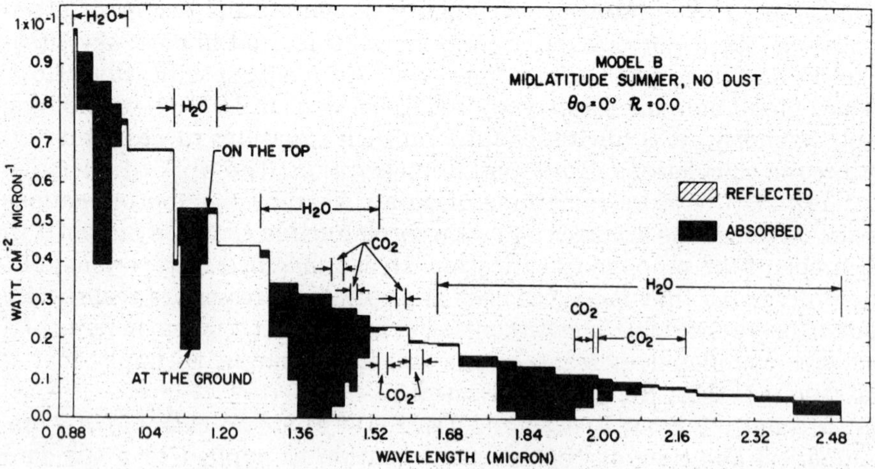

Fig. 5.2. Same as Fig. 5.1 but for the 0.88–2.5 μm region.

figures refer to the effect of the complete atmosphere overlying a non-reflecting ground and to a sun vertically overhead. They serve to illustrate the relative magnitudes of absorption within the atmosphere, scatter back to space and transmission to the surface — all as a function of wavelength. They illustrate that ozone absorption occurs primarily in the ultra-violet (Hartley and Huggins bands) to 0.36 μm, and in the visible regions from about 0.42 to 0.74 μm (the Chappuis bands); that Rayleigh scatter back to space becomes insignificant at wavelengths longer than about 0.8 or 0.9 μm; and that absorption due to various water vapour bands is confined primarily to wavelengths longer than about 0.8 μm.

2. OZONE ABSORPTION

We follow here the work of Lacis and Hansen (1974) which is a classic example of the derivation of a parameterization scheme whose validity has been tested by comparison with the results of a complex but precise numerical solution to the transfer equation. Their precise comparative solution was obtained by the "adding" method.

Their parameterization is based on the fact that ozone absorption occurs high in the atmosphere where there is little scattering because of the low molecular density. Thus absorption concerns primarily the downward direct solar beam and the diffuse radiation scattered or reflected upwards by the underlying atmosphere and ground. In both cases the absorption is primarily the pure exponential attenuation at each individual wavelength which occurs when the source function in the transfer equation can be neglected. Note that it is still necessary to take some account of Rayleigh scatter by the atmosphere since it determines the upwelling diffuse flux which can be absorbed.

Fig. 5.3 gives the two curves of solar absorption as a function of ozone path appropriate to the absorption regions in the ultraviolet and visible parts of the spectrum. The different behaviour of the curves reflects the fact that the Chappuis (visible) bands are "weak" in that absorption continues to increase almost in proportion to ozone amount. The ultraviolet absorption bands are "strong" and saturate very quickly. In fact, the weak bands of the visible region absorb the larger fraction of solar energy for path lengths longer than 1 cm (the global mean vertical path length of ozone is about 0.3 cm — Stergis et al., 1965). The solid curves of the figure are the best-fit relations for absorptance a (referred to the complete solar spectrum) as a function of ozone path x in centimetres:

$$a_{\mathrm{oz}}^{\mathrm{vis}}(x) = \frac{0.02118x}{1 + 0.042x + 0.000323x^2} \qquad [5.1]$$

and:

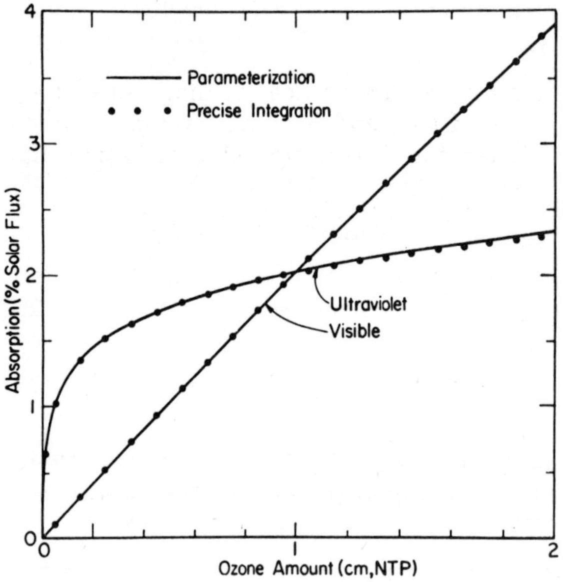

Fig. 5.3. Percent of total solar flux absorbed as a function of ozone amount for the ozone absorption coefficients of Howard et al. (1961) at -44°C. The dotted curves are the results of integrations over wavelength using the wavelengths at which Howard et al. tabulate the absorption coefficient. The parameterization refers to equations [5.1] and [5.2]. (From Lacis and Hansen, 1974.)

$$a_{oz}^{uv}(x) = \frac{1.082x}{(1+138.6x)^{0.805}} + \frac{0.0658x}{1+(103.6x)^3} \qquad [5.2]$$

for which the maximum "error" occurs in the second equation and is less than 0.5% over the range 10^{-4} to 1.0 cm. Such an error is much less than that involved in the precise calculations as a result of inaccuracy in the specification of the absorption coefficients. The total absorptance is the simple sum of equations [5.1] and [5.2].

The ozone amount traversed to level i (counting downwards) in the slant path of the downward direct beam of the sun is given by:

$$x_i = u_i m_r \qquad [5.3]$$

where u_i is the ozone amount in a vertical column above the ith layer and m_r is the relative air mass or magnification factor. m_r needs to be an air mass where earth curvature and the average scale height of ozone is taken into account (see equation [3.16]). Rodgers (1967b) uses a form of equation [3.16] appropriate to a scale height of 8 km. Namely:

$$m_r = 35\mu_0/(1224\mu_0^2+1)^{1/2} \qquad [5.4]$$

where μ_0 is the cosine of the solar zenith angle θ.

The ozone path traversed to level i by the upwelling diffuse radiation is given by:

$$x_i^* = u_t m_r + \bar{m}_r(u_t - u_i) \qquad [5.5]$$

where u_t is the total amount of ozone in the vertical column above the reflecting layer (e.g. the ground or cloud top) and $\bar{m}_r$ is an appropriate and approximate mean magnification factor for diffuse upwelling radiation. In this particular context, Lacis and Hansen (1974) calculated $\bar{m}_r$ to be 1.9 — a value obtained by empirical tuning of $\bar{m}_r$ so that the overall parameterization agreed with the precise solution.

Thus the final expression for ozone absorptance in the ith layer (i.e. the fraction of the total solar constant I_0 which is absorbed in the ith layer) becomes:

$$a_{oz,i} = \mu_0\{a_{oz}(x_{i+1}) - a_{oz}(x_i) + \bar{\alpha}(\mu_0)[a_{oz}(x_i^*) - a_{oz}(x_{i+1}^*)]\} \qquad [5.6]$$

and the equation can be applied to successively lower levels to build a vertical profile of the radiation absorption.

The albedo $\bar{\alpha}(\mu_0)$ contains the contribution of the underlying atmosphere and of the underlying surface which may be ground or cloud. Thus:

$$\bar{\alpha}(\mu_0) = \alpha_R(\mu_0) + [1 - \alpha_R(\mu_0)](1 - \alpha_R^*)\alpha_g/(1 - \alpha_R^*\alpha_g) \qquad [5.7]$$

an expression which takes account of multiple reflections between atmosphere and the surface which must be assumed to be a Lambertian diffuser. $\alpha_R(\mu_0)$ is the zenith angle dependent albedo resulting from Rayleigh scatter by the atmosphere underlying the ozone layer. α_R^* is the corresponding albedo of this scattering atmosphere presented to upwelling diffuse radiation which has been reflected from the ground. For purposes of the present calculation this last albedo is not dependent on zenith angle and can be put equal to the mean value of $\alpha_R(\mu_0)$ defined as follows:

$$\alpha_R^* = 2\int_0^1 \alpha_R(\mu_0)\mu_0 d\mu_0 \qquad [5.8]$$

For clear skies, Lacis and Hansen (1974) have parameterized $\alpha_R(\mu_0)$ as a function of μ_0 according to the formula:

$$\alpha_R(\mu_0) = 0.219/(1 + 0.816\mu_0) \qquad [5.9]$$

which in turn gives α_R^* equal to 0.144 when inserted in equation [5.8]. Note, however, that although this expression is based on a physical model of a pure Rayleigh scattering atmosphere, it has been tuned slightly to give the best comparison between the parameterized ozone absorption and the profile calculated by the precise adding method. It is therefore strictly applicable as a measure of Rayleigh scatter only in the present context (cf. equation [6.21] of Chapter 6). For cloudy skies, the simplest procedure is to neglect Rayleigh scatter altogether on the basis that the amount of atmosphere above average clouds is small. Calculation of cloud reflectivity is a notoriously inaccurate business so that the approximation is not significant.

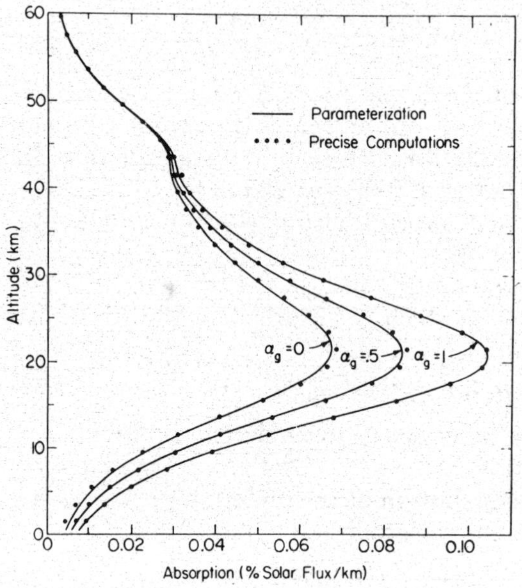

Fig. 5.4. Absorption of solar radiation by ozone, for the solar zenith angle = 60°, three values of the ground albedo α_g, and a standard atmosphere. Comparison between parameterization (equation [5.6]) and precise calculations. (From Lacis and Hansen, 1974.)

Fig. 5.4 gives a comparison of absorption profiles calculated by the parameterization scheme with those profiles obtained by the precise adding method. The comparison is for three surface albedos underlying a standard atmosphere with the sun at a zenith angle of 60°. The "accuracy" probably exceeds that with which the ozone concentration itself is likely to be known except in the most refined situations. Thus the agreement between curves is typically within 1% but may reduce to 10% at altitudes below 15 km when the zenith angle becomes large but when, on the other hand, the energies involved are small.

A major lesson from Fig. 5.4 is the great significance of ground albedo in determining an absorption process in the upper atmosphere. A simple treatment which dealt only with the downward direct beam — for example by reference to equation [5.6], $a_{oz,i} = \mu_0[a_{oz}(x_{i+1}) - a_{oz}(x_i)]$ — would be appropriate only in the grossest calculations.

3. WATER VAPOUR ABSORPTION IN CLEAR SKIES

Absorption of solar radiation by water vapour is the major source of solar radiative heating within the atmosphere itself. At first glance it should be a fairly easy process to parameterize since the absorption occurs at near-infrared wavelengths where Rayleigh scatter is negligible. Unfortunately, although this is true in clear skies, the presence of clouds and their associated Mie scatter at the same levels in the atmosphere as the water vapour re-introduces the necessity to deal with scattering source functions. Furthermore, the absorption coefficient of water vapour is highly dependent on wavelength in a manner which is not as yet known accurately over all parts of the spectrum. These coefficients have also a significant dependence on pressure and temperature which must be accounted for when dealing with a substance which can exist at all atmospheric levels. All estimates of the functional dependence of the absorption of solar radiation by water vapour of the free atmosphere are based on one of two sets of laboratory data.

First, there are the early measurements by Fowle (1912, 1915) of water vapour absorption at various wavelengths. These were assembled by Kimball (1927, 1930) into the simple curve of total absorptance a versus total water path u which is quoted in the Smithsonian Meteorological Tables. Subsequently, a variety of different curves were derived from the same data (Mügge and Möller, 1932; Yamamoto and Onishi, 1952; Houghton, 1954). This unhappy state of affairs was discussed in a classic paper by McDonald (1960) who produced his own $a(u)$ relation, but who admitted that, because of the poor documentation of Fowle's data, he could not claim an absolute accuracy of better than 30%.

Second, there are the relatively modern laboratory measurements by Howard et al. (1955). These were used by Roach (1961) and by Yamamoto (1962) to derive further relations between total absorptance and total water path. Yamamoto's relation is perhaps the most authoritative to date and although his results differ from those of McDonald, he points out that they do so in a manner which supports McDonald's method of treating the data available to him.

Fig. 5.5 is taken from Yamamoto's (1962) paper and shows the various spectral and total absorptances at STP defined as in the caption to the figure. The overall absorption curve (curve *6* in the figure which includes the slight additional CO_2 and O_2 absorption — refer to Fig. 5.2) can be approximated to better than 1% accuracy over the range of water paths 10^{-2} cm $< y < 10$ cm by:

$$a_{wv} = 2.9y/[(1 + 141.5y)^{0.635} + 5.925y] \qquad [5.10]$$

where the absorptance is with respect to the entire solar spectrum (refer to Lacis and Hansen, 1974).

The monochromatic absorptances are functions of temperature and pressure in a manner dependent on whether the individual absorption lines are weak or strong. For example, in a homogeneous medium the pressure dependence ranges from nil for weak lines to a square root dependence for very strong lines as a result of line saturation (refer to Chapter 7). Temperature affects both line width and line intensity but the usual practice is to assume an overall square root temperature dependence of the effective optical path

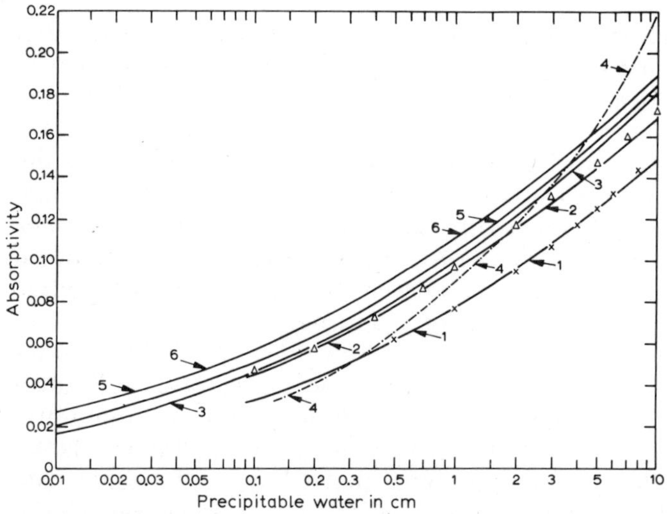

Fig. 5.5. Absorptances of the near-infrared bands. Curve *1* = absorptance summed over the 0.72-, 0.8-, 0.94-, 1.1-, 1.38-, 1.87-μm H_2O bands (× = absorptances estimated by McDonald (1960) over the same bands). Curve *2* = absorptances over the 0.94-, 1.1-, 1.38-, 1.87-, 2.7-, 3.2- and 6.3-μm H_2O bands (Δ = absorptances estimated by Möller (1957) over the same bands). Curve *3* = absorptances over all the H_2O bands considered. Curve *4* = absorptances estimated by Houghton (1954). Curve *5* = absorptances over the H_2O and CO_2 bands. Curve *6* = absorptances over the H_2O, CO_2 and O_2 bands. (From Yamamoto, 1962.)

on the grounds that, since the range of absolute temperature is not great within the atmosphere, the error will not be significant.

A completely accurate calculation of integrated absorptance over all wavelengths and for each individual atmospheric profile (which is generally far from homogeneous) is as yet quite impossible. Furthermore it would probably never be warranted, although there is still some room for improvement over the presently adopted practice of correcting the actual element of water path Δy to an effective element Δy_e according to:

$$\Delta y_e = \Delta y(p/p_s)^n (T_s/T)^{1/2} \quad [5.11]$$

Here p_s and T_s are standard pressure and temperature respectively and n is some number between 0.5 and 1.0. There is little theoretical justification for using a single such pressure correction for the entire spectrum of the water vapour bands. Paltridge (1973) has shown from experimental aircraft measurements that a linear overall pressure correction is at least better than none at all (cf. Kaplan, 1952). His experimental and linearly pressure corrected absorptance values are compared with Yamamoto's calculation in Fig. 5.6.

Thus in a form analogous to that of equation [5.6], absorptance referred to the total solar constant by the *i*th layer in clear skies is given by:

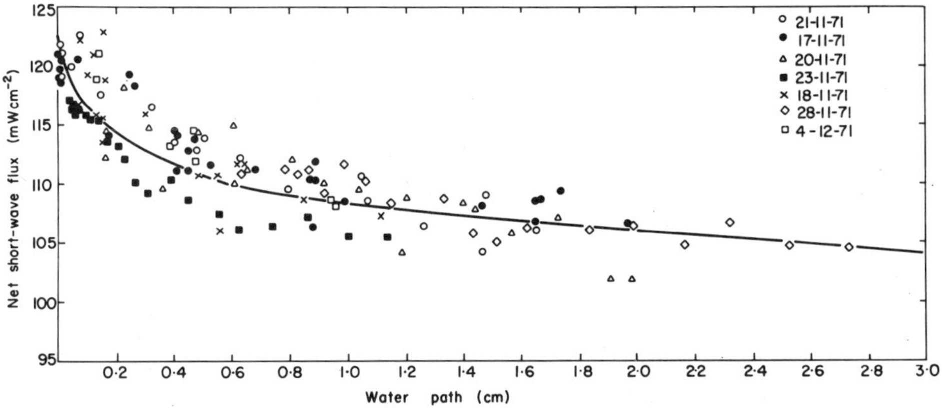

Fig. 5.6. Normalized net short-wave flux as a function of water path (linearly pressure corrected) as measured from an aircraft on the seven clear-sky days. The solid curve is calculated from Yamamoto's theoretical relation (Paltridge, 1973). The net flux at zero water path is 122.5 mW cm^{-2}, and is the value to be expected at heights above the atmospheric water vapour when the sun is vertically overhead. The fractional depletion below this value with increasing water path is the absorptance.

$$a_{wv,i} = \mu_0\{a_{wv}(y_{i+1}) - a_{wv}(y_i) + \alpha_g[a_{wv}(y_i^*) - a_{wv}(y_{i+1}^*)]\} \qquad [5.12]$$

where, in terms of vertical water path u and in turn the usually measured or computed parameters pressure p, temperature T and specific humidity q:

$$y_i = m_r u_i = \frac{m_r}{g}\int_0^{p_i} q\left(\frac{p}{p_s}\right)^n\left(\frac{T_s}{T}\right)^{1/2} dp \qquad [5.13]$$

and:

$$y_i^* = \frac{m_r}{g}\int_0^{p_g} q\left(\frac{p}{p_s}\right)^n\left(\frac{T_s}{T}\right)^{1/2} dp + \frac{\bar{m}_r}{g}\int_{p_i+1}^{p_g} q\left(\frac{p}{p_s}\right)^n\left(\frac{T_s}{T}\right)^{1/2} dp \qquad [5.14]$$

Here p_i, p_{i+1} and p_g are the pressures at the top of the ith layer, at the bottom of the ith layer and at the ground respectively. The magnification factor $\bar{m}_r$ for the upwelling diffuse radiation could be subject to tuning if a more accurate comparative model was available (cf. the previous section). Meanwhile the Elsasser figure of 1.66 suffices (Chapter 7). The factor g is the acceleration due to gravity.

Calculated water vapour absorption profiles in terms of heating rate are given in Fig. 5.7 for a standard atmosphere, solar zenith angle of 60°, a surface albedo of 0.07 (≈ sea surface), and a typical mid-latitude winter water vapour profile where the total precipitable water is equal to 0.8 cm. The three curves correspond to different degrees of pressure correction and the difference between them is as good a measure as any of the order of likely error in any such calculation based on present knowledge. The significant

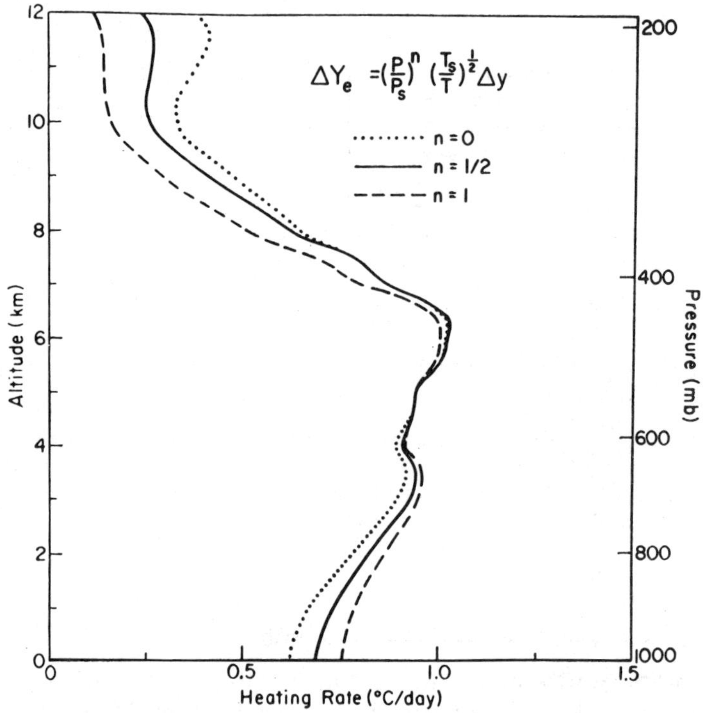

Fig. 5.7. The effect of pressure scaling on the heating rate in a clear sky. The solar zenith angle is 60° and the standard atmosphere and water vapour distribution of McClatchey et al. (1972) are assumed. The computations use Yamamoto's (1962) absorptance (equation [5.10]). The ground albedo is 0.07. (From Lacis and Hansen, 1974.)

qualitative feature is the dominant absorption at high altitude because of the rapid saturation of the absorbing lines.

It should be noted that surface albedo varies very significantly with wavelength (see Chapter 6) so that α_g for the above water vapour computations should strictly be an albedo appropriate to near-infrared wavelengths. Similarly, the α_g associated with the earlier ozone absorption discussion should be some mean value appropriate to ultraviolet and visible wavelengths.

4. THE EFFECT OF CLOUD

Introduction of clouds into an atmospheric profile vastly complicates the radiative calculations because of the necessity to re-introduce the scattering source function in the transfer equation — a source function furthermore which involves Mie scatter and which in turn is a complex function of cloud microphysics. It is of course possible to nominate some ideal cloud of arbitrarily specified geometry and microphysics and perform very precise computations to solve the transfer equation. However, it is rarely possible to

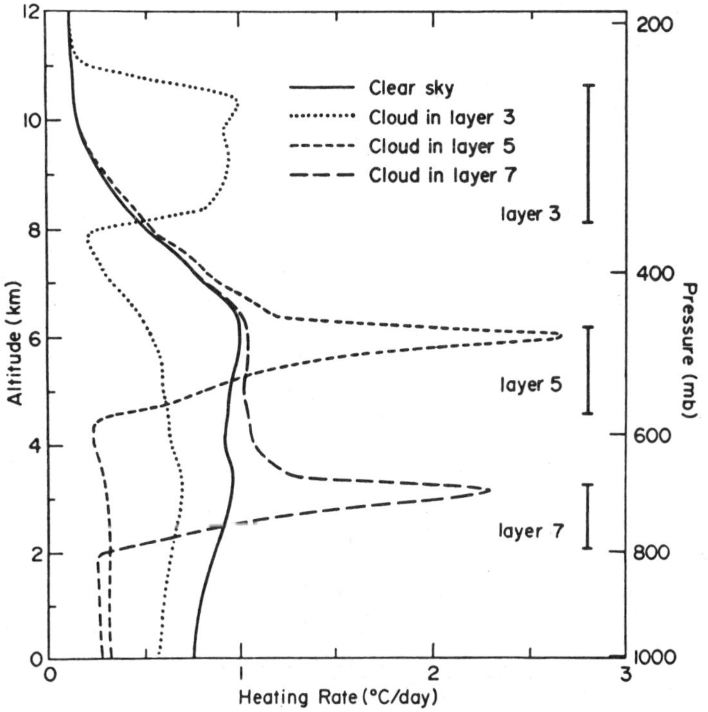

Fig. 5.8. The effect of cloud scatter on the heating rates in the troposphere computed for water vapour absorption when clouds are inserted in the layers shown. Solar zenith angle of 60° and 100% relative humidity in cloud layer. (From Lacis and Hansen, 1974.)

measure, let alone predict in a model, the relevant characteristics required to determine the gross parameters of total reflectance and transmittance of a cloud. Thus the accuracy (as opposed to the precision) of the associated radiative calculations may be reduced by an order of magnitude.

That cloud scatter plays a dominant role in determining solar absorption by water vapour within the atmosphere itself — quite apart from its major influence on the radiant energy available at the ground — can be appreciated by reference to Fig. 5.8. The profile of the heating rate of a typical clear-sky atmosphere (cf. Fig. 5.7) is compared with the profiles obtained by insertion of "typical" cloud at three different levels. The difference in each case results entirely from the effect of cloud scatter on the effective water vapour path length. No account has been taken of absorption by liquid water itself — a process which may also be important (see later). It is significant that the differences are primarily in the *distribution* of absorption with height. Owing to saturation of the absorption bands, the effect of cloud on *total* water vapour absorption by the atmosphere is not very large.

The literature on the subject is surfeited with precise numerical solutions of the transfer equation in specific cloud and specifically cloudy atmospheres. Their major value is as test-beds for any parameterization scheme for use in an operational model. Almost invariably such a scheme is some version of the two-stream or Eddington approximation discussed in the previous chapter.

The two-stream approximation is inherently inaccurate but is the most commonly used on the grounds of computational simplicity and a compatibility with the very low accuracy of cloud specification. It is based on the assumption of completely diffuse scatter within all parts of the cloud and in its unmodified form automatically excludes consideration of albedo, say, as a function of solar zenith angle. Referring to the relevant discussion in Chapter 4, the transfer equation reduces to consideration of only the upward and downward "stream" radiances $N\uparrow$ and $N\downarrow$ respectively and becomes:

$$(1/3^{1/2})\frac{dN\uparrow}{d\tau_e} = -N\uparrow + N\uparrow\bar{\omega}_0(1-bf) + N\downarrow\bar{\omega}_0 bf \quad [5.15]$$

$$-(1/3^{1/2})\frac{dN\downarrow}{d\tau_e} = -N\downarrow + N\downarrow\bar{\omega}_0(1-bf) + N\uparrow\bar{\omega}_0 bf \quad [5.16]$$

where $\bar{\omega}_0$ is the single scattering albedo equal to $d\tau_{sc}/d\tau_e$, which in turn is equal to 1 for a pure scattering medium. τ_{sc} and τ_e are the scattering and total extinction optical depths respectively. bf is the fraction of the radiation scattered into the backward hemisphere on a single scatter. The $3^{1/2}$ is again a magnification factor which takes account of the fact that the mean path length of diffuse radiation is longer than that of a direct vertical beam. Since path length enters non-linearly into solutions of the transfer equation, the adoption of a fixed magnification factor representative of all wavelengths is an approximation adopted for convenience and its value can be justifiably tuned to improve a particular parameterization within the likely range of 1.3–2.0 (Goody, 1964, p. 244). In the present application, Neiburger (1954) used the factor (1/2.0) instead of the factor $(1/3^{1/2})$ adopted by Sagan and Pollack (1967).

Application of appropriate boundary conditions allows analytic solution (from Sagan and Pollack, 1967) of equations [5.15] and [5.16] for the fractional transmission Tr, reflection Re and absorption a of a complete and homogeneous cloud layer:

$$Re = \frac{(u+1)(u-1)[\exp(\tau_{ef}) - \exp(-\tau_{ef})]}{(u+1)^2\exp(\tau_{ef}) - (u-1)^2\exp(-\tau_{ef})} \quad [5.17]$$

$$Tr = 4u/[(u+1)^2\exp(\tau_{ef}) - (u-1)^2\exp(-\tau_{ef})] \quad [5.18]$$

and:

$$a = 1 - Re - Tr \quad [5.19]$$

where $u^2 = (1-\bar{\omega}_0 + 2bf\bar{\omega}_0)/(1-\bar{\omega}_0)$ and $\tau_{ef} = 3^{1/2}u(1-\bar{\omega}_0)\tau^*$.

Layer depth enters the equations via the total vertical extinction optical depth τ^*. This equals $\tau_{sc} + \kappa_a u$ where κ_a is the absorption coefficient of water vapour and u is the effective water vapour amount in the vertical depth of the layer.

These exact solutions of an approximate problem can be used as the basis of any number of numerical techniques for investigating the radiative properties of a layered atmosphere where some layers are pure cloud and some are cloud free. Specific allowance for the direct beam (see Chapter 4) is probably not worthwhile in multilayer cloud models so that one deals only with the total vertical global flux densities. Lacis and Hansen (1974) adopt a parameterization scheme which is simply a very coarse application of the adding method using only nine atmospheric layers and a coarse numerical stratification of the water vapour absorption coefficient. The "recipe" for adding layers involves calculating the reflection and transmission of composite layers according to the scheme of equations [4.46] to [4.51]. In summary of the method, if the optical depth τ_{ef} and the single scattering albedo $\bar{\omega}_0$ of the individual layers are specified, the reflectance and transmittance of each layer can be calculated using equations [5.17] and [5.18]. The "recipe" then leads to the complete atmospheric profiles of the downward and upward flux densities once the flux density $I_{0\lambda} \cos\theta$ impinging on the top of the atmosphere is specified since the adding method allows one to calculate the reflectance and transmittance of the atmosphere between *any* two levels. However, it should be noted that equations [5.17] and [5.18] apply for a specific wavelength — or for a band of wavelengths over which τ_{ef} and $\bar{\omega}_0$ are constant. If this is not the case the spectrum must be divided into separate regions where τ_{ef} and $\bar{\omega}_0$ can be assumed constant and the process repeated for the individual bands. The total flux densities can then be obtained by a final summation over all the bands. Cloud drops have very little short-wave absorption (although see later) so that for most applications $\bar{\omega}_0$ can be set at a constant value very close to unity. Nevertheless their size distribution can certainly ensure that τ_{ef} varies with wavelength and therefore that a final wavelength integration is necessary. However, it is often a fairly good assumption that cloud drops are sufficiently large (relative to short wavelengths) that their scattering cross-section is independent of wavelength and is twice their geometrical cross-section. Mie theory shows that diffraction introduces the factor of 2. In such situations the reflectance and transmittance equations can be applied to the total spectrum.

Fritz (1954) added a degree of refinement to the two-stream approximation in that he devised a model which deals specifically with both the direct and diffuse radiation fluxes within a cloud and which is therefore capable of handling zenith angle dependence of the reflection function or albedo. However, as far as the present authors are aware, the model has been applied only to this specific aspect and has not been developed to the extent of formally deriving the associated transmission (and therefore absorption) function. The model is strictly applicable only to clouds of "large drops" where the size parameter $2\pi r/\lambda$ is 40 or more and where the single scatter pattern is strongly forward peaked.

His model envisages the "generation" of diffuse and isotropic radiant energy by the direct beam on its path down through the cloud. His definition of the direct beam includes the radiation which after a single-scattering event is scattered forward within 60° of the true direct beam. The diffuse photons are defined as those scattered at angles greater than 60°. Fritz derived an empirical formula for the rate of diffuse energy generation per unit volume n_g as a function of depth in the cloud:

$$n_g = 0.02(4m_r/L)^2 z \exp(-0.5656 z m_r/L) \qquad [5.20]$$

where m_r is relative air mass sec θ and L is a mean free path defined in terms of drop radius r and drop number density n as $1/\sum_i n_i \pi r_i^2$. Thus $1/L$ is in fact a volume scattering coefficient and z/L is an optical depth which refers only to the *geometrical* cross-section for scattering. (The actual scattering cross-section for large drops is twice the geometrical cross-section as a result of diffraction.)

The diffuse generated energy is governed by a typical diffusion equation of the form:

$$-\Gamma \mathrm{d}^2 n/\mathrm{d}z^2 = n_g \qquad [5.21]$$

where n is the number of photons per unit volume and Γ is a diffusion coefficient. By various machinations involving boundary conditions and semi-empirical relations between Γ and L and the velocity of the photons, Fritz (1954) obtains an expression for the diffuse upward flux emerging from the top of the cloud of depth h. This is also the cloud albedo α_c when the down-coming flux is normalized to unity. That is:

$$\alpha_c = -\left(\frac{\mathrm{d}n}{\mathrm{d}z}\right)_{z=0} = 1 - \{\exp(-0.5656 m_r h/L)[(h/L)(1-1.48m_r) + 3.54/m_r - 2.62] - 3.54/m_r - 2.62\}/[(h/L)+5.2] \qquad [5.22]$$

where h is the total thickness of the cloud.

As it stands the formula takes no account of the asymmetry of scatter towards the top of the cloud where a fraction of the forward scattered energy after single scatter escapes upwards out of the cloud top for any non-zero zenith angle. This fraction f_s will increase strongly with increasing zenith angle. Fritz made one form of correction for this. Here we adopt the arbitrary procedure of defining albedo as $\alpha_c + f_s$ where α_c remains as the albedo according to equation [5.22]. f_s was defined simply as $k_1/\cos\theta$ where k_1 is a constant of order 0.02.

Fig. 5.9 is a comparison (G.W. Paltridge and S. Twomey, unpublished) of the albedos predicted by the Fritz model corrected for asymmetry as above and those predicted by the precise doubling method of Twomey et al.

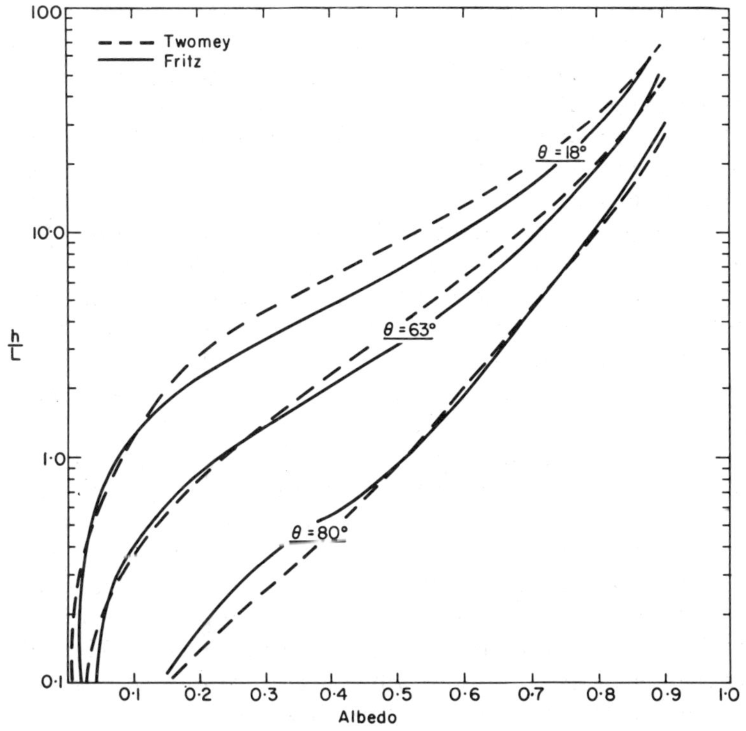

Fig. 5.9. Comparison of the Fritz (1954) model for cloud albedo with precise calculations by Twomey et al. (1967) using the adding method — for three solar zenith angles. The ordinate h/L (cloud depth over mean free path) is equal to one-half the vertical optical depth, since the definition of optical depth for large drops involves a drop cross-section equal to twice the geometric cross-section.

(1967) for a standard cloud of fixed drop size distribution (see case 6 of their paper). The constant k_1 has been adjusted (in fact to 0.02) to yield a minimum mean square deviation between the three pairs of curves in the figure where each comparative pair corresponds to a different solar zenith angle. The mean "error" over the range displayed is about 6%.

The major lessons from the exercise are that cloud albedo is enormously variable with solar zenith angle and that the accuracy of very simple cloud models which are capable of analytic solutions is not very good.

4.1. *Liquid water absorption*

The dominant effect of cloud on solar radiation *absorption* within the atmosphere is via the increase in the optical path of water vapour as a result of multiple scatter. Thus clouds tend to concentrate water vapour absorption within or near their boundaries, and in this sense are certainly absorbers of short-wave radiation. Usually this is at the expense of water vapour absorption below the clouds because of the tendency of the bands to saturate at optical paths greater than about 1 cm. The increase in the absorption of the whole atmosphere may therefore be small.

The question arises as to whether absorption by the liquid water (or ice) of clouds contributes a significant addition to the total solar absorption by virtue of the fact that ice and water have absorption bands different from the vapour. Experimental measurements of the difference between the net short-wave flux density at the top and bottom of clouds have virtually no hope of answering this question as anyone who has attempted the exercise will admit. Neiburger (1949) and Griggs (1968) have tried and by averaging their results for low-altitude stratiform clouds obtained short-wave absorptions of 7% and 4% respectively. The standard deviations were two or three times the averages. Robinson (1958) obtained values ranging between 13% and 29%. Even accepting any of these figures at face value, the problem still remains to separate the vapour absorption from the liquid absorption. It seems the question can be answered only by theoretical analysis. Such an analysis was attempted by Fritz (1958), but he relied upon absorption coefficients appropriate to bulk liquid water — not on the coefficients appropriate to water dispersed in the form of droplets. The absorption by drops may bear no obvious relation to that of bulk water. The full Mie theory treatment is necessary. As far as the present authors are aware, this particular problem has not been examined in detail.

Rough calculations show that the extra liquid water absorption is certainly small because the relevant bands occur only in the "tail" of the solar spectrum at wavelengths longer than about 1 μm — and also because there is some overlap between the vapour and liquid absorption bands. For instance, consider a typical cloud of 0.001 g cm^{-2} total liquid water in the vertical column. Consider also two versions of such a cloud: in one case let all the drops have a radius of 3 μm and in the second case let all the drops have a radius of 10 μm. The two cases correspond to vertical optical depths for scatter τ_{sc} of 50 and 15 respectively on the basis of the large-drop approximation where the scattering cross-section is twice the geometrical cross-section. (Note that for a given amount of liquid water the optical depth decreases with increasing drop size because the volume of a drop is proportional to r^3 while its cross-section is proportional to r^2.)

Irvine and Pollack (1968) have tabulated the single scattering albedo $\bar{\omega}_0$ as a function of wavelength for drops of these and other sizes (see also Fig. 8.5). Now $1-\bar{\omega}_0$ is a measure of the fractional absorptance of a drop. That is, $1-\bar{\omega}_0 = \kappa_a/\kappa_e \approx d\tau_a/d\tau_{sc} \approx \tau_a/\tau_{sc}$ where it is assumed that the scattering is very much larger than the absorption. Thus the fractional absorption $a_{CL}(\lambda)$ by a cloud of a vertical beam of wavelength λ is given by:

$$a_{CL}(\lambda) \approx 1 - \exp\{-[1-\bar{\omega}_0(\lambda)]\tau_{sc}\} \qquad [5.23]$$

Considering only the direct beam of a sun vertically overhead, and assuming the cloud to be above all the atmospheric water vapour so that the impinging irradiance is that of the extra-terrestrial spectrum, one can calculate the total absorption by the liquid water in the direct beam (expressed as a fraction of the solar constant) as:

$$a_{\mathrm{CL}} \approx (1/I_0) \int_0^\infty I_{0\lambda}(1 - \exp\{-[1 - \bar{\omega}_0(\lambda)]\tau_{\mathrm{sc}\lambda}\})\,\mathrm{d}\lambda \qquad [5.24]$$

where the values of $I_{0\lambda}$ can be obtained from Table 3.1 and the integration need be performed only over the wavelengths of significant absorption greater than 1 μm. Such calculation yields an a_{CL} of 3.0% for the cloud of 3-μm drops and of 4.7% for the cloud of 10-μm drops.

If the cloud is put at the bottom of an atmosphere containing, say, 1 cm of precipitable water vapour, the same calculations can be done with $I_{0\lambda}$ replaced by I_λ — that is, by the solar spectrum after the beam has passed through the water vapour. This spectrum is given by:

$$I_\lambda = I_{0\lambda} \exp(-\kappa_{\mathrm{a}\lambda} u) \qquad [5.25]$$

where $\kappa_{\mathrm{a}\lambda}$ here refers to the mass absorption coefficient of water *vapour* at various wavelengths given, say, by Fowle (1912) and reported by Fritz (1958); and the u is the 1 cm optical path of the vapour. For this situation the liquid water absorptions by the cloud reduce to 0.4% and 1.4% for the cloud of 3-μm drops and the cloud of 10-μm drops respectively.

These rough calculations serve to show that cloud drop absorption is small even when the cloud is not protected by overlying water vapour. However, it is significant. Multiple scatter within a cloud can increase the effective optical path by almost an order of magnitude and may double the absorption figures above. As a comparison, the net long-wave loss of energy from clouds — which certainly may have a controlling influence on the character and dynamics of certain clouds as is discussed in Chapter 10 — has a maximum value of about 8 mW cm^{-2}. Daytime solar absorption of a few percent of the 136 mW cm^{-2} of the solar constant may go a long way towards neutralizing the effect of this long-wave loss.

Finally it should be emphasized that these calculations assume pure water drops. It is possible that aerosol particles may attach to drops and alter their absorbing properties. The subject has been discussed by Twomey (1970) among others and raises the possibility of a subtle, and perhaps highly significant, mode of climate modification.

5. RAYLEIGH SCATTER

The phase function for Rayleigh scatter (i.e. by particles very much smaller than the wavelength of light) is a known and simple analytic function

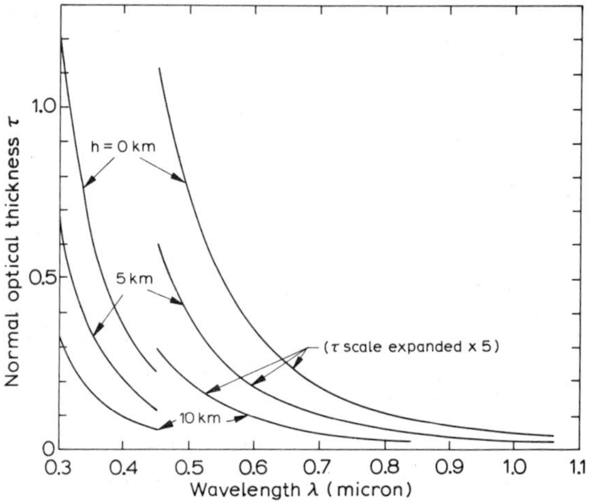

Fig. 5.10. Rayleigh normal optical thickness between the top of the atmosphere and the altitudes given on the figure. (From Marggraf and Griggs, 1969.)

of the scattering angle ψ:

$$\tilde{P}(\psi) = \tfrac{3}{4}(1+\cos^2\psi) \qquad [5.26]$$

and as a result the computations of such scatter within the atmosphere can be, and are generally, carried out to very high accuracy. The impetus for such accurate analysis comes mainly from activities associated with remote sensing. In most meteorological applications the subject can be treated somewhat cursorily since despite the very complicated dependence of scattered intensity on wavelength, solar zenith angle and on observing angle, this dependence is not highly variable with the state of the atmosphere itself. Nevertheless, there is at least one situation where a detailed consideration is necessary — namely in the analysis of clear but dusty sky situations where the Rayleigh scatter and the scatter by dust occur together in both wavelength and spatial position and are of comparable magnitude.

Rayleigh scatter decreases with increasing wavelength (roughly in proportion to λ^{-4}). Fig. 5.10 shows that the total vertical optical depth of the clear atmosphere decreases rapidly with wavelength from about unity in the ultraviolet. The figure is from Marggraf and Griggs (1969) and gives also the vertical optical depth to the top of the atmosphere from various heights h above the surface. They supply an approximate analytic expression for τ_R as a function of λ in microns and h in kilometres.

$$\tau_R(\lambda, h) = 0.0088\lambda^{(-4.15+0.2\lambda)}[\exp(-0.1188h-0.00116h^2)] \qquad [5.27]$$

Accurately computed values of τ_R are tabulated by Elterman and Toolin (1965).

The optical thicknesses are not sufficiently small to ignore the effect of

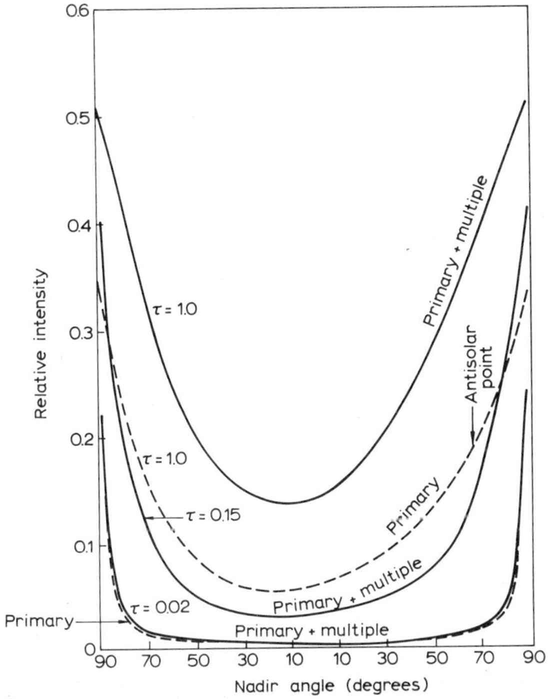

Fig. 5.11. Relative intensity of sunlight Rayleigh scattered back to space from an atmosphere overlying a "black" surface. The comparison is with single and multiple scatter for three optical depths. (From Coulson, 1959.)

multiple scatter which greatly complicates the analysis. Thus Fig. 5.11 gives the relative intensity (or relative radiance) of sunlight Rayleigh scattered back to space from an atmosphere where the underlying surface albedo is zero. The figure compares, for several optical depths, the intensities calculated when all orders of magnitude are considered and those calculated on the basis of single scatter according to the formula (derivable from the work of Hammad, 1948, and Hammad and Chapman, 1939, for instance):

$$N = \frac{F_0}{4} \tilde{P}(\psi)[\mu_0/(\mu + \mu_0)] \left\{1 - \exp\left[-\tau_R\left(\frac{1}{\mu} + \frac{1}{\mu_0}\right)\right]\right\} \qquad [5.28]$$

where μ and μ_0 are the cosines of the nadir and zenith angles respectively, and πF_0 is the incident parallel beam irradiance perpendicular to the beam which would be the spectral value of the solar constant $I_{0\lambda}$ when the source is the sun. The calculations in the figure assume F_0 to be unity.

The basic meteorological problem is to compute the total downward and upward flux densities F at various levels h in the atmosphere. The direct beam component in the downward direction is simply:

$$F{\downarrow}_{\mathrm{DIR}}(\tau_{\mathrm{R}}, h) = I_{0\lambda}\mu_0 \exp(-\tau_{\mathrm{R}}/\mu_0) \qquad [5.29]$$

and must of course be integrated over wavelength to give the corresponding total flux density. Evaluation of the monochromatic diffuse flux density or sky radiation $D{\downarrow}_\lambda$ requires calculation of the radiance at each zenith and azimuth angle $N_\lambda(\theta, \phi)$ which must then be integrated over the hemisphere according to:

$$D{\downarrow}_\lambda = \int N_\lambda(\theta, \phi) \cos\theta \, \mathrm{d}\omega \qquad [5.30]$$

Again the total sky radiation involves an integration of $D{\downarrow}_\lambda$ over all wavelengths.

Chandrasekhar (1950) has manipulated the problem so as to derive an expression for $D{\downarrow}_\lambda$ in terms of two functions γ_{e} and γ_{r} of the vertical optical depth τ_{R} and the cosine of solar zenith angle μ_0. Thus, for an atmosphere for which the underlying surface albedo is zero:

$$D{\downarrow}_\lambda = I_{0\lambda}\mu_0[\tfrac{1}{2}\gamma_{\mathrm{e}}(\mu_0, \tau_{\mathrm{R}}) + \tfrac{1}{2}\gamma_{\mathrm{r}}(\mu_0, \tau_{\mathrm{R}}) - \exp(-\tau_{\mathrm{R}}/\mu_0)] \qquad [5.31]$$

Physically, the functions $\gamma_{\mathrm{e}}/2$ and $\gamma_{\mathrm{r}}/2$ are the monochromatic global radiation flux densities resulting from the components polarized parallel and normal respectively to the vertical plane, for unit incident flux of unpolarized radiation when there is no reflection by the surface and when the source or sun is directly overhead. Ignoring polarization the quantity of interest is the function $Z(\mu_0, \tau_{\mathrm{R}}) = (\gamma_{\mathrm{e}} + \gamma_{\mathrm{r}})/2.0$.

Invariably the surface albedo is not zero so that the downcoming diffuse flux is augmented by radiation which has been reflected by the ground and is subsequently scattered so as to contribute to the downward flux density. Assuming diffuse reflection by the surface of albedo α_{g}, the augmented radiation can be included in the above equation by use of a function S which is dependent only on τ_{R} and not on μ_0, but which is derived in terms of the γ_{e} and γ_{r} above (see Chandrasekhar, 1950, or Diermendjian and Sekera, 1954). Then:

$$D{\downarrow}_\lambda = I_{0\lambda}\mu_0 \left[\frac{Z(\tau_{\mathrm{R}}, \mu_0)}{1 - S(\tau_{\mathrm{R}})\alpha_{\mathrm{g}}} - \exp(\tau_{\mathrm{R}}/\mu_0)\right] \qquad [5.32]$$

for the diffuse flux density at the bottom of the atmosphere where τ_{R} is the vertical optical depth.

The functions Z and S are not simple to evaluate and must be calculated by successive iteration in terms of τ_{R} and μ_0. Such calculations have been tabulated by Chandrasekhar and Elbert (1954). The functions are presented graphically in Fig. 5.12. The accuracy of a specific calculation of $D{\downarrow}_\lambda$ is determined by the effort expended in computing their values. Perhaps the simplest approach is that of Marggraf and Griggs (1969) which is to fit empirically a multivariate polynomial to the curves of Fig. 5.12 and use this polynomial directly.

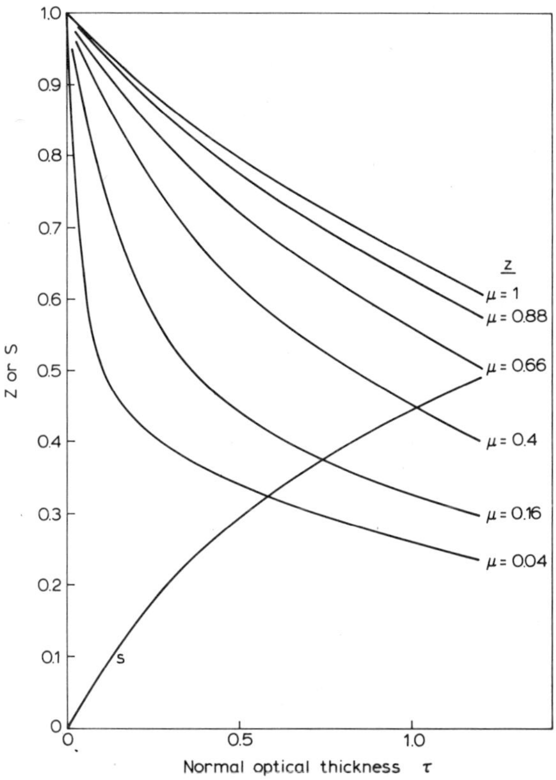

Fig. 5.12. $Z(\tau, \mu)$ and $S(\tau)$ functions versus normal optical thickness. (From Marggraf and Griggs, 1969.)

An equivalent expression to equation [5.32] exists for the total monochromatic diffuse flux $D{\uparrow}_\lambda$ emerging from the top of the atmosphere (Coulson, 1959). It involves the same Z and S functions and is:

$$D{\uparrow}_\lambda = I_{0\lambda}\mu_0 \left[1 - \frac{Z(\tau_R, \mu_0)(1-\alpha_g)}{1-\alpha_g S(\tau_R)}\right] \qquad [5.33]$$

Fig. 5.13 and Table 5.1 are from Coulson's paper. Fig. 5.13 shows that with increasing zenith angle the peak of the spectrum of scattered light tends towards longer wavelengths. This is true also for light scattered downwards. The figure contains also a comparison of the spectra of upward and downward scattered radiation for a large zenith angle of 84.3°. It is notable that the upscattered radiation is greater than the downscattered radiation. The difference is an increasing function of optical depth. Physically this is an interesting phenomenon in a medium where the scattering phase function is symmetrical. It arises because the radiation *available* for scatter

TABLE 5.1.

Values and comparisons of the total absolute flux density for 3200 Å < λ < 8000 Å at six different sun elevations (units of flux density: $W\,m^{-2}$ — after Coulson, 1959)

Sun elevation	1.1°	5.7°	11.5°	23.6°	53.1°	90.0°
F_0	15.4	77.0	154.0	307.9	615.8	769.9
$(F\uparrow)_0$	8.0	25.0	34.3	44.0	48.9	51.8
$(F\uparrow)_{25}$	9.7	37.1	62.4	103.6	179.9	214.8
$(F\uparrow)_{80}$	13.7	65.9	128.3	249.6	488.6	609.6
$(\alpha_0)_0$	0.519	0.325	0.233	0.143	0.079	0.067
$(\alpha_0)_{25}$	0.630	0.482	0.405	0.336	0.292	0.279
$(\alpha_0)_{80}$	0.890	0.856	0.833	0.811	0.793	0.792
$(F\downarrow)_0$	7.4	52.0	119.0	263.9	566.9	718.1
$(F\downarrow)_{25}$	7.6	53.2	122.1	272.4	581.2	740.1
$(F\downarrow)_{80}$	8.5	55.5	128.5	291.5	636.0	801.5

F_0 = incident flux on horizontal surface at top of atmosphere.
$(F\uparrow)_A$ = emergent flux through horizontal surface at top of atmosphere (subscript A represents value of surface reflectivity: 0, 0.25 or 0.80).
$(\alpha_0)_A$ = $(F\uparrow)/F_0$ = albedo: ratio of emergent to incident flux.
$(F\downarrow)_A$ = flux incident on horizontal surface at ground.

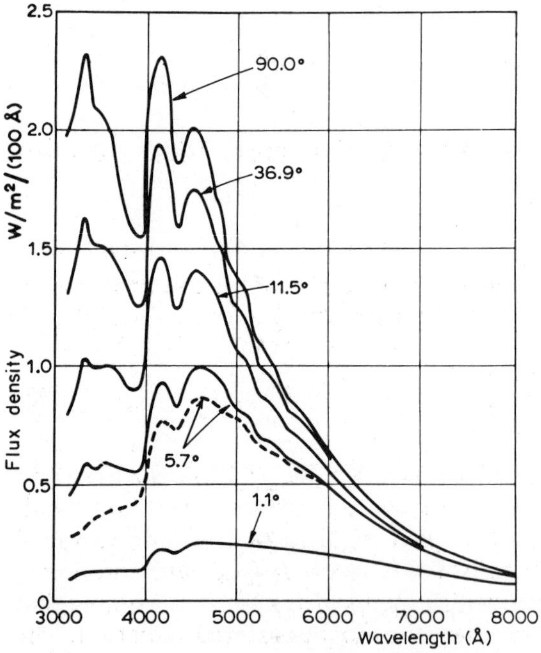

Fig. 5.13. Absolute monochromatic flux from the top of a plane-parallel model atmosphere as a function of wavelength for zero surface reflectivity and five different solar elevations. Dashed curve is for downcoming diffuse flux for solar elevation 5.7°. (Coulson, 1959.)

downwards decreases towards the bottom of the atmosphere. Thus for very high optical thickness the reflection coefficient of a pure scattering medium tends to unity.

Table 5.1 is the result of integration over wavelength of the curves of Fig. 5.13 and equivalent curves from Diermendjian and Sekera (1954). It gives the total flux density $F\uparrow$ emerging from the top of the atmosphere, the total flux density $F\downarrow$ out of the bottom, and the planetary albedo α_0 — each for the wavelength range 0.32–0.8 μm and for a pure Rayleigh atmosphere, and each for three different values of surface albedo and for various solar zenith angles. Note that it is quite possible for the surface albedo to exceed the planetary albedo.

Coulson (1959) derived a curve of total planetary albedo (taking account of the fact that the incident light extends over the whole spectrum) as a function of zenith angle θ. He integrated it appropriately so as to derive a total planetary albedo for the Rayleigh atmosphere itself $\bar{\alpha}_R$ when the effect of the earth's curvature is taken into account. His figure for $\bar{\alpha}_R$ (= 6.9%) is slightly larger than more modern calculations such as that of Lacis and Hansen (1974) who arrive at a figure of 6.3%. Coulson assumed an ozone absorption which was slightly low.

CHAPTER 6

Radiation at the Ground

1. INTRODUCTION

Referring back to Fig. 1.4 of Chapter 1, the major input of solar energy to the overall global system occurs at the ground. Compared with the total fractional solar absorption $[(1-\alpha_0) \approx 0.72]$ only about 0.18 is absorbed within the atmosphere itself partly by ozone but mostly by water vapour. Although it may be possible to treat atmospheric absorption of solar radiation in a cursory manner, most meteorological problems involve consideration of surface absorption in some detail. This entails the prediction of the downcoming solar radiation flux density $F_S\downarrow$ at ground level since surface albedo α_g is normally an "external parameter" of the system which is known (cf. the planetary albedo α_0 which is not) and the net solar input to the ground is then $F_S\downarrow(1-\alpha_g)$.

Atmospheric scatter ensures that the downcoming flux density has a diffuse as well as a direct beam component. At any given solar zenith angle θ the total vertical flux density (global radiation) is given by:

$$F_S\downarrow \; = F_{DIR} \cos\theta + \int N_{DIFF}\, d\omega \qquad [6.1]$$

where F_{DIR} is the irradiance of the direct beam on a surface perpendicular to the beam and N_{DIFF} is the radiance of the diffuse radiation and the integral is over all solid angles of the upper hemisphere. N_{DIFF} is in fact a strongly varying function of position in the sky as can be seen from the lower half of Fig. 6.1 which are Coulson's (1959) calculations of the zenith angle dependence of skylight in a pure Rayleigh atmosphere for various values of surface albedo and at a particular solar zenith angle and optical depth; or from Fig. 6.2 which shows measured values by C. Dorno (see N. Robinson, 1966) of sky luminance (a photometric measure roughly proportional to radiance) as a function of both zenith and azimuth angles — again for a particular solar zenith angle in a clear sky. In all practical meteorological problems it is always assumed that diffuse radiation is isotropic on the basis that in clear skies the vertical component of diffuse flux density is an order of magnitude less than the total global radiation $F_S\downarrow$ (refer to Figs. 6.3, 6.4 and 6.5) so that the error is insignificant; and that in cloudy skies, where $F_S\downarrow$ is made up entirely of diffuse radiation, any calculation of the

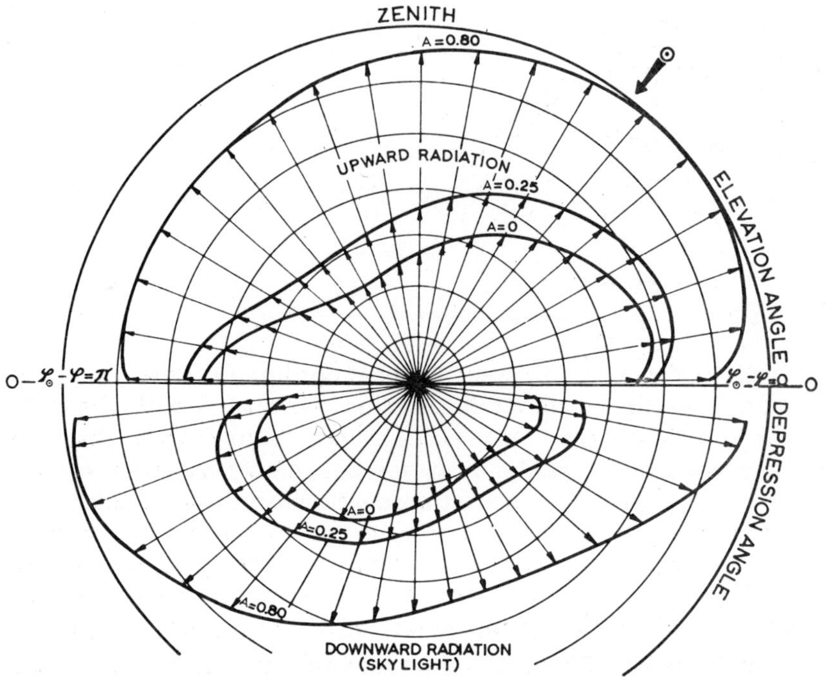

Fig. 6.1. Distribution of the intensity of upward radiation emerging from the top of the atmosphere and of the skylight. The sun's elevation is 53° and its position is indicated by ← ⊙. Large optical depth ($\tau = 1.0$); earth's surface albedo $A = 0$, 0,25, 0,80. (From Coulson, 1959.)

directional properties of the radiance would be a pure waste of time in view of the poor accuracy of cloud specification. In any event the large multiple scatter within clouds tends to induce isotropy. Thus equation [6.1] reduces to:

$$F_S\downarrow \ = I \cos\theta + D\downarrow \qquad [6.2]$$

where the symbol I is used here and in the rest of the chapter in place of F_{DIR} to be compatible with the solar constant I_0. $D\downarrow$ is the vertical diffuse flux density measured or computed as a single parameter.

All three components are a function of solar angle and as a frame of reference the measured clear-sky values obtained at Aspendale over several years are presented in Figs. 6.3, 6.4 and 6.5. The measurement accuracy of an individual point each of which is an hourly average is about 2 mW cm^{-2}. The spread of points outside this limit is an indication of the natural variability induced by atmospheric changes over a typical mid-latitude station (Aspendale has an annual mean atmospheric water content of 1.5 cm and an annual mean Ångström turbidity of 0.04 ± 20%) and of the difficulty of

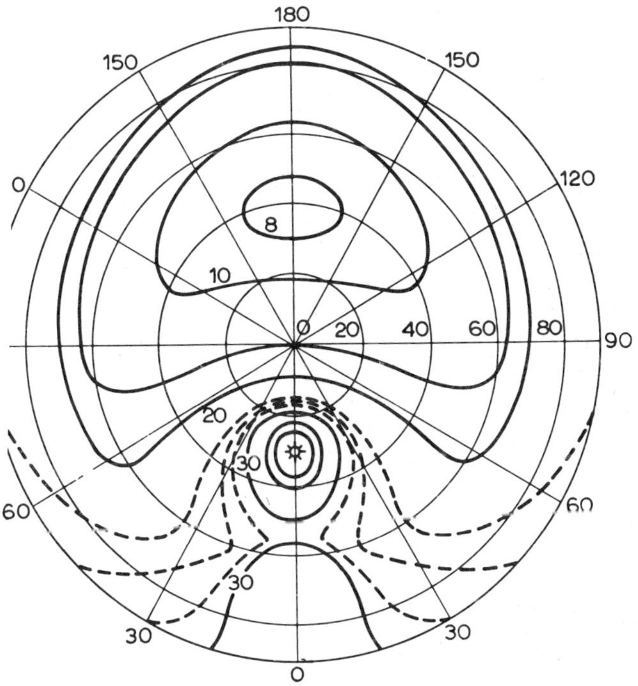

Fig. 6.2. Sky luminance distribution from C. Dorno. (Quoted by N. Robinson, 1966.)

defining a clear-sky from radiation records alone. Here the criteria were based on practical experience and were that, for the period of the day between 10.00 and 15.00 hours LAT, the direct radiance was greater than 60 mW cm^{-2}, global radiation was greater than twice D, and D was less than 8 mW cm^{-2} in winter and was less than 13 mW cm^{-2} in summer. Analytical curves fitted to the data by least squares analysis are given in the captions. For many meteorological applications and models *which do not involve feedback* to atmospheric transparency, one could do far worse than simply take these analytical curves as generally applicable input data with perhaps some correction for regions of excessive dust pollution and for extremes of total precipitable water.

Deacon (1969) has calculated charts of the clear-sky daily total global radiation input to the surface as a function of latitude and time of year. He used the daily inputs to the top of the atmosphere which are quoted in the Smithsonian Meteorological Tables, and assumed atmospheric transmissivities of 0.7 and 0.8. The first corresponds roughly to dusty and industrial localities and to moist tropical atmospheres, and the second is appropriate for a fairly clear atmosphere. Comparison with the observed totals at Aspendale (for instance) are good to 5% or better.

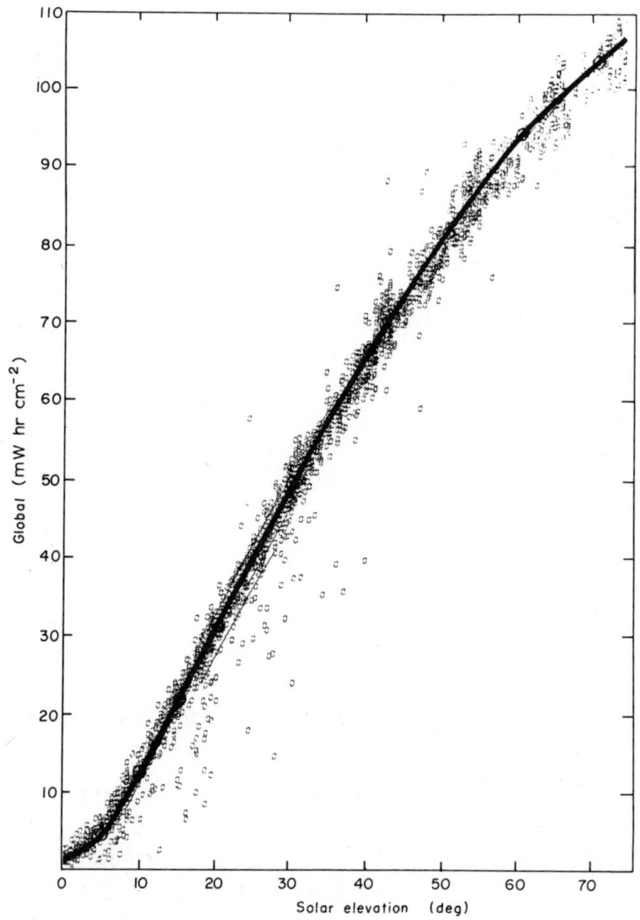

Fig. 6.3. Global solar radiation (hourly total) in clear skies as a function of solar elevation γ at Aspendale for the period 1967–1972. Solid curve is $1.0 + 141.1 \sin\gamma - 31.0 (\sin\gamma)^{1/2}$.

2. SOLAR DIRECT BEAM

Calculation of the direct beam irradiance I in clear skies (it is usually zero in cloudy skies) is simple in principle since it involves only the solution of the transfer equation applied to parallel radiation where the source function is zero. That is:

$$I = \int_0^\infty I_{0\lambda} \exp(-\tau_{e\lambda} \sec\theta) d\lambda \qquad [6.3]$$

where $I_{0\lambda}$ is the irradiance of the solar beam at wavelength λ outside the earth atmosphere and $\tau_{e\lambda} \sec\theta$ is the total extinction optical depth of the

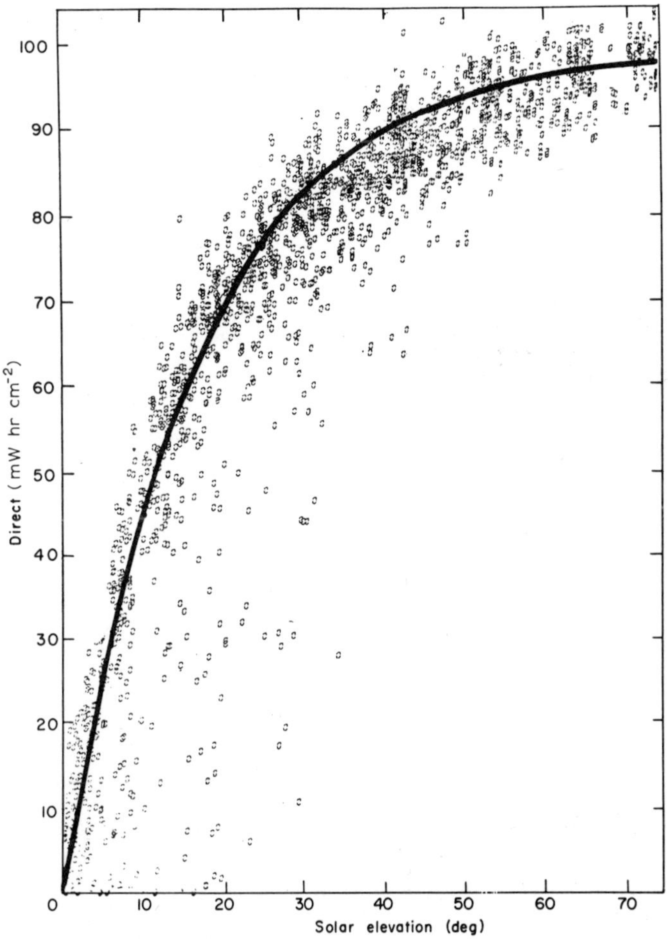

Fig. 6.4. Direct radiation (hourly total) in clear skies as a function of solar elevation γ at Aspendale for the period 1967–1972. Solid curve is least square fit: $I = 100.0\,[1 - \exp(-0.06\gamma)]$.

atmosphere in the direct path of the solar beam. The $\sec\theta$ is the approximate value of relative air mass m_r. $\tau_{e\lambda}$ is the sum of the extinctions due to scatter and absorption. Thus

$$I = \int_0^\infty I_{0\lambda} \exp[-(\tau_{R\lambda} + \tau_{oz\lambda} + \tau_{wv\lambda} + \tau_{D\lambda})m_r]\,d\lambda \qquad [6.4]$$

where the subscripts R, oz, wv and D refer to Rayleigh scatter, ozone absorption, water vapour absorption, and extinction (both scatter and absorption) by dust. The optical depths in each case are the *vertical* optical depths from the ground to infinity and are tabulated for a standard atmos-

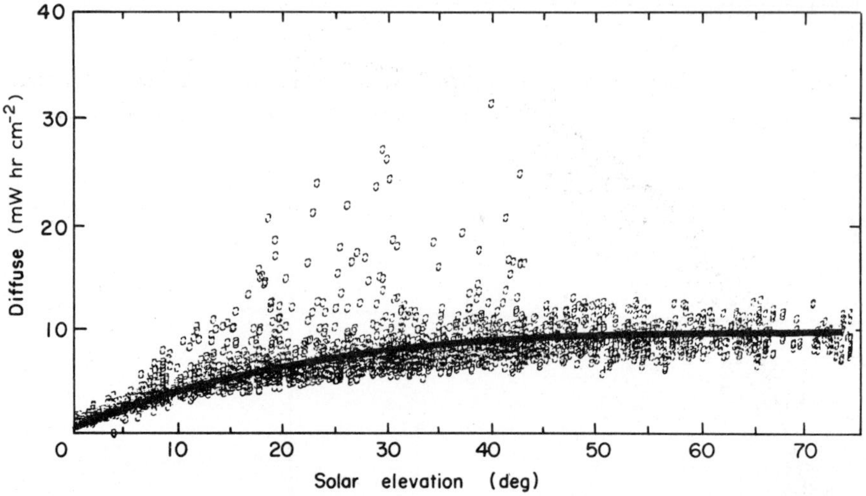

Fig. 6.5. Diffuse radiation (hourly total) in clear skies as a function of solar zenith angle γ at Aspendale, 1967–1972. Solid curve is least square fit: $D\downarrow = 0.5 + 9.6[1 - \exp(-0.05\gamma)]$. Note the definition of "clear sky" in the text.

phere by Elterman and Toolin (1965). In terms of transmittances Tr, the equation is:

$$I = \int_0^\infty I_{0\lambda}(Tr_{R\lambda} \cdot Tr_{oz\lambda} \cdot Tr_{wv\lambda} \cdot Tr_{D\lambda}) m_r \mathrm{d}\lambda \qquad [6.5]$$

Equation [6.3] is simply an expression of Beer's law of exponential extinction applied to each wavelength individually. The law is not applicable over a band of wavelengths where the extinction coefficient varies with wavelength, and all the above extinction processes are in fact wavelength dependent. Taking water vapour absorption on its own purely as an example, Fig. 5.5 of the previous chapter shows that the total water vapour absorptance a_{wv} is far from a single exponential function of water path and a "nice" expression such as $I = I_0 \exp(-\tau_{wv} m_r)$ cannot be obtained. One is forced to the more general form $I = I_0\, Tr_{wv}(um_r)$ where $Tr_{wv}(um_r)$ is the average transmission of the water path um_r. Further, it is not simply multiplicative with the transmission functions appropriate to the other processes. That is, it is not true as a general statement that $I = I_0(Tr_R \cdot Tr_{wv} \cdot Tr_{oz} \cdot Tr_D)$, for if we imagine the spectrum divided into two separate wavelength bands of different average transmittances Tr_1 and Tr_2 and imagine further an impinging irradiance I_0 on each band:

$$I = I_0(Tr_1 + Tr_2) \qquad [6.6]$$

There are, however, a number of fortunate circumstances which allow good approximations without the necessity to do the complete wavelength integration.

2.1. *Rayleigh scatter*

The extinction by Rayleigh scatter varies with wavelength in a smooth and calculable manner. For the whole atmosphere the vertical optical depth is given by:

$$\tau_{R\lambda} = \int_0^\infty \kappa_{R\lambda}\,\rho\,dz = \int_0^\infty \beta_{R\lambda}\,dz \qquad [6.7]$$

where $\kappa_{R\lambda}$ is the Rayleigh mass scattering coefficient and $\beta_{R\lambda}$ is the more usually quoted volume scattering coefficient. If $\hat{N}$ is the number density of molecules at the particular pressure p and temperature $T(\hat{N} = p/kT$ where k is the gas constant per molecule):

$$\beta_{R\lambda} = J_{R\lambda}\hat{N} \qquad [6.8]$$

where $J_{R\lambda}$ is the Rayleigh scattering cross-section. For air at standard pressure and temperature:

$$(J_{R\lambda})_s = \frac{8\pi^3}{3}\,\frac{(n_s^2-1)^2}{\lambda^4\hat{N}_s^2}\left(\frac{6+3\rho_n}{6-7\rho_n}\right) \qquad [6.9]$$

where n_s and $\hat{N}_s$ are the refractive index and molecular number density of air respectively at STP, and the factor in large brackets takes account of the optical anisotropy of air molecules. Penndorf (1957) took ρ_n to be 0.035 which is a fair mean of various experimental and theoretical evaluations of the term.

A common technique is to assume that $(J_{R\lambda})_s = J_{R\lambda}$ = constant on the basis of the Gladstone and Dale law (see Longhurst, 1960, p. 417) that for a gas $(n-1)/\hat{N}$ is a constant. On this basis, by expanding $(n_s^2-1)^2$ and putting $(n_s+1) = 2$ because n_s is very close to unity:

$$\beta_{R\lambda} = \frac{32\pi^3(n-1)^2}{3\lambda^4\hat{N}}\left(\frac{6+3\rho_n}{6-7\rho_n}\right) \qquad [6.10]$$

or:

$$\beta_{R\lambda} = (J_{R\lambda})_s\cdot\hat{N} \qquad [6.11]$$

Equation [6.10] is the usually quoted formula for $\beta_{R\lambda}$. However, it should be realized that both equations [6.10] and [6.11] are approximations valid to an accuracy of only ±10% over the atmospheric range of temperatures. The error arises because $(n-1)/\hat{N}$ is in fact a function of temperature. The correct formula for $J_{R\lambda}$ in equation [6.8] is:

$$J_{R\lambda} = \frac{8\pi^3}{3}\,\frac{(n^2-1)^2}{\lambda^4\hat{N}^2}\left(\frac{6+3\rho_n}{6-7\rho_n}\right) \qquad [6.12]$$

where both n and $\hat{N}$ must be computed as their separate functions of temperature. For details of such computations refer again to Penndorf (1957).

In the present context of total atmospheric extinction, where most of the scatter occurs in the lower atmosphere because the density or $\hat{N}$ is largest there, the effect of temperature on the overall optical depth is reduced to the 1% level. Equation [6.10] is sufficiently accurate except for the more refined remote sensing applications such as the determination of monochromatic turbidities. Thus by making use of Edlén's (1953) formula for the refractive index of air at STP:

$$(n_s - 1)10^{-8} = 6432.8 + \frac{2949810}{146 - 1/\lambda^2} + \frac{25540}{41 - 1/\lambda^2} \qquad [6.13]$$

where the wavelength λ is in microns, one can perform the calculations indicated by equations [6.9], [6.10] and [6.7] to produce the vertical optical depth $\tau_{R\lambda}$ of the whole atmosphere as a function of wavelength. Various simplifications of this function are available of which the most commonly used is probably that quoted by N. Robinson (1966) for standard pressure:

$$\tau_{R\lambda} = 0.00888\lambda^{-4.05} \qquad [6.14]$$

where λ is in microns.

Since the surface pressure p_g is a measure of the total mass of the atmosphere, optical depth is closely proportional to p_g for the relatively small changes which do occur. Thus the final equation for the irradiance of the direct beam of the sun at the bottom of a clean dry atmosphere becomes:

$$I = \int_0^\infty I_{0\lambda} \exp[-(p_g/p_s)\tau_{R\lambda} m_r]\,d\lambda \qquad [6.15]$$

where $\tau_{R\lambda}$ is given by equation [6.14] and p_s is the standard surface pressure of 1013 mb.

Despite the fact that Beer's law is not valid when applied to overall bands, it is sometimes convenient for computational purposes to perform the integration over wavelength and force the final result into the exponential form. That is, to set by definition:

$$I = I_0 \exp[-\bar{\tau}_R(m_r)\cdot(p_g/p_s)\cdot m_r] = \int_0^\infty I_{0\lambda} \exp[-\tau_{R\lambda}(p_g/p_s)m_r]\,d\lambda \qquad [6.16]$$

where the "mean optical depth" $\bar{\tau}_R(m_r)$ is no longer a constant (as is the case for Beer's law) but is itself a function of relative air mass m_r. Table 6.1 (after N. Robinson, 1966) gives $\bar{\tau}_R(m_r)$ and the corresponding values of direct beam irradiance as a function of m_r, where the value for solar constant has been chosen as 135.6 mW cm^{-2}.

TABLE 6.1

Rayleigh scatter "mean optical depth" and corresponding direct beam irradiance as a function of relative air mass

m_r	$\bar{\tau}_R(m_r)$	$I(\text{mW cm}^{-2})$
0.5	0.105	128.6
1.0	0.099	122.8
1.5	0.094	117.7
2.0	0.089	113.5
3.0	0.082	106.0
4.0	0.076	100.0
6.0	0.067	90.7
8.0	0.060	83.9
10.0	0.055	78.2

2.2. *Ozone and water vapour absorption*

The transmission or absorption functions of these two gases have already been discussed in the previous chapter and the parameterized versions of the wavelength integrated absorptances are given there by equations [5.1] and [5.2] for ozone and by equation [5.10] for water vapour. In the case of ozone one can take advantage of the fact that its absorption takes place mostly above the main bulk of the atmosphere and of the fact that the total ozone absorption in the direct beam is in any event small (a few percent of the solar constant) compared with total Rayleigh extinction. Further, the ozone absorption occurs at wavelengths in the ultraviolet and visible where also the Rayleigh scatter is dominant so that to a first approximation the two transmittances are multiplicative. Thus the surface value of the direct beam irradiance I in a clean dry atmosphere containing ozone might be expressed:

$$I = I_0[1 - a_{oz}(u_{oz}m_r)] \exp[-\bar{\tau}_R(m_r)\cdot(p_g/p_s)\cdot m_r] \qquad [6.17]$$

where u_{oz} is the vertical path of ozone as in the previous chapter.

Similarly the water vapour absorption occurs at wavelengths outside those affected by ozone and Rayleigh scatter so that the relevant transmittances are roughly additive. Thus for a clean "wet" atmosphere containing ozone:

$$I = I_0\{[1 - a_{oz}(u_{oz}m_r)] \exp[-\bar{\tau}_R(m_r)(p_g/p_s)m_r] - a_{wv}(u_{wv}m_r)\} \qquad [6.18]$$

where u_{wv} is the total water path in the vertical column of the atmosphere. Preferably it should be the pressure and temperature corrected *effective* water path according to equation [5.13].

2.3. *The effect of atmospheric dust*

Atmospheric aerosols are a tremendous nuisance from the point of view of radiative calculations for much the same reasons as are clouds. First, they consist of particles large compared with the wavelengths of solar radiation and their scattering properties must be dealt with in terms of the inherently complicated Mie theory. Second, their scatter and absorption occur at the same levels in the atmosphere as the dominant Rayleigh scatter and water vapour absorption as well as over the entire spectrum. Thus one cannot take advantage of either spatial or spectral separation. Third, the size distribution of the particles, the vertical profile of their number density and the amount of their absorption relative to their scatter are all highly variable from place to place and time to time; and it is rarely possible to specify these characteristics with any accuracy. This last problem is the fundamental stumbling block of the three because once the aerosol characteristics are specified it is at least possible in principle to carry out the various integrations necessary to evaluate for each wavelength and each volume of the atmosphere the fundamental radiative parameters necessary to calculate the absorption and scatter by that volume — and then to solve the complete transfer equation.

In the foreseeable future it is probable that in anything other than very specialized circumstances the only *predictable* parameter of aerosols will be the total number density n_T or total mass per unit volume as a function of height or geographical position. It is unlikely that size distribution or the absorption to scatter ratio will be predictable *a priori*, and meteorologists will be forced to assign specific and fixed size distributions at a particular position in the atmosphere based on experimental experience. The tools of the meteorologist will therefore include a set of typical aerosols (see the Appendix) for which at least the following parameters have been pre-calculated for each wavelength: (1) the mean effective radius r_e; (2) the mean (over all particle sizes) of the extinction efficiency per particle $\bar{Q}_e$; (3) the mean (over all particle sizes) of the single-scattering albedo $\bar{\omega}_0$ which specifies the amount of absorption; and (4) the mean (over all particle sizes) of the scattering pattern — or at the very minimum the mean asymmetry factor $\langle\overline{\cos\psi}\rangle$. Examples of such data sets are given by Diermendjian (1969).

Assuming such information is available together with the predicted number density $n_T(h)$ as a function of height h, in the present context of calculating direct beam irradiance at the surface one could have:

$$\tau_{D\lambda} = \int_0^\infty \beta_{D\lambda}\,dh = \int_0^\infty n_T(h)\bar{Q}_e(h)\cdot\pi[r_e(h)]^2\,dh \qquad [6.19]$$

where $\beta_{D\lambda}$ is the volume extinction coefficient due to aerosols. $\tau_{D\lambda}$ can then be inserted in the full wavelength integration of equation [6.4].

As an example of the magnitudes involved, the standard atmosphere dust

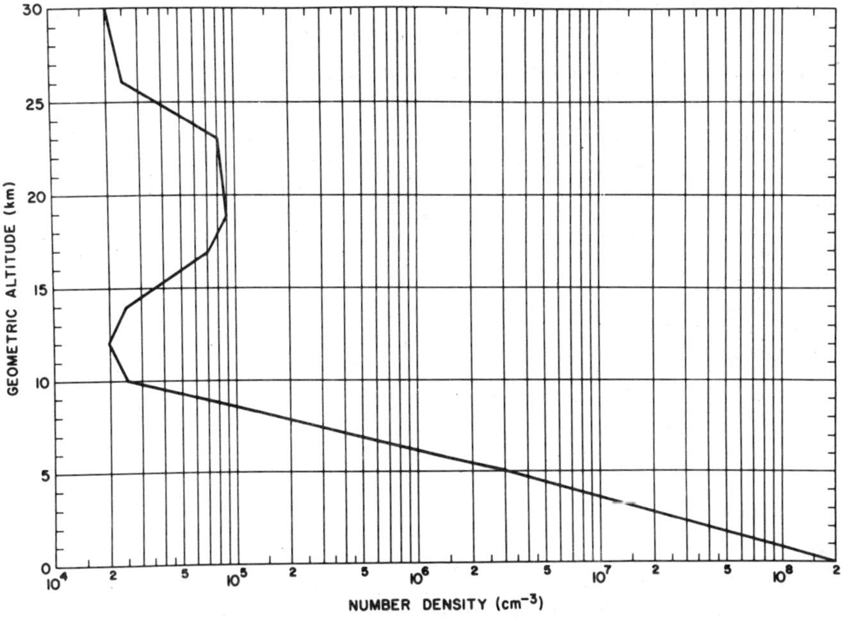

Fig. 6.6. A "standard atmosphere" aerosol number density versus altitude profile. (From Elterman and Toolin, 1965.)

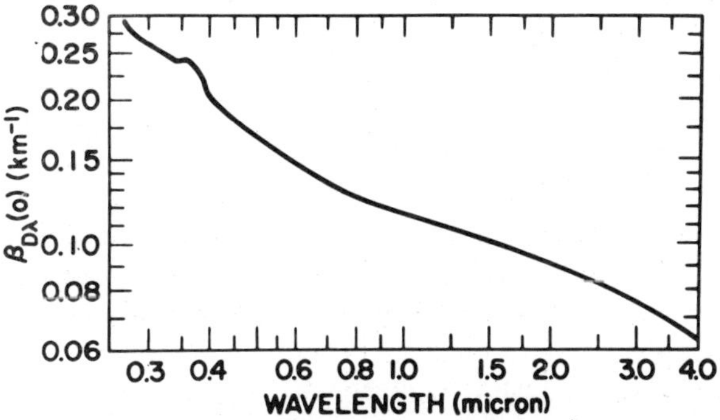

Fig. 6.7. Aerosol volume extinction coefficient at sea level, $\beta_{D\lambda}(0)$, versus wavelength, for an average clear day (meteorological range about 25 km). (From Elterman and Toolin, 1965.)

loading given by Elterman and Toolin (1965) (Fig. 6.6) and the standard atmosphere aerosol volume extinction coefficient as a function of wavelength [$\beta_{D\lambda}(h) = \beta_{D\lambda}(0)n_T(h)/n_T(0)$, where $\beta_{D\lambda}(0)$ is given in Fig. 6.7] were used to compute I for a dry but dusty atmosphere according to:

$$I = \int_0^\infty I_{0\lambda} \exp[-(\tau_{R\lambda} + \tau_{D\lambda})]\, d\lambda \qquad [6.20]$$

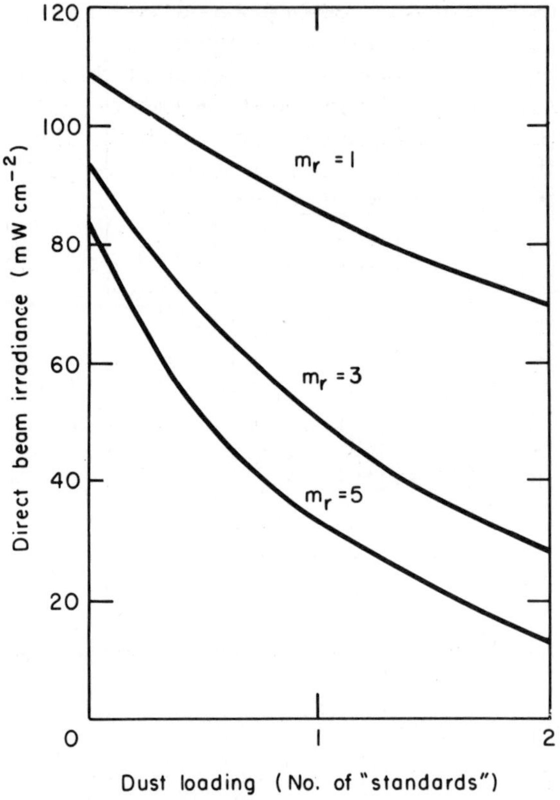

Fig. 6.8. Direct beam irradiance in a dry atmosphere as a function of atmospheric dust loading (relative to standard dust loading of Figs. 6.6 and 6.7) for three relative air masses.

The results are plotted in Fig. 6.8 where the abscissa is the fraction by which $n_T(h)$ at all levels has been multiplied relative to the standard. The parameter is air mass. For a vertical sun ($m_r = 1$) and the standard atmosphere, the reduction of I by dust is of order 20%. This standard atmosphere corresponds to a fairly high turbidity as defined by the Ångström turbidity coefficient (which is the aerosol vertical optical depth at a wavelength of 1 μm — see Chapter 9). For this model $\tau_{D, 1\,\mu m} \approx 0.15$ which is two or three times the commonly measured value over Australia for example. Nevertheless, even low turbidities represent a significant reduction of I which is accentuated at the high air masses.

It is possible to avoid the wavelength integration under certain limited circumstances. First, it may be that the aerosol consists of large and non-absorbing particles — where $\bar{\omega}_0 = 1$ and where the scatter pattern is very strongly forward peaked. For such a case the aerosol can be ignored on the basis that any scattered radiation still remains within the reasonable definition of direct beam. This assumption can rarely be used for clouds because of their large optical depth and multiple scatter. Second, the aerosol may be non-absorbing and have a broad scatter pattern in the forward direction but have little backscatter ($\langle\cos\psi\rangle \to 1.0$). To somewhat less accuracy than the first example one can assume that the overall effect of the aerosol on *global* radiation is negligible on the basis that radiation removed from the direct beam re-appears as diffuse radiation. A comparison of the spread of points in Fig. 6.3 as compared with Figs. 6.4 and 6.5 bears out this point. Third, for a pure absorbing aerosol one could pre-calculate a set of curves similar to those of Fig. 6.8 and subtract the indicated absorptions at any particular m_r and normalized $n_T(h)$ from the radiance calculated according to equation [6.18]. For a part-absorbing aerosol one could reduce that absorption in proportion to the specified $1 - \bar{\omega}_0$ and , if appropriate, disregard the scatter in the same manner as the first or second examples.

It should be apparent that experimentally measured turbidities which indicate only direct beam depletion are not by themselves necessarily useful in the meteorological context where the interest is in total global flux density at the surface. One must still make some assumption about aerosol type in order to specify $\bar{\omega}_0$ and the scattering pattern if the direct beam turbidity measurements are to be used as input data to a model requiring global radiation.

3. DIFFUSE RADIATION IN CLEAR SKIES

Because diffuse radiation is a small fraction of the vertical component of the direct beam it is natural to assume that almost any simple parameterization would be sufficient — such as, for instance, using a single, universal $D\downarrow$ versus θ function such as that of Fig. 6.5. However, this is true only for modelling efforts which contain no feedback via ground interaction to atmospheric transparency. The absolute *change* in diffuse flux at the surface as a result of scatter can be of the same order as the absolute change in direct radiation. Therefore, if one is concerned with changes in global flux, the two components must be computed with equal accuracy and in a manner which preserves the obvious relation between them. This is the reason that aerosol scatter is such a nuisance. It is the major changeable factor in clear-sky scatter and cannot often be ignored.

Accurate calculation of diffuse flux in a dusty atmosphere requires the full transfer theory treatment such as that of Yamamoto and Tanaka (1972) involving finally an integration over wavelength. It is not possible to shorten this process by going to the Rayleigh atmosphere formula (equation [5.32]) and inserting $\tau_{R\lambda} + \tau_{D\lambda}$ in place of $\tau_{R\lambda}$ since the formula is based on the pure Rayleigh polar pattern. Any approximations which remove the wavelength integrations must be fairly drastic. Perhaps the most obvious is to take advantage of the fact that over most wavelengths the optical depths of both

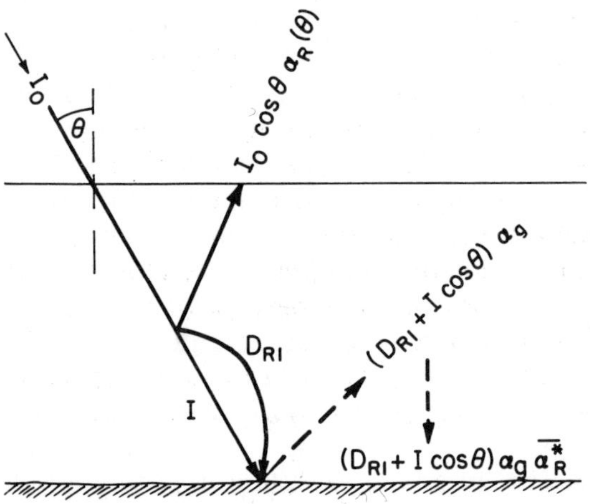

Fig. 6.9. Schematic diagram for simple calculation of Rayleigh scattered diffuse flux density — see text.

Rayleigh and aerosol scatter are fairly small. On this basis one might deal with the two processes separately and simply add the total flux densities attributable to each. For instance, the following basic treatment may be inaccurate in the absolute sense but it preserves a degree of consistency between direct and diffuse radiation in the presence of aerosols. It can be expanded for greater accuracy as desired.

A least square fit to the accurate calculation of Rayleigh scatter by the total standard atmosphere above a black earth leads to:

$$\alpha_R(\theta) = 0.28/(1 + 6.43 \cos\theta) \qquad [6.21]$$

where $\alpha_R(\theta)$ is the Rayleigh albedo as a function of solar zenith angle θ (Lacis and Hansen, 1974) and the formula is accurate to 1%. (While on the subject, their equivalent planetary albedo $\bar{\alpha}_R$ integrated over all sun angles for a clean atmosphere with standard ozone and to the complete solar spectrum is 6.22%. Coulson's (1959) value of 6.9% underestimated ozone absorption.) Referring to Fig. 6.9, the diffuse flux density scattered upwards from the Rayleigh atmosphere is $I_0 \cos\theta\, \alpha_R(\theta)$. The diffuse flux density $D{\downarrow}_{RI}$ scattered downwards is given by the difference:

$$D{\downarrow}_{RI} = (I_0 - I)\cos\theta - I_0 \cos\theta\, \alpha_R(\theta) \qquad [6.22]$$

where I might be calculated using equation [6.16]. For a non-black surface of albedo α_g, one must take into account also the surface reflected radiation which in turn is Rayleigh scattered backwards with a reflection coefficient $\bar{\alpha}_R^*$ more or less equal to $\bar{\alpha}_R$. According to Lacis and Hansen (1974), $\bar{\alpha}_R^*$ is actually somewhat larger than $\bar{\alpha}_R$ since there is no intervening ozone absorption. They calculate $\bar{\alpha}_R^* = 0.0685$. Thus the total Rayleigh-scattered downcoming diffuse flux density $D{\downarrow}_R$ becomes:

$$D{\downarrow}_R = D{\downarrow}_{RI} + (D{\downarrow}_{RI} + I\cos\theta)\alpha_g\bar{\alpha}_R^* \qquad [6.23]$$

where for simplicity here the increased atmospheric albedo due to dust scattering of upwelling radiation is ignored. Dust scatter is strongly forward peaked.

To this must be added the diffuse flux density $D{\downarrow}_D$ associated with aerosol scatter. The amount of radiation dI_D aerosol-scattered out of the direct beam must be calculated somehow according to the previous section. For typical aerosols of little back-scatter, $D{\downarrow}_D$ will be some function of dI_D and θ whose value ranges from dI_D at $\theta = 0°$ to approximately $0.5dI_D$ at $\theta = 90°$. Thus:

$$D{\downarrow}_D = dI_D f_1(\theta) \qquad [6.24]$$

where multiple scatter and the back-scatter of ground-reflected aerosol scatter is ignored. For a "spherical" forward scatter pattern of the aerosol, $f_1(\theta) = (1 - \cos\theta)^2 (2 + \cos\theta)/4$. Finally:

$$D{\downarrow} = D{\downarrow}_R + D{\downarrow}_D \qquad [6.25]$$

The process can be refined in a number of ways but whether the additional complication is worth the trouble will depend on the accuracy of the parameterization of the wavelength-integrated aerosol scatter from the direct beam.

4. RADIATION ON A TILTED SURFACE

The solar radiation quantities normally calculated or measured at the ground surface are the direct beam irradiance I and the vertical component of the diffuse flux density $D{\downarrow}$. These are sufficient for the usual meteorological problem of computing the total solar radiation input (the global radiation $F{\downarrow}_S$) to the horizontal surface according to $F{\downarrow}_S = I\cos\theta + D{\downarrow}$. They are sufficient also to calculate the input G' per unit area to a sloping surface of known tilt angle ζ of the surface normal to the vertical — but to an accuracy limited by the assumption of diffuse radiation isotropy and by the knowledge of the albedo α_g of the ground surface facing the tilted plane. The calculation involves a pure geometrical exercise to establish the angle θ' between the direct beam and the normal to the plane and to establish the fractions f' and $1 - f'$ of sky and ground surface respectively which can be seen by the plane. Then:

$$G' = I\cos\theta' + D{\downarrow}f' + F{\downarrow}_S(1 - f')\alpha_g \qquad [6.26]$$

where, in addition to specifying α_g, it is implied also that the surface is a diffuse reflector. This last is the common assumption in all meteorological work but is not always a very good one.

Referred to a system where the z-axis is vertical, the y-axis is in the direction of north and the x-axis is to the east, the direction cosines (x, y, z) of the sun are:

$$\sin\theta \sin\phi, \quad \sin\theta \cos\phi, \quad \cos\theta \qquad [6.27]$$

where ϕ is the solar azimuth angle measured east of north. Both θ and ϕ are predictable from equations in Chapter 3 and are tabulated in the Smithsonian Meteorological Tables. The corresponding direction cosines of the normal to the plane are:

$$\sin\zeta \sin\phi', \quad \sin\zeta \cos\phi', \quad \cos\zeta \qquad [6.28]$$

where ϕ' is the azimuth angle of the normal. The angle θ' is the angle between the directions specified by equations [6.27] and [6.28] and is therefore given by:

$$\cos\theta' = \sin\theta \sin\phi \sin\zeta \sin\phi' + \sin\theta \cos\phi \sin\zeta \cos\phi' + \cos\theta \cos\zeta \qquad [6.29]$$

Appropriate integration shows also that:

$$f' = \cos^2\zeta/2 \qquad [6.30]$$

The dominant error in any calculation involving equation [6.26] will normally arise from the assumption of a diffusely reflecting ground. However, since the polar pattern of ground reflection is rarely known with any significant accuracy, there is very little one can do to improve the situation in practical models even if one was prepared to go to the trouble of a detailed geometrical exercise. If it is known that the surface is close to being a pure specular reflector (water at large solar zenith angles?) it may be worth modifying equation [6.26] to:

$$G' = I\cos\theta' + D{\downarrow}f' + D{\downarrow}\alpha_G(1-f') + I\alpha_{GS}\sin\theta'' \qquad [6.31]$$

where α_G is the albedo of the surface to diffuse radiation, α_{GS} is the albedo appropriate to the direct beam at its zenith angle θ and $\sin\theta'' = \cos\zeta\cos\theta - \cos\theta\sin\theta\cos(\phi-\phi')$. The reader is referred to N. Robinson (1966) or Kondratyev (1969) for a more detailed discussion of measurement versus the flux densities computed according to the above.

5. RADIATIVE PROPERTIES OF THE SURFACE

A fundamental lower boundary condition of the atmosphere in all meteorological and climate problems is the requirement for surface energy balance. The upward flux densities of latent and sensible heats $\widetilde{LE}$ and $\widetilde{H}$ and the downward heat flux $\widetilde{G}$ into the ground are balanced by the net radiative inputs of both short- and long-wave radiation. That is:

$$\widetilde{LE} + \widetilde{H} + \widetilde{G} = F{\downarrow}_S(1-\alpha_g) + \epsilon_g(F_L{\downarrow} - \sigma T_g^4) \qquad [6.32]$$

where the first term on the right-hand side is the net short-wave input involving the available global radiation $F{\downarrow}_S$ and the short-wave albedo α_g. The last term is the net long-wave input involving the downcoming flux density $F_L{\downarrow}$, the black-body radiance of the surface at its temperature T_g and the surface emissivity or absorptivity ϵ_g. Since the spectrum of $F_L{\downarrow}$ is not that of a black body the last term is an approximation which assumes ϵ_g to be spectrally independent.

In typical climate models α_g and ϵ_g are treated as fundamental pieces of input data like the solar constant. They are the external parameters which determine all the other variables of the system such as the $\widetilde{L}\widetilde{E}$, $\widetilde{H}$, T_g and $F{\downarrow}_S$. However, neither α_g nor ϵ_g are necessarily independent variables and indeed there are marked feedback processes which produce changes from one time to another. Perhaps the most obvious is the very large increase in albedo accompanying snowfall and, say, the advance of the polar ice caps. Less spectacular seasonal changes occur in concert with the change in vegetative cover and the change in soil moisture content. In the ultimate *a priori* model both α_g and ϵ_g will be computed parameters which vary with the climate itself.

It is often forgotten that the measurement of regional albedo and emissivity is never likely to attain an accuracy, even a relative accuracy, anywhere near that of the present measurement of the solar constant. Bearing in mind that a change of 0.01 in surface albedo is almost as effective in altering the energy input to the system as a 1% change in solar constant (other things being equal), the ultimate limit to the accuracy of climate change prediction will probably be set by the relative error in measuring α_g and ϵ_g. In this restricted sense as opposed to the requirements of remote sensing, it is somewhat futile to demand ever more accurate measurements of the solar constant. Further, it is much easier to envisage all sorts of natural and *climate independent* processes which could produce changes in surface albedo of several percent than to envisage equivalent changes in solar output.

5.1. *Surface albedo*

There are vast numbers of measurements available in the literature which pertain to particular surfaces at particular times. Very few of these are in a format useful to the meteorologist.

First, the meteorologist is generally concerned with the average albedo over an area which may be several hundred kilometres on a side. Overland and sometimes over the sea such areas are made up of a mosaic of individual patches each of which has its own albedo and albedo behaviour. The problem is to obtain a mosaic average which if necessary is a function of independent variables which a model can predict or obtain elsewhere. Further, since α_g varies strongly with solar zenith angle θ — especially at the higher zenith angles characteristic of the daily average — the meteorologist needs the *function* $\alpha_g(\theta)$. Given $\alpha_g(\theta)$ for clear skies it is always possible to work out an approximate albedo appropriate to diffuse radiation such as exists under cloudy skies. Under the usual assumption of isotropic diffuse radiation:

$$\alpha_g(\text{diff}) \approx 2 \int_0^{\pi/2} \alpha_g(\theta) \cos\theta \sin\theta \, d\theta \qquad [6.33]$$

where the "approximately equal" sign arises because it is assumed in this equation that in clear skies $\alpha_g(\theta)$ is appropriate for direct radiation only.

Second, the meteorologist needs information on the spectral albedo. At the very least he needs the mean albedo appropriate to the 300–400-nanometre (nm) band, the 300–800-nm band and the 0.8–2-μm band. These are the spectral regions where surface reflected radiation affects ozone absorption, Rayleigh scatter and water vapour absorption respectively. Fig. 6.10 from Kondratyev (1973) shows that the relevant albedos are generally very different. For most surfaces (snow is an obvious exception) the ultraviolet albedo is down by a factor of 2 or 3 on that for visible radiation. The near-infrared albedo may be up by a factor of 3 or 4 on that for visible radiation. This last is especially relevant for vegetation-covered surfaces. Leaves become very transparent beyond the chlorophyll absorption bands which extend to 700 nm. At the longer wavelengths the albedo tends to that of the underlying soil rather than that of the vegetative cover.

The fact that α_g is so spectrally dependent introduces a further complication with regard to the interpretation of actual measurements. The standard technique is to measure the total (wavelength-integrated) upward and downward flux densities and define albedo as their ratio. Such a measured value may not be appropriate for another set of atmospheric conditions which yield a different spectrum of solar radiation. The problem is usually ignored on the basis that on any reasonable time average the error will smooth out to a level well below that associated with the actual measurement of α_g. This is reasonable with state-of-the-art models where the least of the problems is the absolute accuracy of surface albedo.

The Russian school has been particularly active in producing measured albedos of different surface types in order that regional-average figures can be derived. Their work is summarized in some detail by Kondratyev (1969, 1973). Similarly Posey and Clapp (1964) have summarized available data and have produced seasonal maps of the single-valued (i.e. time-averaged) surface albedo for both hemispheres.

Future development in cataloguing measured surface albedos will come from the use of satellite data. Such data are ideal from the point of view of obtaining the mosaic average but suffer slightly from the necessity to take account of the intervening atmosphere. Initial work of this type was performed by Conover (1965) using the TV scanner data from the Tiros VII satellite. Recent satellites (e.g. Nimbus) incorporate medium resolution

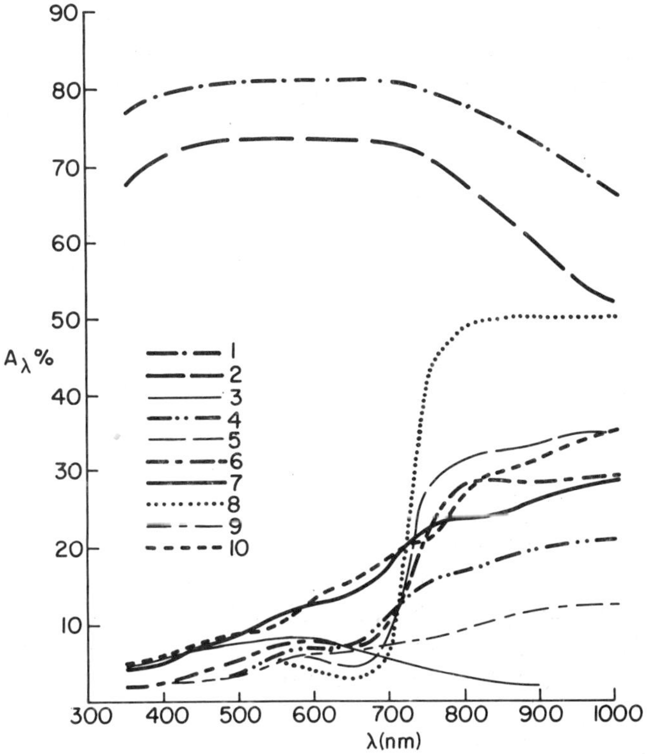

Fig. 6.10. Spectral albedo of different natural underlying surfaces at various solar elevations γ. *1* = snow with ice crust, $\gamma = 38^\circ$; *2* = large-grained wet snow, $\gamma = 37^\circ$; *3* = water surface of a lake, $\gamma = 56^\circ$; *4* = soil after thawed snow, $\gamma = 24^\circ 30'$; *5* = silage corn, $\gamma = 54^\circ$; *6* = tall greencorn, $\gamma = 56^\circ$; *7* = yellow corn, $\gamma = 46^\circ$; *8* = sudan grass, $\gamma = 52^\circ$; *9* = chernozem, $\gamma = 40^\circ$; *10* = stubble of cereals, $\gamma = 35^\circ$.

radiometers which measure the upward radiance over the entire short-wave region (0.2–4 μm) from ground areas of the order of 500 km a side. Thus Vonder Haar and his colleagues (e.g. Raschke et al., 1973) have produced data on the geographical distribution of minimum planetary albedo which, to fair accuracy, corresponds to the planetary albedo of clear skies. It is easy enough to derive the actual surface albedo α_g to an accuracy of perhaps 0.05 as follows. Suppose at a given solar zenith angle θ the planetary albedo is $\alpha_0(\theta)$, the atmospheric reflectance to solar radiation is $\alpha_A(\theta)$ and its absorptance is $\hat{m}(\theta)$, and its reflectance to upwelling diffuse radiation is α_A^*. Then considering only a single ground reflection:

$$\alpha_0(\theta) = \alpha_A(\theta) + [1 - \alpha_A(\theta) - \hat{m}(\theta)]\,\alpha_g(\theta)\cdot(1 - \alpha_A^*) \qquad [6.34]$$

On the basis that most short-wave absorption tends to saturate in the downward direct beam (refer to Fig. 5.6 for instance) one can adopt a value $\bar{u}$ of

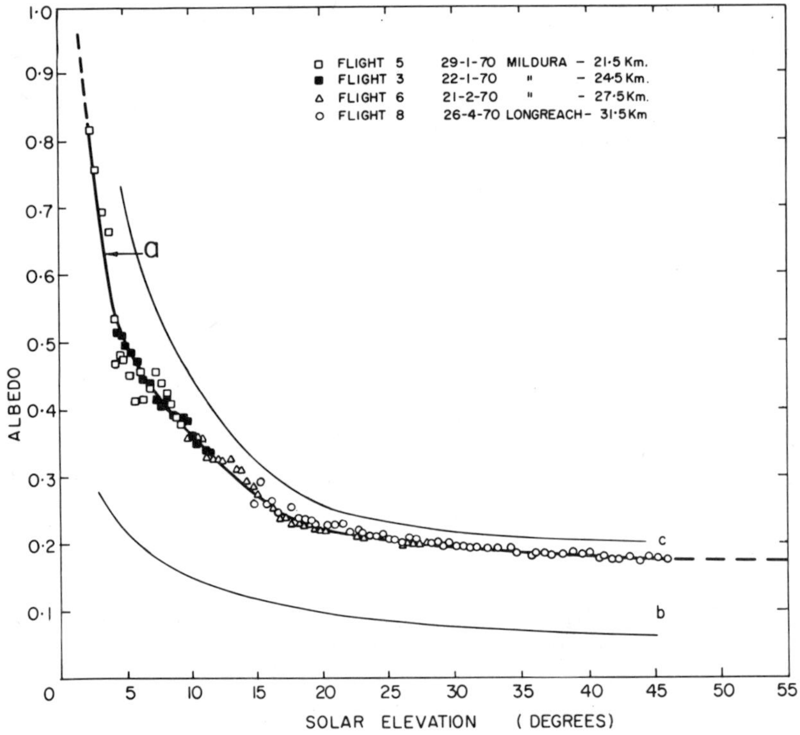

Fig. 6.11. Short-wave planetary albedo versus solar elevation. Curve *a* is the hand-drawn line of best fit through the measured values obtained on four constant level balloon flights, curve *b* is the theoretical albedo for a pure Rayleigh scattering atmosphere, and curve *c* is the computed surface albedo. (After Paltridge and Sargeant, 1971.)

the total precipitable water in the vertical column of the atmosphere appropriate to the region. Neglecting ozone absorption, $\hat{m}(\theta)$ then equals the $a_{wv}(\bar{u}\sec\theta)$ which can be calculated from equation [5.10]. Then:

$$\alpha_g(\theta) = \frac{\alpha_0(\theta) - \alpha_A(\theta)}{[1 - \alpha_A(\theta) - \hat{m}(\theta)](1 - \alpha_A^*)} \qquad [6.35]$$

It is obvious physically that planetary albedo is not the simple sum of the surface and atmospheric albedos. It is not so obvious but can be seen by rearranging equation [6.34] that planetary albedo may be considerably less than surface albedo — particularly for high surface albedos such as that of snow.

Equation [6.35] can be refined considerably but has been used here in its raw form on balloon-measured planetary albedo data taken over the semi-desert and dry grassland area which characterizes much of the Australian continent. Fig. 6.11 shows the observed $\alpha_0(\theta)$ from several balloon flights, the Rayleigh scatter atmospheric reflectance $\alpha_R(\theta)$ from Coulson (1959), and the surface $\alpha_g(\theta)$ calculated from equation [6.35] on the basis that $\bar{u} = 1.5$ cm and $\alpha_A^* = 0.10$ and $\alpha_A(\theta)$ equals $\alpha_R(\theta)$. This value of α_A^* is slightly larger than the 0.06 appropriate for pure Rayleigh scatter in order to take account of back-scatter due to dust.

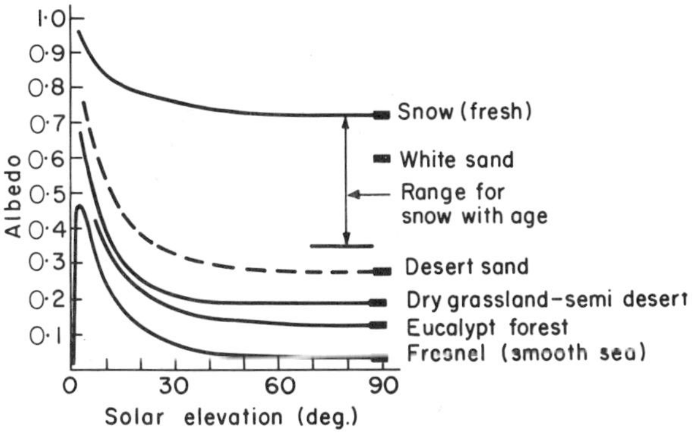

Fig. 6.12. Albedo versus solar elevation γ for a number of surfaces. The values at $\gamma = 90°$ correspond to the α_1 of equation [6.36]. The dashed curve for desert was drawn using this equation.

Parameterizing mosaic average surface albedo will remain a fairly inaccurate business for a long while even with extensive work on presently available data. In the meantime and within the accuracy of most models, it is probably sufficient to assume that the form of α_g versus θ is the same for all surfaces. Reference to Fig. 6.12 shows that to a first approximation most surfaces retain the dominant Fresnel reflection characteristic — namely, a fairly constant value for θ up to about 50° and a rapidly increasing value at higher θ. Thus for operational models which require the solar angle dependence of albedo it is probably quite justifiable to assume a single formula of one disposable parameter which represents the albedo α_1 at high solar elevations. That is, a formula of a form similar to:

$$\alpha_g(\theta) = \alpha_1 + (1 - \alpha_1)\exp[-k(90° - \theta)] \qquad [6.36]$$

where k is of order 0.1. For instance, the dashed curve of Fig. 6.12 representing $\alpha_g(\theta)$ for desert has been drawn using equation [6.36] and an α_1 typical of deserts.

Apart from the similarity of shape of the various $\alpha_g(\theta)$ curves, the obvious point is the extreme effect of snow cover. Snow cover is therefore a variable which figures largely in climate models since it can profoundly alter the planetary energy balance. Note that snow albedo decreases markedly with "age" and the corresponding α_1 in equation [6.36] can be as low as 0.35 for slushy snow at the end of winter.

It is an interesting scientific problem as to why most rough surfaces display an α_g versus θ character similar to that of pure Fresnel reflection — which applies specifically to specular reflection from a smooth surface. There can be exceptions where, for instance, the shadowing by roughness elements increases with θ faster than the increase of albedo with θ of those portions of the surface which are sunlit. Kondratyev (1973) quotes certain cases where ice albedo has shown a decrease with θ. A certain amount of theoretical work on the subject has been done by the Russian school but the basic physics still needs to be clarified.

It is particularly important to specify accurately the albedo of the sea since it covers two-thirds of the earth's surface. In the terminology of equation [6.36], its α_1 can range from about 0.03 (the Fresnel value for smooth water) to about 0.1 for very rough seas. The basic problem from the modeller's point of view is to parameterize sea-surface albedo in terms of such things as windspeed. This will need an analysis similar to that of Cox and Munk (1954) which involves specifying the reflectivity in terms of the probability distribution of wave slope. Cox and Munk's work was concerned specifically with the measurement of sunglint and its relation to windspeed.

5.2. *Surface emissivity*

Mainly because modelling efforts to date have been concerned with other problems, it has almost invariably been assumed that the emissivity of all surfaces is unity at all wavelengths. In fact emissivity is highly variable with wavelength and is significantly variable with surface character. The variation with wavelength is not too serious since the dominant *net* long-wave exchange with the surface is via the atmospheric window from 8 to 14 μm (see Chapter 1) and the measured values are usually derived from instruments tuned to this wavelength interval. The variation with surface character is far more serious. Suppose, for instance, that the surface energy balance was maintained purely by radiative gains and losses. That is:

$$(1-\alpha_g)F{\downarrow}_S = \epsilon_g(1-\epsilon)\sigma T_g^4 \qquad [6.37]$$

where in terms of the model in Chapter 1 the net long-wave loss from the surface of emissivity ϵ_g can occur only within the window region. This introduces also the emissivity of the atmosphere ϵ on the right-hand side. (The model assumes the downward atmospheric emission from the opaque regions of the spectrum varies directly with the upward surface emission — see next section.) Thus if the original energy balance gives rise to a surface temperature of about 300 K, a drop in ϵ_g of 0.01 requires a rise of 1.8 K in T_g to maintain the balance. In most practical situations the balance is maintained as well by upward fluxes of latent and sensible heats $\widetilde{LE}$ and $\widetilde{H}$ and the downward ground heat flux $\widetilde{G}$. In this case the effect of ϵ_g on T_g is more obscure since the relation between $\widetilde{LE}$ or $\widetilde{H}$ or $\widetilde{G}$ and T_g is not necessarily

TABLE 6.2

Vertical emissivity ϵ_g of various materials determined from the Barnes IT-2 infrared thermometer and an "emissivity box" (from Beuttner and Kern, 1965)

Material	ϵ_g
Quartz (agate)	0.712
Granite	0.815
Feldspar	0.870
Obsidian	0.862
Basalt	0.904
Dunite	0.856
Granite, rough side	0.898
Obsidian, rough side, broken glass appearance	0.837
Basalt, rough side, shiny	0.934
Dunite, rough side	0.892
Silicon sandstone, polished side	0.909
Silicon sandstone, rough side	0.935
Dolomite, polished side, $CaMg(CO_3)_2$	0.929
Dolomite, rough side	0.958
Dolomite gravel, 0.5-cm rocks	0.959
Plate silicon glass	0.865
Parson's black on Al sheet (no grooves)	0.988
Plywood sheet	0.962
Human skin	0.980
Sand, quartz large grain	0.914
Sand, quartz large grain wet with water (nearly saturated)	0.936
Sand, Monterey, quartz small grain	0.928
Concrete walkway, dry	0.966
Asphalt paving	0.956
Water, pure	0.993
Water, plus thin film of petroleum oil	0.972
Water, plus thin film of corn oil	0.966
Water, covered by a thin sheet of polyethylene	0.961

known. However, assuming for the sake of argument that a rise in T_g gives an increase in the sum of $\tilde{L}\tilde{E}$ and $\tilde{H}$ and $\tilde{G}$ in direct proportion to the increase in σT_g^4, and assuming the global mean ratio between $\tilde{L}\tilde{E} + \tilde{H} + \tilde{G}$ and $(1-\epsilon)\sigma T_g^4$ (≈ 3 to 1) is held constant, the change of 0.01 in ϵ_g would change T_g by 0.8 K.

Since typical surfaces have ϵ_g ranging from 0.997 (wet snow) down to 0.712 for quartz (see Table 6.2 after Beuttner and Kern, 1965) the problem is not minor. Indeed it is far worse than these particular figures would indicate since they refer to vertical emissivities appropriate to remote measurement of temperature by vertically oriented radiometers on spacecraft. The meteorologically significant quantity is the global or flux emissivity which for natural surfaces may be much less. In common with short-wave reflectance typical surfaces have an infrared reflectance *Re* which increases with angle of incidence. Since $\epsilon_g = 1 - Re$, the actual flux emissivity (that is, the integral of $1 - Re$ over all solid angles of the hemisphere) will in general be less than the "vertical" ϵ_g. For instance, in the case of water (Pontier and Dechambenoy, 1966):

$$\epsilon_g(\text{flux})_{8-14\ \mu m} \approx 0.95 \qquad [6.38]$$

In the absence of detailed measurements, it is necessary to make some assumption about the shape of infrared reflectivity as a function of θ (cf. equation [6.36] for short-wave radiation) and perform the relevant integration to give ϵ_g (flux) in terms of ϵ_g (vertical). The overall effect may be minimized because the atmospheric window tends to "close" at large zenith angles.

6. PARAMETERIZED COMPONENTS OF NET RADIATION AT THE GROUND

A very common requirement is the calculation of the net radiation input to the ground — or the components of the net input — in terms of cloud cover and very little else. The accuracy of cloud specification is so poor that there is often no point in accounting for the input variations which arise from changes in water vapour, dust and ozone.

Referring first to short-wave input, many authors have devised empirical regression relations between daily total or daily mean global irradiance to the observed fractional cloud cover or, more commonly, to the hours of direct sunshine as recorded at weather stations (e.g. Hamon et al., 1954). Such formulae may yield an individual daily mean global irradiance which is in error by 50% or more, but the error must reduce roughly in proportion to the square root of the number of days considered down to the limit set by other variable factors.

Typical meteorological cloud observation includes reports of the amount of cloud at various levels. The accuracy of an individual report may be horribly low, but presumably the error is more or less random and a sufficient number of such observations should give a reliable measure of the cloud climatology of a particular site. Further, there are a large number of radiation-driven surface processes which are non-linear to the extent that a time average of the process bears no relation to the time average of the radiation input. The instantaneous input must be used to calculate the instantaneous process response and any time average which is performed to reduce the effect of random error in the input data must be done last. A typical example might be calculation of the seasonal average efficiency of solar-powered machines. For such cases it may be essential to make use of the available detail in cloud observation and to calculate both the direct and diffuse components of the global flux density.

Direct beam irradiance $I(\theta)$ may be given as a function of zenith angle θ

by some empirical relation such as in Fig. 6.4. Calculation of the daily mean is simply a matter of calculating θ as a function of time via equation [3.4], integrating $I(\theta)$ over the day and reducing the integrated value by the observed total fractional cloud cover CC since the presence of cloud in the direct beam normally reduces $I(\theta)$ to zero. Mathematically the appropriate formula is:

$$\bar{I} = \sum_{t_1}^{t_2} (1 - CC) I(\theta)/t \qquad [6.39]$$

where t_1 and t_2 are the times of sunrise and sunset and t is the 24-hour period in units appropriate to the time step of the summation.

The calculation of global flux is more difficult since one must take into account not only the fraction of cloud cover but also the contribution of the clouds to the diffuse flux. In cloudy skies the diffuse flux may be the dominant or only radiant energy flux. The global flux $F_s\downarrow(\theta)$ at a particular time (i.e. a particular θ) may be expressed as:

$$F_s\downarrow(\theta) = I_0 \left(1 - \sum_i \alpha_{ci} CC_i\right) [1 - \hat{m} - \alpha_A (1 - CC)] \cos\theta \qquad [6.40]$$

where I_0 is the solar constant, $\hat{m}$ is the absorption due to water vapour and α_A is the albedo of the atmosphere in the fraction of clear sky $1 - CC$. The α_{ci} are the albedos of the clouds plus the atmosphere above them at level i, and the CC_i are the fractional amounts of cloud at level i which are visible from above when cloud overlap is taken into account. By analogy with equation [6.39], the daily mean global flux density $\bar{F}_s\downarrow$ is given by:

$$\bar{F}_s\downarrow = \sum_{t_1}^{t_2} F_s\downarrow(\theta)/t \qquad [6.41]$$

Paltridge (1974b) has used the above formulae to produce contour maps of seasonal average direct and global radiation over Australia plus other derivable quantities such as the diffuse flux density and the flux densities on inclined planes. The fundamental input data were 5 years of cloud amounts at each of four levels recorded every 3 hours at 45 meteorological stations over the continent. In order to calculate the CC_i required in equation [6.40], a simple "overlapping" geometrical exercise leads to the formula:

$$CC_i = CC_{Ti} \left(1 - \sum_{j=1}^{i-1} CC_j\right) \qquad [6.42]$$

where CC_{Ti} is the recorded total cloud amount at level i and the layer count is downwards. There are objections to this procedure but it is a good check that, averaged over a few days, the quantity $\sum_1^4 CC_i$ is generally very close to the average total cloud amount actually recorded by the observers.

The major difficulty in calculating $F_s\downarrow$ is to assign accurate values to the α_{ci}. For this sort of problem one is not really justified in going to the enormous trouble of including a dependence of α_{ci} on solar angle since the error is swamped by the error in cloud amount specification. For this reason also, Paltridge (1974b) assigned a fixed value of 0.18 to $\hat{m}$ on the basis that water vapour absorption tends to saturate at about this value, and ignored any possible additional absorption by clouds on the grounds that the absorption bands of water vapour and of liquid water tend to overlap (see Chapter 5). He assigned the cloud albedo values α_{ci} according to Table 6.3, which are mean values obtained from (among other sources) experience with actual radiometric measurements from high-altitude balloons and aircraft.

TABLE 6.3

Mean short-wave planetary albedo α_{ci} and long-wave flux emissivity ϵ_{ci} assumed to be associated with various cloud types. These values have led to good agreement between predicted and observed monthly average global flux densities of short- and long-wave radiation at Aspendale

Cloud level	Cloud type	α_{ci}	ϵ_{ci}
1	Cirrus	0.35	0.3
2	Altocumulus-altostratus	0.55	0.9
3	Low cloud 2	0.60	1.0
4	Low cloud 1	0.50	1.0

Despite the approximations the predicted *monthly* average values of global flux density are remarkably close to the observed amounts recorded at a number of sites about the continent. The maximum error is rarely more than 10% on any month at any site and is generally much less. The agreement is much better than it deserves to be. Very probably the situation is helped by the fact that cloud cover is one of the few "U-shaped" distributions in nature. That is, when there is cloud it will probably be total cloud cover rather than some intermediate fraction. The *seasonal* mean absolute error in the predicted value of direct, global and diffuse radiation was about 1–2 mW cm^{-2} for the one station in Australia (Aspendale) where long-term records of all these parameters are available.

Referring now to the net long-wave flux at the ground, the upward component is the black-body radiation σT_g^4 reduced to $\epsilon_g \sigma T_g^4$ if the surface emissivity is less than 1.0. In clear skies the downward component $F_L\downarrow$ arises from the emission bands of water vapour and carbon dioxide, and accurate values can be obtained only by detailed computations involving the total profiles of temperature and water vapour (see Chapter 7). However, various empirical formulae have been derived which relate $F_L\downarrow$ to measured surface (screen) values of temperature T_g and water vapour pressure e. Their validity depends on the correlations which exist between surface values and the profiles in the free atmosphere. Thus Brunt (1932) proposed a formula:

$$F_L\downarrow/\sigma T_g^4 = a + be^{1/2} \quad [6.43]$$

where a and b are constants which are not, however, universally applicable. He proposed $a = 0.52$ and $b = 0.065$, values which gave a correlation coefficient of 0.97 between $F_L\downarrow/\sigma T_g^4$ and $e^{1/2}$ for one particular site. The agreement degenerates at other localities. Swinbank (1963) examined the problem and came to the conclusion that the apparent correlation between $F_L\downarrow$ and e derived from the fact that both $F_L\downarrow$ and e are correlated to temperature. On this argument there should be a direct relation between $F_L\downarrow$ and T_g alone and Swinbank derived the empirical formula

$$F_L\downarrow = 5.31 \times 10^{-14}\,T^6 \quad \{\text{mW cm}^{-2}\} \quad [6.44]$$

$F_L\downarrow$ is so strongly correlated with screen temperature because the intensities of the water vapour and carbon dioxide absorption bands at lower atmospheric mixing ratios are sufficiently large to produce effectively full black-body emission over the wavelengths they cover — that is, at wavelengths outside the atmospheric window from 8 to 14 μm. The screen temperature is therefore a good approximation to that of the radiating medium. However, the concept is only an approximation and a more representative temperature of the "centre of gravity" of emission might be that of somewhat higher in the atmosphere, perhaps at 200 or 300 m. The exact height would depend on the humidity distribution. Swinbank's regression equation is based on night-time data and is therefore biased towards inversion conditions. During day-time lapse conditions it is to be expected that the effective centre of gravity temperature would be lower relative to screen temperature and the night-time regression would over-estimate $F_L\downarrow$ during the day. Experimentally this seems to be true (Paltridge, 1970) and on average during the day $F_L\downarrow$ from equation [6.44] should be reduced by about 2 mW cm^{-2}. There are valid objections to the actual physics of Swinbank's model (Deacon, 1970) but for normal clear-sky conditions (apart from dry atmospheres over deserts and on tops of mountains) equation [6.44] with a suitable daytime correction estimates $F_L\downarrow$ with the quite remarkable accuracy of ± 2 mW cm^{-2}.

The model above is a specific application of the "opaque and clear" model of atmospheric long-wave emission devised originally by Simpson (1928). Roughly 0.7 of the total energy in the black-body spectrum exists at wavelengths outside the 8–14-μm window at normal terrestrial temperatures. The fraction varies slightly with temperature as the spectrum shifts relative to the window which is one of the reasons why the Swinbank formula involves a power of T different from 4. However, the next simplification is to assume the atmospheric emissivity as viewed from the ground (i.e. $F_L\downarrow/\sigma T_g^4$) to be a constant 0.7 independent of temperature. This is the basis of the long-wave model used in equation [1.2] of Chapter 1. Note that because the lower atmospheric air temperature is highly correlated with, and is not greatly different from, surface temperature (as a result of sensible heat exchange if nothing else), the *net* long-wave exchange over those wavelengths outside the window is always small. To a first approximation, net exchange occurs only in the window region, and in clear skies consists only of the upward component from the ground. That is $F_{\mathrm{L,net}}\uparrow \approx (1-0.7)\epsilon_{\mathrm{g}}\sigma T_{\mathrm{g}}^{4}$.

On this model the presence of cloud affects the net flux at the surface only via the window region. Thus cloud of temperature T_{c} and emissivity ϵ_{c} and amount CC increases the downcoming long-wave flux from the $F_{\mathrm{L}}\downarrow$ predicted by the Swinbank formula to:

$$F_{\mathrm{LT}}\downarrow \;=\; F_{\mathrm{L}}\downarrow + (1-0.7)\epsilon_{\mathrm{c}}\sigma T_{\mathrm{c}}^{4}\cdot CC \qquad [6.45]$$

Knowledge of cloud height will give T_{c} either from an observed radiosonde profile or, to a not very significant error, from a standard atmosphere profile. Cloud emissivity ϵ_{c} is usually very close to unity for low cloud. As a first guess the emissivities of middle and high cloud might be assumed close to the values quoted in Table 6.3 (see also Chapter 8).

The next stage of simplification is to assign a single average value for the term $(1-0.7)\epsilon_{\mathrm{c}}\sigma T_{\mathrm{c}}^{4}$ obtained from experimental data on the deviation of $F_{\mathrm{L}}\downarrow$ from the clear-sky formula [6.44] as a function of total observed cloud cover. Such an average may be different from place to place but for Aspendale which is a fairly typical mid-latitude station:

$$\overline{(1-0.7)\epsilon_{\mathrm{c}}\sigma T_{\mathrm{c}}^{4}} \;=\; 6\,\mathrm{mW\,cm^{-2}} \qquad [6.46]$$

This value can perhaps be used generally in equation [6.45].

7. RADIATION IN THE OCEAN

Table 6.4 gives the observed extinction coefficient of "pure" sea water as a function of wavelength together with the theoretical scattering coefficient. The data are effectively those of distilled water also since the presence of dissolved salts apparently makes very little difference except perhaps in the extreme ultraviolet. The scatter has the typical Rayleigh characteristic of decreasing rapidly with increasing wavelength. It becomes insignificant at wavelengths longer than 800 nm. The absorption varies rapidly with wavelength, but unlike the atmosphere it dominates the scatter over the whole of the spectrum. The overall effect is for the spectrum of downward flux density to be narrowed with increasing depth such that infrared energy (>800 nm) is removed within the first metre. At a depth of 100 m the spectrum is reduced to the band between 400 and 500 nm. As a general and very approximate rule of thumb, 50% of the short-wave energy is absorbed within the first metre, 80% within the first 10 and 99% within the first 100 m when the sun is vertically overhead.

TABLE 6.4

Observed volume extinction coefficient β_e and theoretical volume scattering coefficient β_{sc} as a function of wavelength for pure water (after Jerlov, 1968). Values at $\lambda > 1.0\ \mu m$ are approximate means over a band $\pm 0.1\ \mu m$ either side of quoted wavelength

λ (nm)	β_e (m^{-1})	β_{sc} (m^{-1})	λ (μm)	β_e (m^{-1})	β_{sc} (m^{-1})
375	4.5×10^{-2}	6.6×10^{-3}	1.0	20	—
400	4.3×10^{-2}	5.0×10^{-3}	1.2	60	—
425	3.3×10^{-2}	3.9×10^{-3}	1.4	400	—
450	1.9×10^{-2}	3.0×10^{-3}	1.6	500	—
475	1.8×10^{-2}	2.4×10^{-3}	1.8	700	—
500	3.6×10^{-2}	1.9×10^{-3}	2.0	3000	—
525	4.1×10^{-2}	1.6×10^{-3}	2.2	2000	—
550	6.9×10^{-2}	1.3×10^{-3}	2.4	3000	—
575	9.1×10^{-2}	1.1×10^{-3}			
600	1.8×10^{-1}	9.3×10^{-4}			
625	2.3×10^{-1}	7.8×10^{-1}			
650	2.9×10^{-1}	6.7×10^{-4}			
675	3.7×10^{-1}	5.8×10^{-4}			
700	5.0×10^{-1}	4.9×10^{-4}			
725	1.2×10^{-0}	4.3×10^{-4}			
750	2.4×10^{-0}	3.8×10^{-4}			
775	2.4×10^{-0}	3.3×10^{-4}			
800	2.0×10^{-0}	2.9×10^{-4}			

Thus the climatologist whose interest is with energy flow has not been greatly concerned with the detail of oceanic radiative processes since on the scale of interest to him the absorption is effectively surface absorption. However, this view is somewhat short-sighted. Far more than the atmosphere, the ocean surface layers are subject to great variations in both organic and inorganic suspended particulate matter. This drastically affects the scattering coefficient which in turn can alter the surface albedo and hence the general energy budget of the globe. No significant work on this subject has been done and, until it has, speculation on possible climate change will remain an exercise involving only sensitivity tests or partial differential solutions. In principle, the oceanic problem can be examined by the same theoretical techniques as are used in the atmosphere. In practice the difficulty is also the same as that of the atmosphere — namely, to specify the distribution of particulate matter and to separate the scattering and absorption coefficients.

CHAPTER 7

Long-Wave Radiation Transfer in the Clear Atmosphere

1. INTRODUCTION

Long-wave infrared radiation is absorbed strongly by certain minor atmospheric gases. This absorption is very spectrally dependent and the spectral region where absorption occurs is determined by the gaseous molecular structure. The radiation is absorbed in spectral *bands.* These absorption bands consist of many individual narrow spectral *lines.* The line absorption is caused by changes in the molecular vibrational or rotational energies. At wavelengths in the centre of molecular bands, the infrared radiation can be absorbed within a few metres of atmosphere.

An absorbing gas is also an emitter of infrared radiation (at the absorbing wavelengths) as a result of its finite temperature. Thus a given volume of the atmosphere both receives and emits infrared radiation, and will cool or warm according as to whether the *net* radiative input is negative or positive. The meteorologist is primarily concerned with the cooling or heating of a given layer in the atmosphere which results from the net emission or absorption of energy from the layer. That is, he is concerned with the change in the vertical net flux density with height and with the associated cooling rates. The computation of vertical infrared flux densities and of the associated cooling rates is the prime concern of this and the following chapter.

2. DISTRIBUTION AND NATURE OF ATMOSPHERIC ABSORPTION

Water vapour (H_2O), carbon dioxide (CO_2) and ozone (O_3) have the greatest effect on infrared transfer in the atmosphere. Because its mixing ratio is constant with altitude, CO_2 has a dominant role in the cooling of the stratosphere where the water vapour mixing ratios are very small. H_2O absorption is predominant in the troposphere, particularly the lower troposphere. O_3 absorption is significant mainly in the stratosphere. Water vapour has strong vibration-rotation absorption bands at 1.4, 1.9, 2.7 and 6.3 μm. The first three are important only for solar radiation transfer and will not be considered further in this chapter. The bands of importance to terrestrial infrared transfer are the 6.3-μm band and the pure rotation bands between 20 μm and 1 mm. The most important carbon dioxide band is a

TABLE 7.1

Wavelengths at the centres of absorption bands of the various atmospheric gases of importance to infrared transfer. Corresponding wavenumbers are in brackets

		"Strong" (μm)	"Weak" (μm)
Water vapour	H_2O	1.4 (7142) 1.9 (5263) 2.7 (3704) 6.3 (5787) 13.0 − 1000	0.9 (11111) 1.1 (9091)
Carbon dioxide	CO_2	2.7 (3704) 4.3 (2320) 14.7 (680)	1.4 (7142) 1.6 (6250) 2.0 (5000) 5.0 (2000) 9.4 (1064) 10.4 (962)
Ozone	O_3	4.7 (787) 9.6 (1042) 14.1 (709)	3.3 (3030) 3.6 (2778) 5.7 (1754)
Nitrous oxide	N_2O	4.5 (2222) 7.8 (1282)	3.9 (2564) 4.1 (2439) 9.6 (1042) 17.0 (588)
Methane	CH_4	3.3 (3030) 3.8 (2632) 7.7 (1299)	
Carbon monoxide	CO	4.7 (2128)	2.3 (4348)

vibration-rotation band at 14.7 μm which causes intense absorption. Bands of less importance occur at 2.7 and 4.3 μm. Radiation is completely absorbed within a short distance by the 14.7-μm CO_2 band at altitudes up to 20 km or more. Ozone absorbs intensely in the region of 9.6 μm and has other bands at 4.7 and 14.1 μm. Absorption occurs also in several other minor atmospheric constituents such as carbon monoxide, nitrous oxide, methane and nitric acid.

In the regions between the molecular absorption bands atmospheric gases are relatively transparent. However, water vapour has a continuum absorption which manifests itself most markedly in the so-called "atmospheric window" in the region between about 8 and 13 or 14 μm, and between the rotational H_2O lines. The origin of this absorption is still a subject of controversy. Undoubtedly, a part of the absorption is due to the wings of strong H_2O lines in the rotation band and part is due to weak

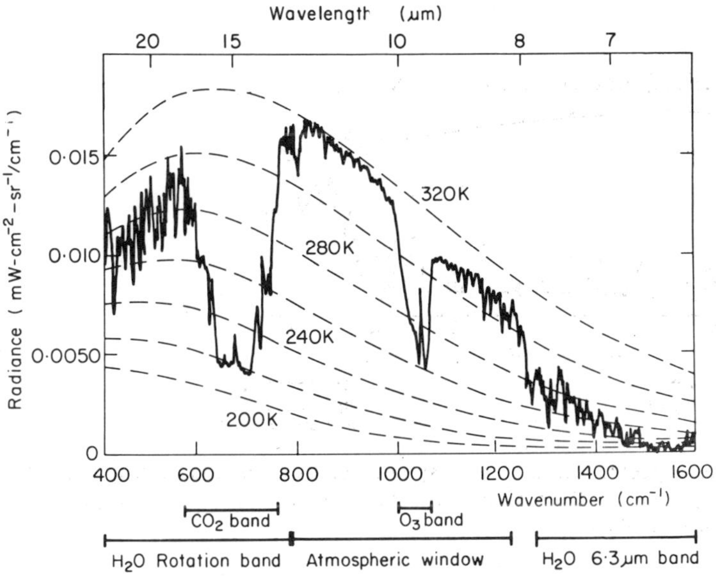

Fig. 7.1. Atmospheric spectrum obtained with a scanning interferometer on board the Nimbus 4 satellite. The interferometer viewed the earth vertically and the satellite was passing over the North African desert. (After Hanel et al., 1972.)

rotational lines in the window itself. Appreciable absorption is also thought to be due to phenomena associated with the clustering of two or more water vapour molecules. The continuum absorption is thus most marked in moist atmospheres in the middle and lower troposphere.

The absorption properties of the various gases are listed in Table 7.1. Fig. 7.1 shows the infrared terrestrial spectrum as taken by a scanning interferometer looking downward from a satellite over a desert region. Also shown are black-body radiances at various temperatures. The CO_2 band causes emission at the stratospheric temperature of 220 K. In the window region, the radiance corresponds closely to terrestrial temperatures except at about 8–9 μm where the temperature is depressed due to surface emissivity variations of sand. The influence of the ozone band is clearly visible and the water vapour bands cause emission at mid-tropospheric temperatures. Fig. 7.2 shows for the 8–13-μm window the inverse case of the zenith infrared radiance of the sky observed from the surface. The atmosphere has a low radiance in this spectral region due to its small absorption but the effects of weak water vapour lines, the continuum emission and ozone emission are clearly identifiable. Fig. 7.3 shows a similar spectrum from 15 to 22 μm. The emission corresponds quite closely to that of a black body at ground temperature except for a relatively transparent region between 17 and 19 μm. The radiance in this region is considerably lessened on top of a high mountain as shown in Fig. 7.4. In this case the atmosphere is quite

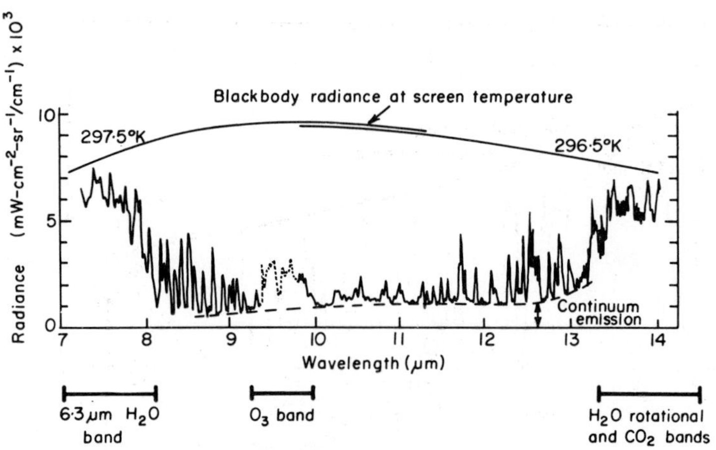

Fig. 7.2. Zenith emission spectrum in the "atmospheric window" measured with a grating spectrometer on the ground at Beer Sheva. The total water vapour optical path was 1.35 g cm^{-2}. (After Bolle., 1965.)

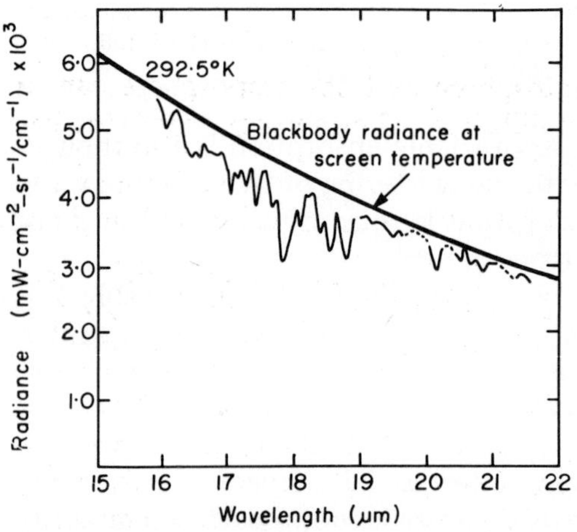

Fig. 7.3. Zenith emission spectrum in the 15–22-μm region measured with a grating spectrometer on the ground at South Agata. The total water vapour optical path was about 1.0 g cm^{-2}. (After Bolle et al., 1963.)

transparent between the water vapour lines. Beyond 30 μm wavelength and extending to about 1,000 μm, the atmosphere is much more absorbing and lines are only identifiable in stratospheric spectra (Harries and Burroughs, 1971).

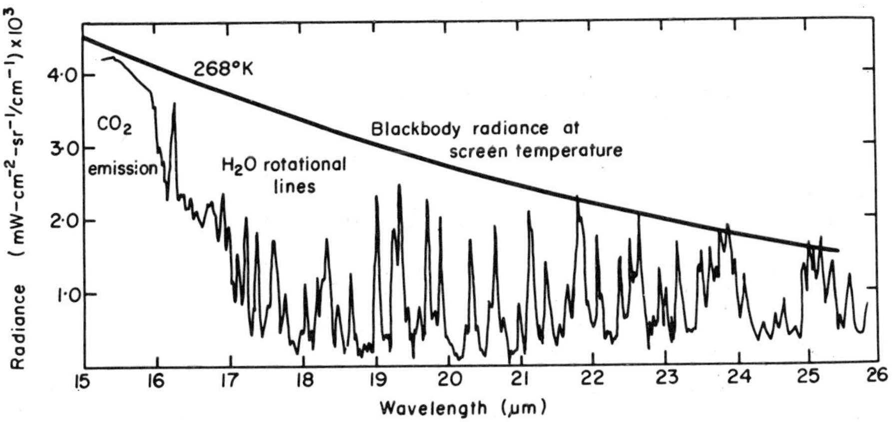

Fig. 7.4. Zenith emission spectrum in the 15–26-μm region measured with a grating spectrometer on the ground at Jungfraujoch. Total water optical path of about 0.1 g cm^{-2}. (After Bolle et al., 1963.)

3. THE EQUATION OF RADIATIVE TRANSFER

The atmosphere in the lower stratosphere and the troposphere can be considered in local thermodynamic equilibrium. Then the equation of radiative transfer takes the form:

$$\cos\theta \, dN_\nu(z, \theta)/\kappa_\nu \rho \, dz = -N_\nu(z, \theta) + \tilde{J}_\nu(z, \theta) \qquad [7.1]$$

where in this case the source function $\tilde{J}_\nu(z, \theta)$ is simply the Planck blackbody radiance $B_\nu(T_z)$ at wavenumber ν. The factor κ_ν is the mass absorption coefficient of a gas of density ρ. This expression can be integrated to yield:

$$N_\nu(z, \theta) = B_\nu(T_g) \exp[-\tau_\nu(z, z_0)\sec\theta]$$
$$+ \int_{z_0}^{z} B_\nu(T_{z'}) \exp[-\tau_\nu(z, z')\sec\theta] \, \kappa_\nu \rho \sec\theta \, dz' \qquad [7.2]$$

where $\tau_\nu(z, z') = \int_z^{z'} \kappa_\nu \rho \, dz''$. This equation can readily be interpreted with reference to Fig. 7.5. Consider the upward radiance at z at an angle θ to the zenith of a horizontally homogeneous atmosphere, bounded by a surface below. If the surface emits at the Planck radiance $B_\nu(T_g)$, then the radiance at z from the surface is $B_\nu(T_g) \exp[-\tau_\nu(z, z_0)\sec\theta]$. The second term on the right-hand side of equation [7.2] thus represents the integrated contribution at z from each slab along the path $(z - z_0)\sec\theta$.

Now the exponentials of equation [7.2] are obviously transmittances. That is:

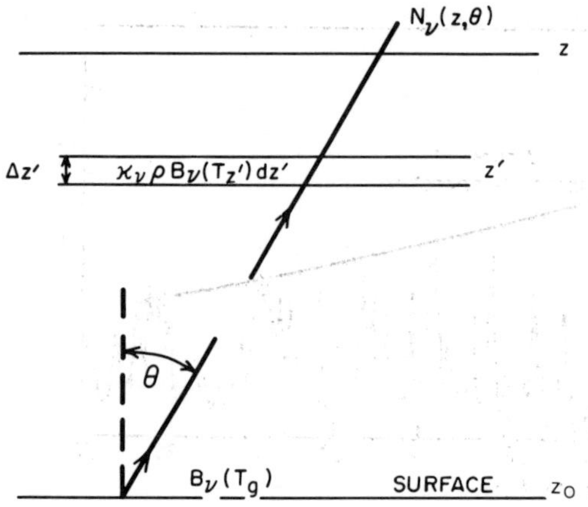

Fig. 7.5. Schematic diagram of radiance $N_\nu(z, \theta)$ at an altitude z.

$$Tr_\nu(z, z', \theta) = \exp[-\tau_\nu(z, z')\sec\theta] \quad [7.3]$$

which has the derivative:

$$dTr_\nu(z, z', \theta)/dz' = \exp[-\tau_\nu(z, z')\sec\theta]\kappa_\nu \rho \sec\theta \quad [7.4]$$

Substituting back into equation [7.2]:

$$N_\nu(z, \theta) = B_\nu(T_g)\exp[-\tau_\nu(z, z_0)\sec\theta] + \int_{z_0}^{z} B_\nu(T_{z'})[dTr_\nu(z, z', \theta)/dz']dz' \quad [7.5]$$

This integro-differential equation enables the radiance $N_\nu(z, \theta)$ to be calculated if both temperature and transmission dependence with z' are known. As both $B_\nu(T_{z'})$ and $Tr_\nu(z, z', \theta)$ are variables, a second form of equation [7.5] can be obtained from an integration by parts. Considering only vertical transfer, so that $\sec\theta = 1$, we have:

$$d[B_\nu(T_{z'})Tr_\nu(z, z')]/dz' = B_\nu(T_{z'})[dTr_\nu(z, z')/dz']dz' + Tr_\nu(z, z')[dB_\nu(T_{z'})/dz']dz' \quad [7.6]$$

from which:

$$\int_{z_0}^{z} B_\nu(T_{z'})[dTr_\nu(z, z')/dz']dz' = [B_\nu(T_{z'})Tr_\nu(z, z')]_{z_0}^{z} - \int_{z_0}^{z} Tr_\nu(z, z')[dB_\nu(T_{z'})/dz']dz' = -B_\nu(T_{z_0})Tr_\nu(z, z_0) + B_\nu(T_z) - \int_{z_0}^{z} Tr_\nu(z, z')[dB_\nu(T_{z'})/dz']dz' \quad [7.7]$$

Hence, the vertical upward radiance is given by:

$$N_\nu(z) = Tr_\nu(z, z_0)[B_\nu(T_g) - B_\nu(T_{z_0})] + B_\nu(T_z) - \int_{z_0}^{z} Tr_\nu(z, z')[dB_\nu(T_{z'})/dz']dz' \qquad [7.8]$$

$B_\nu(T_g) - B_\nu(T_{z_0})$ is only different from zero if the ground *surface* temperature is different from the ground air temperature. Similar equations exist for the downward radiance where the equivalent boundary condition at the top of the atmosphere is that of zero downward radiance.

4. THE EQUATION FOR VERTICAL FLUX DENSITY

Both equations [7.5] and [7.8], as well as other of their transformations, have been used in computations of the atmospheric spectral radiance. However, for heat budget and cooling calculations it is the spectral radiance integrated over all angles of the relevant hemisphere (that is, the flux density) which is required. Calculation of the flux density F_ν at a point involves the integration of the radiance over both zenith and azimuth angles θ and ϕ. However, since the concern is generally with plane-parallel atmospheres and since the source function at least is isotropic, it is far simpler to go immediately to a form of two-stream model whereby the equations of transfer are reduced to only the two appropriate to the upward and downward flux densities $F_\nu\uparrow$ and $F_\nu\downarrow$. The relevant transmittances must be *flux* transmittances, and it is in these that the angular integration is taken into account by various devices or approximations. The final total flux densities $F\uparrow$ and $F\downarrow$ are obtained by summing the $F_\nu\uparrow$ or $F_\nu\downarrow$ over the spectrum. The equations for the flux density at a height z in the atmosphere are of exactly the same form as the vertical radiance equations. This can be shown (at least for the case where the upward and downward fields are individually isotropic) as follows.

The *flux density* transmittance $\widetilde{Tr}_\nu(z)$ across a slab of thickness z is given by:

$$\widetilde{Tr}_\nu(z) = (1/\pi)\int_0^{2\pi} d\phi \int_0^{\pi/2} \exp(-\kappa_\nu \rho z \sec\theta)\cos\theta \sin\theta \, d\theta \qquad [7.9]$$

from which we obtain:

$$\widetilde{Tr}_\nu(z) = 2\int_1^{\infty} \exp(-\kappa_\nu \rho z \eta) d\eta/\eta^3 \qquad [7.10]$$

where $\eta = \sec\theta$. This integral is termed the third exponential integral $\mathrm{Ei}_3(\kappa_\nu \rho z)$ (Elsasser, 1942), so that:

$$\widetilde{Tr}_\nu(z) = 2\mathrm{Ei}_3(\kappa_\nu \rho z) \quad [7.11]$$

Similarly, the flux density emitted from a slab of thickness dz' at z' and received at a distance z is given by

$$dF_\nu = \kappa_\nu \rho dz' \int_0^{2\pi} d\phi \int_0^{\pi/2} B_\nu(T_{z'}) \exp(-\kappa_\nu \rho z \sec\theta) \sec\theta \cos\theta \sin\theta \, d\theta \quad [7.12]$$

After carrying out the integration over θ and ϕ we obtain:

$$dF_\nu = 2\rho\pi\kappa_\nu B_\nu(T_{z'}) \mathrm{Ei}_2(\kappa_\nu \rho z) dz' \quad [7.13]$$

where $\mathrm{Ei}_2(\kappa_\nu \rho z)$ is the second exponential integral. As it can be shown that $d\mathrm{Ei}_n(\tilde{\chi})/d\tilde{\chi}$ equals $-\mathrm{Ei}_{n-1}(\tilde{\chi})$, then in the terminology of the last section:

$$F_\nu(z) = -2 \int_{z_0}^{z} d[\mathrm{Ei}_3(z, z')/dz'] \tilde{B}_\nu(T_{z'}) dz' \quad [7.14]$$

It follows that equations of the *form* of [7.5] and [7.8] will hold where $Tr_\nu(z, z')$ is replaced by the *flux density* transmittance $\widetilde{Tr}_\nu(z, z')$, and $B_\nu(T_{z'})$ is in *flux density* units — that is, the radiance function multiplied by π. Thus equations for the upward and downward flux densities at a level z are given by (Rodgers and Walshaw, 1966):

$$F_\nu(z)\uparrow = \widetilde{Tr}_\nu(z, z_0)[\tilde{B}_\nu(T_g) - \tilde{B}_\nu(T_{z_0})] + \tilde{B}_\nu(T_z) - \int_{z_0}^{z} \widetilde{Tr}_\nu(z, z')[d\tilde{B}_\nu(T_{z'})/dz'] dz' \quad [7.15]$$

$$F_\nu(z)\downarrow = \widetilde{Tr}_\nu(z, Z)[\tilde{B}_\nu(T_Z) - \tilde{B}_\nu(\mathrm{top})] - \tilde{B}_\nu(T_z) - \int_{z}^{Z} \widetilde{Tr}_\nu(z, z')[d\tilde{B}_\nu(T_{z'})/dz'] dz' \quad [7.16]$$

The level Z is the top boundary of integration and $\tilde{B}_\nu(\mathrm{top})$ represents flux that is incident from above Z.

The radiation field in the atmosphere is not generally isotropic. However, with suitable treatment of the flux transmission $Tr(z, z')$, equations [7.15] and [7.16] give acceptable accuracy for many purposes and have been used almost exclusively for computations of atmospheric cooling.

To make the flux computation possible, the transmission must be averaged over finite spectral intervals $\Delta\nu$ so that:

$$\widetilde{Tr}(z, z') = (1/\Delta\nu) \int_{\Delta\nu} \widetilde{Tr}_\nu(z, z') \quad [7.17]$$

In the following, the flux and the transmittance and Planck functions will normally be quoted without the formal subscript to indicate integration over a finite spectral interval.

The net loss of energy from a layer in the atmosphere due to a spectral interval $\Delta\nu$ is given by the rate of change of net flux density F_N where $F_N(z)$ is $F(z)\downarrow - F(z)\uparrow$. The time rate of change of the atmospheric temperature is:

$$\mathrm{d}T(z)/\mathrm{d}t = (1/\rho_a c_p)[\mathrm{d}F_N(z)/\mathrm{d}z] \qquad [7.18]$$

where ρ_a is the density of air and c_p is its specific heat at constant pressure. Integrating equations [7.15] and [7.16] over a spectral interval $\Delta\nu$ and subtracting:

$$\mathrm{d}F_N(z)/\mathrm{d}z = [\mathrm{d}\tilde{T}r(z, Z)/\mathrm{d}z][\tilde{B}(T_Z) - \tilde{B}(\mathrm{top})] - [\mathrm{d}\tilde{T}r(z, z_0)/\mathrm{d}z][\tilde{B}(T_{z_0}) - \tilde{B}(T_g)] - \int_0^Z [\mathrm{d}\tilde{T}r(z, z')/\mathrm{d}z][\mathrm{d}\tilde{B}(T_{z'})/\mathrm{d}z']\mathrm{d}z' \qquad [7.19]$$

where $\tilde{B}(T_{z'})$ is the Planck function at the centre of $\Delta\nu$. The total time rate of change of temperature is obtained by summing $\mathrm{d}T(z)/\mathrm{d}t$ over all spectral intervals $\Delta\nu$ in every atmospheric absorption band.

The basic problem in atmospheric radiation transfer is the computation of $\tilde{T}r(z, z')$ between all levels z and z'. This requires integrations over wavenumber, height and angle. The wavenumber integration involves the averaging of the effects of hundreds of spectral absorption lines in a spectral interval $\Delta\nu$ in a molecular band. The height integration involves an atmosphere in which the temperature and pressure are continuously changing. The absorption of molecular bands is dependent on pressure and temperature so that this dependence has to be incorporated in the transmission function. Another problem is the permissible spectral width $\Delta\nu$ over which the Planck function can be considered to be constant as this has a direct bearing on the economy of computing time. The simplest models use a "grey-body" approximation with $\int \tilde{B}_\nu(T)\mathrm{d}\nu = \sigma T^4$. The most detailed models to date divide the various molecular bands into from 2 to 10 spectral intervals, depending on the accuracy required.

As the transmittance is normally averaged over finite spectral intervals it is necessary to do this *before* making the integrations over angle. When using the flux equations the angular integration is accomplished by simply multiplying the vertical optical path u by a constant "diffusivity factor".

A great deal of work has been done over the last three decades on the problems involved in the integration of the transmittance function and many ingenious solutions have been devised. As the transmittance is central to the theory of radiative cooling, the next section is devoted to this topic before some actual methods which have been used to compute flux densities and cooling rates are discussed. In what follows it is convenient sometimes to discuss the absorptance, rather than the transmittance. In a non-scattering gas the sum of absorptance and transmittance is unity.

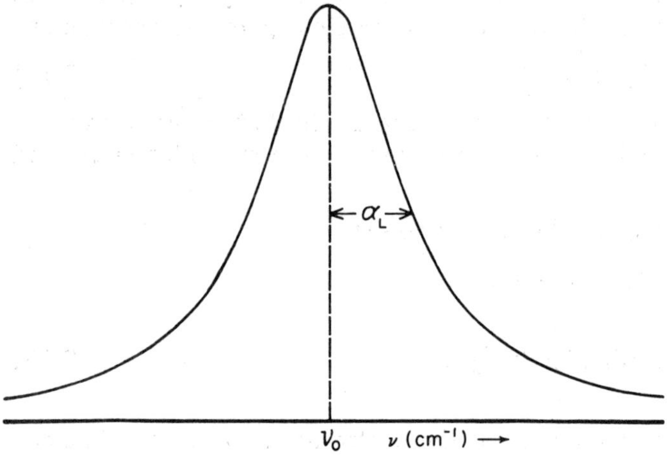

Fig. 7.6. The shape of a spectral line, where α_L is the (Lorentz) line width at the "½ power point".

5. TRANSMITTANCE MODELLING

5.1. *Absorption by a single line*

The mass absorption coefficient κ_ν over a single molecular absorption line varies in a manner illustrated in Fig. 7.6. The line width α is a function of the gas pressure and temperature since the spread in wavenumber is in fact caused by molecular motions and collisions. Following Houghton and Smith (1966), the integral of the absorption coefficient over the line, $\int \kappa_\nu d\nu$, is called the line strength S. The line strength is a function of the type of molecular transition and also the temperature. In a gas at tropospheric pressures, the dependence of κ_ν on the distance from the centre frequency ν_0 is given quite well by the Lorentz line shape for pressure broadening:

$$\kappa_\nu = (S\alpha_L/\pi)/[(\nu-\nu_0)^2 + \alpha_L^2] \qquad [7.20]$$

which describes a line which is broadened by molecular collisions. Although this shape is not quite accurate, particularly for $(\nu - \nu_0) \gg \alpha_L$, it is adequate for most atmospheric problems. On simple kinetic theory the line width $\alpha_L(p, T)$ at pressure p and temperature T is given by:

$$\alpha_L(p, T) = \alpha_L(p_s, T_s)(p/p_s)(T_s/T)^{1/2} \qquad [7.21]$$

The integrated absorptance a of a given optical path u over a single line is given by:

$$a = \int [1-\exp(-\kappa_\nu u)]\,d\nu \qquad [7.22]$$

where $1-\exp(-\kappa_\nu u)$ is the absorptance a_ν, and $u = \rho z$ over a path z of gas density ρ. If a can be calculated for one line and if the mean line spacings and the distribution of line strengths in a gaseous absorption band are both known there is the possibility of calculating the *mean* absorptance due to all the lines. The integration of equation [7.22] over the single line gives:

$$a = 2\pi\alpha_L \chi \exp(-\chi)[JJ_0(i\chi) - iJJ_1(i\chi)] \qquad [7.23]$$

where $\chi = Su/2\pi\alpha_L$ and JJ_0 and JJ_1 are Bessel functions. Equation [7.23] can be simplified both when $\chi \to 0$ and when $\chi \to \infty$ and these approximations are very important to radiative transfer in the atmosphere.

For small χ ("weak-line" approximation):

$$a = Su \qquad [7.24]$$

and for large χ ("strong-line" approximation):

$$a = 2(S\alpha_L u)^{1/2} \qquad [7.25]$$

A plot of log a versus log u thus has a slope of 1 for small χ and $\frac{1}{2}$ for large χ. Such a curve is known as a "curve of growth". The strong-line approximation requires a large value of $Su/2\pi\alpha_L$. For large line strengths this approximation can still apply even for small absorber amounts. It is still valid, for instance, for absorption of water vapour in the stratosphere at wavelengths near the centre of the rotation band.

Interesting features of equations [7.24] and [7.25] are the dependence of a only on line strength for a given absorber mass in weak lines, but an equal dependence on line width α_L in strong lines. The line strength S is dependent on temperature but not on pressure. However, α_L, and therefore a in equation [7.25], is a function of pressure. Fig. 7.7 illustrates the absorptance a_ν for "weak-line", intermediate and "strong-line" absorption. For the strong-line case, the line centres are completely absorbed and changes in a are determined by the wings of the lines.

At low atmospheric pressure, pressure broadening becomes small and Doppler broadening becomes significant. Doppler broadening is a consequence of the random velocities of molecules, so that the emitted radiation suffers random frequency shifts. The critical height above which Doppler broadening should be considered ranges from 44 km for H_2O to 52 km for CO_2 (Rodgers and Walshaw, 1966). The mass absorption coefficient for pure Doppler broadening is given by:

$$\kappa_\nu = [(S/\alpha_D(\pi)^{1/2}]\exp\{-[(\nu-\nu_0)/\alpha_D]^2\} \qquad [7.26]$$

where α_D, the Doppler half-width, is given by:

$$\alpha_D = (\nu_0/c)(2\hat{k}T/m)^{1/2} \qquad [7.27]$$

m being the mass of the molecule. α_D is directly dependent on ν_0 and therefore often becomes dominant in the visible and ultraviolet regions of the spectrum. A line of Doppler shape is more intense at the centre and weaker in the wings compared to a Lorentz line. If a line is fully absorbed at the centre, then the effect of Doppler broadening is minimal.

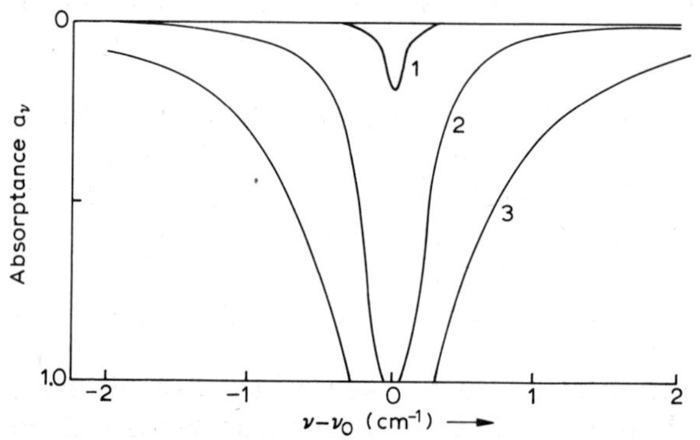

Fig. 7.7 Fractional absorptance by a spectral line for three different absorption categories. *1* = weak absorption (weak-line approximation applies), *2* = intermediate absorption (equation [7.23] applies), *3* = strong absorption (strong-line approximation applies).

5.2. *Absorption over a band*

The basis of all transmittance functions is the consideration of the absorptance of many lines in a given spectral interval $\Delta\nu$. The simplest case is when the lines are of equal strength and are spaced at a mean distance δ which is always much greater than α, so that no overlapping of lines occurs. Suppose there are n lines in the interval $\Delta\nu$. Then the average absorptance $\bar{a}$ is given by

$$\bar{a} = na/n\delta = a/\delta \qquad [7.28]$$

Then the "weak"- and "strong"-line approximations for a Lorentz line shape are:

$$\bar{a} = Su/\delta \qquad [7.29]$$

$$\text{and } \bar{a} = (2/\delta)(S\alpha_L u)^{1/2} \qquad [7.30]$$

respectively.

To go a stage further, suppose there are n lines of unequal strength. If the integrated absorptance of the ith line is a_i, and δ is the mean spacing, then

$$\bar{a} = (1/n\delta) \sum^{n} a_i = \bar{\bar{a}}/\delta \qquad [7.31]$$

provided that δ and the a_i are independent. $\bar{\bar{a}}$ is the arithmetic mean of the a_i.

As the absorber mass increases, equation [7.30] cannot hold indefinitely as absorption in neighbouring lines must eventually overlap and the transmission must reach zero. Several attempts have been made to consider overlap between single lines, but better progress has been made in considering a molecular absorption band as one single entity which has certain statistical properties. The most well-known, and indeed the most used and accepted, band models are those of Elsasser (1942) and Goody (1952), and only these will be considered here. A variation of Goody's model will be considered in section 7. For a detailed discussion of band models, the interested reader is referred to Goody (1964).

Historically, Elsasser (1942) was the first to consider an ensemble of lines in a band. His band model consists of an infinite array of Lorentz lines of equal intensity, spaced at equal intervals. This type of array is most applicable to bands appropriate to linear molecules. Although no molecular band corresponds exactly to Elsasser's concept, gases such as CO_2 can be represented fairly well by a series of separate Elsasser bands of different spacing and strength. For the derivation of the average absorptance $\bar{a}$ of an Elsasser band, we follow Elsasser and Culbertson (1960). The mass absorption coefficient κ_ν is given by:

$$\kappa_\nu = \sum_{n=-\infty}^{n=+\infty} (S\alpha_L/\pi)/[(\nu - n\delta)^2 + \alpha_L^2] \qquad [7.32]$$

where it is assumed that the band stretches to infinity in wave number. In actual fact, contributions from lines which are distant by more than a few line spacings δ from ν become very small. The transmittance of such a band turns out to be:

$$\overline{Tr} = 1 - \mathrm{erf}\,(\pi u S\alpha_L)^{1/2}/\delta \qquad [7.33]$$

with the usual definition that:

$$\mathrm{erf}(x) = 2/\sqrt{\pi} \int_0^x \exp(-x^2)\mathrm{d}x \qquad [7.34]$$

For small values of x, we have $\mathrm{erf}(x) = 2x/\pi^{1/2}$ and equation [7.33] reduces to the square root absorption law of equation [7.30] since transmittance is equal to 1 minus the absorptance. In practice, for the Elsasser model the square root absorption holds for a rather limited range of optical thicknesses.

Although the Elsasser model might be suitable for CO_2 under certain circumstances, the water vapour rotational band whose absorption is dominant in the troposphere has a very irregular appearance, as shown in Fig. 7.4. This fact suggested to Goody (1952) that a statistical treatment of the water vapour molecular absorption bands might be fruitful. In Goody's model he considers an interval $\Delta\nu$ with n lines of mean spacing δ, and that any line has equal probability of being anywhere in the interval $\Delta\nu$. When n becomes large the expression for Tr becomes an exponential. For an arbitrary distribution of line strengths the transmittance takes the form:

$$Tr = \exp(-\bar{a}) \qquad [7.35]$$

The remarkable point about this equation is that $\bar{a}$, the average absorptance over the spectral interval, appears within an exponential. Equation [7.35] is not derived on the assumption of any particular line shape and is valid for any distribution of line strengths. Various intensity distributions have been discussed by Goody (1964). The value of $\bar{a}$ can also be found from experimental data. The restrictions on the use of the model are that the lines have random spacing, that the line width is constant and that integration is over a sufficiently large spectral interval to include many lines.

The distribution of line strengths which gives good agreement with experimental water vapour data, as well as theoretical computations, is a simple Poisson distribution:

$$P(S) = (1/S_0)\exp(-S/S_0) \qquad [7.36]$$

where S_0 is the *mean* line strength. Then $\bar{a} = (S_0 u/\delta)(1 + S_0 u/\pi\alpha)^{-1/2}$ and the expression for Tr is:

$$Tr = \exp[-(S_0 u/\delta)(1 + S_0 u/\pi\alpha)^{-1/2}] \qquad [7.37]$$

In the strong-line case, $(S_0 u/\pi\alpha) \gg 1$ and equation [7.37] reduces to:

$$Tr = \exp[-(S_0 u\pi\alpha)^{1/2}/\delta] \qquad [7.38]$$

Values of the parameters S_0, δ and α are obtainable from experiment or from quantum mechanical calculations.

Goody (1964) gives tables of the quantities $\sum_i S_i$ and $\sum_i (S_i\alpha_i)^{1/2}$ computed theoretically by Benedict and Kaplan (1959) for intervals of wavenumber equal to 20 cm^{-1} in the rotation band. These quantities are related to the random model parameters in equation [7.37] by $S_0/\delta = \sum_i S_i/\Delta\nu$ and $S_0\pi\alpha/\delta^2 = [2\sum_i (S_i\alpha_i)^{1/2}/\Delta\nu]^2$. The theoretical figures are found to agree well with experimental values for wavelengths greater than 20 μm (see section 6.1). Hence the Goody random model provides an excellent basis for computations of water vapour transmission.

The above theoretical approach has the advantage that pressure and temperature effects can be introduced into the schemes on a sound physical basis. The absorption parameters which are affected are the line width α and

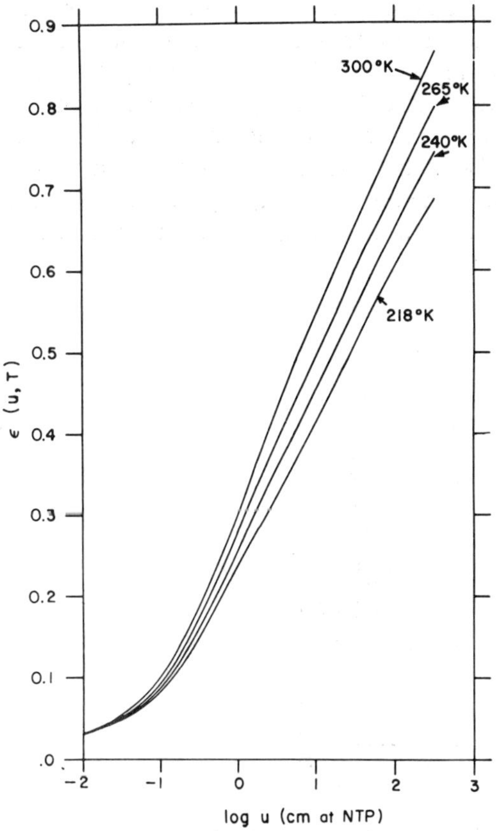

Fig. 7.8. Flux emissivity of carbon dioxide as a function of temperature and optical path. (After Manabe and Möller, 1961.)

the line strength S. The line width α_L for a Lorentz line is dependent on both pressure and temperature through equation [7.21]. The pressure dependence is much stronger than the temperature dependence. The line strength is almost independent of pressure and is dependent only on temperature. Thus in the "weak-line" absorption approximation (equation [7.24]) there is no pressure dependence.

The change of strength S with temperature is due to the relative change of the population of the molecular energy levels. The population of the ith level of energy E_i is given by the Boltzmann formula:

$$n(E_i) = (1/P_a)\exp(-E_i/\hat{k}T) \qquad [7.39]$$

where the partition function P_a is the sum of the Boltzmann factors for all states. The excited states of a complex molecule are very numerous as is evident from the large number of lines in a band. Elaborate calculations are thus needed to obtain accurate temperature dependences and approximations are often used. The temperature dependence of the absorptance or flux emissivity of the 14.7-μm band of CO_2 as estimated by Manabe and Möller (1961) from the computations of Sasamori (1959) is shown in Fig. 7.8. The temperature dependence of the product $S\alpha_L$ for the water vapour rotation band and for spectral intervals of $\Delta\nu = 10\,\text{cm}^{-1}$, is shown in Fig. 7.9. The general trend is for an increasing temperature variation with increasing wave-number (see also Fig. 7.12). Various schemes for simplifying the pressure and temperature dependences in bands will be considered in the following.

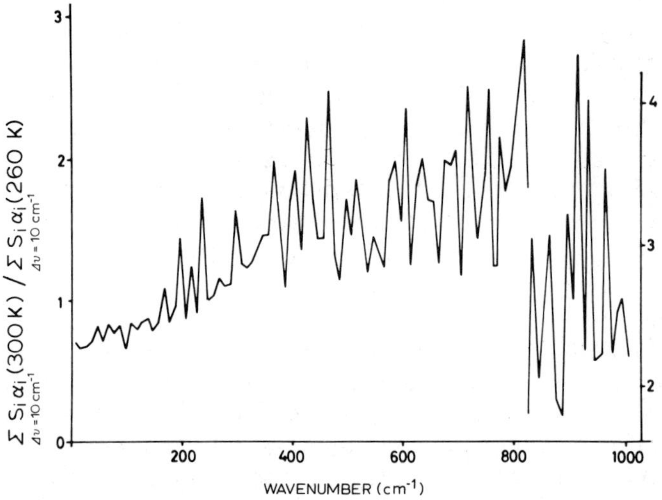

Fig. 7.9. Temperature dependence of water vapour rotation band. The figure gives mean values for $10\,\text{cm}^{-1}$ intervals of $\Sigma\, S_i\alpha_i(300\,\text{K})/\Sigma\, S_i\alpha_i/(260\,\text{K})$. (After Bolle, 1965.)

5.3. *Empirical transmittance models and generalized absorption coefficients*

An alternative approach to the frequency integration is to sidestep the theoretical problem and use empirical transmittance models derived from experimental measurements. Such measurements are made in the laboratory under controlled conditions of temperature and pressure. Although pressure variations corresponding to tropospheric and lower stratospheric conditions are quite easy to simulate, corresponding temperature variations are not. Consequently temperature effects are normally introduced separately on a semi-theoretical basis.

A concept which has been used extensively is that of a "generalized absorption coefficient". This idea was first introduced in order to make possible a simple one-parameter scheme for calculation of fluxes which led to the invention of the radiation chart discussed later. The transmittance function is written in terms of the generalized absorption coefficient l for a spectral interval $\Delta\nu$ and a "scaled" absorber amount u^*:

$$Tr = Tr(u^*, l) \qquad [7.40]$$

The absorber amount is suitably scaled in order to take implicit account of temperature and pressure dependences. The transmittance then depends only on u^* and l, and l is made to vary continuously with wavenumber. The concept of a generalized absorption coefficient is only useful if an experimental plot of transmittance Tr against u^* (or $\log u^*$) has the same shape for different spectral intervals $\Delta\nu$. Then the same empirical formula will hold for each spectral interval except for the substitution of the appropriate l.

Palmer (1960) found that a generalized curve was a good approximation for the rotation band of H_2O with $\Delta\nu \approx 20\,\text{cm}^{-1}$ or greater. Elsasser and Culbertson (1960) (EC) derived generalized absorption coefficients for H_2O (rotation band and 6.3 μm), CO_2 (14.7 μm) and O_3 (9.6 μm), showing that the concept is reasonable for CO_2 and H_2O, but perhaps not so successful for O_3. They scaled the absorber amount or optical path length according to $u^* = u_s(p/p_s)(T_s/T)^{1/2}$. They argued that in the strong-line case, both regular (equation [7.33]) and random (equation [7.38]) models contain the product $u\alpha_L$, and that therefore according to equation [7.21] the above scaling is correct.

The method of including the temperature dependence of S by EC was to consider the molecules as classical rigid rotators. They approximated l as a series of parabolas representing different temperatures. This somewhat arbitrary procedure has been criticized by Sasamori (1968) and will be mentioned again later when the overall effects of temperature dependence on the cooling rate are considered.

Because of the difficulty of representing CO_2 and O_3 accurately with a simple band model or a generalized coefficient, recourse is often made to one empirical transmittance function averaged over a whole absorption band. Howard et al. (1955) made extensive measurements with a CO_2 absorber concentration from 1 to 1,000 atm cm and total pressures from 20 to 740 mm Hg. Walshaw (1957) made measurements on O_3 for amounts up to 1.5 cm STP and for pressures from 10 to 760 mm Hg at a temperature of 20°C. They then obtained empirical equations of transmission.

The linear scaling of the absorber amount with pressure is not strictly accurate. An absorption band is a mixture of strong and weak lines. Because of this an equation such as:

$$u^* = u_s(p/p_s)^i \qquad [7.41]$$

is a more accurate form. When fitting their results, Howard et al. (1955) found $i = 0.86$ gave the best value.

It should be noted that overlap between the H_2O rotational and CO_2 14.7-μm bands has a significant effect on atmospheric cooling. This aspect will be discussed along with other problems of atmospheric models in section 6.

5.4. *Direct integration transmittance models*

The increasing speed of digital computers has brought us nearer to the day when line-by-line calculations of transmission of radiation through a realistic atmosphere, including all molecular absorbers, become possible on a routine basis. Such computations have already been made for the interpretation of high-resolution spectra of the earth obtained by satellites (Hanel et al., 1972) and of similar atmospheric spectra obtained from balloon measurements (Kyle, 1969; Goldman et al., 1973). The results of direct integration will be no better than the input parameters. Kyle considered his results to be of the same order of accuracy as the experimental measurements (≈5%). Kunde and Maguire (1974) found agreement to within 5–10% between direct integration calculations and laboratory measurement which, in their words, "is a number which is probably indicative of the absolute accuracy of present state-of-the-art molecular transmittances". Drayson (1966) came to a similar conclusion by comparison of his model with experiment.

The direct integration model consists of computing the absorption coefficient $\kappa_\nu = \sum_i (S_i/\pi)\{\alpha/[(\nu-\nu_{0i})^2 + \alpha^2]\}$ in very narrow ranges of the spectrum where the index i extends over all lines in the absorption band. Various approximations can be used to decrease computer time.

The absorption coefficient κ_ν varies by orders of magnitude on a scale of 0.001 cm^{-1} near ν_{0i}, while in the wings the absorption coefficient varies much more gradually. Thus to obtain high accuracy and at the same time to minimize computer time, a variable wavenumber interval is desirable ranging from less than 0.001 cm^{-1} near the line centres to 0.1 cm^{-1} in the wings.

Perhaps the most useful aspect of a line-by-line model is its application to parts of the atmosphere where the pressure is lower than in measured laboratory transmittances. Thus models like Drayson's or Kunde and Maguire's now find use for computation of CO_2 cooling up to 100 km. As the CO_2 mixing ratio is assumed constant with altitude, the transmission along any path can be integrated directly with respect to pressure. Thus for a Lorentz line Drayson obtains the transmittance Tr_ν between pressures p_1 and p_2 as:

$$Tr_\nu = \exp\left(-C_1 \int_{p_2}^{p_1} \kappa_\nu \mathrm{d}p\right)$$

$$= \exp\left(-C_1 \sum_i (S_i/2\pi\alpha_{\mathrm{Ls}}) \ln\{[\alpha_{\mathrm{Ls}}^2 p_2^2 + (\nu-\nu_{0i})^2]/[\alpha_{\mathrm{Ls}}^2 p_1^2 + (\nu-\nu_{0i})^2]\}\right) \qquad [7.42]$$

where C_1 is a constant and α_{Ls} is the line width at standard pressure p_s.

For the troposphere and lower stratosphere, it is doubtful whether the direct integration models have any advantage for cooling calculations.

Certainly, the computer time for routine calculations is still prohibitive, particularly as the water vapour *distribution* must be considered. Also the radiance must be calculated at discrete angles between 0 and $\pi/2$ in order to obtain the flux density. As the ultimate accuracy is only 5–10%, normal band models seem justified. However, direct integration could be a useful check on the various approximations made in band models.

5.5. *Water vapour continuum*

Within the "atmospheric window" between wavenumbers of 715 and 1200 cm^{-1} (8 and 14 μm), atmospheric absorption is found to be more continuous in nature than in the band regions, particularly in moist atmospheres. The continuum is overlain by many weak water vapour lines, two ozone absorption bands (one of which exhibits strong absorption in the stratosphere) and two very weak CO_2 bands. Typical atmospheric zenith emission at ground level was illustrated in Fig. 7.2. The continuum absorption appears to extend well into the water vapour rotation band (Bolle, 1965; Bignell, 1970). The atmospheric window is the region where dominant radiative transfer occurs directly between the earth's surface and space, and between clouds and space. The transmittance in the window is thus an important quantity.

Perhaps because of the rather weak absorption in this region the absorption mechanisms have been extraordinarily difficult to separate. Early ideas on the continuum centred around two factors; absorption due to the combined effects of wings of distant strong rotational lines, and absorption due to aerosols in the atmosphere. The difficulty in accounting for the first factor lies in our ignorance regarding line shapes in the wings far from the line centre. The strongest lines lie at about 50 μm wavelength and beyond. Bignell et al. (1963) calculated the continuum on the assumption of a Lorentz line shape for pressure broadening by a foreign gas and obtained not unreasonable agreement with experiment, at least in the 8–13-μm region. However, the discrepancy increased at larger wavelengths and at 20 μm the absorption coefficient was about five times greater than that predicted. Measurements by many other authors give a considerable range of continuum absorption coefficient at one single wavenumber, as indicated by the surveys of Bignell (1970) and Platt (1972). This range has often been attributed in the past to the presence of aerosols.

Burch (1970) and Bignell (1970) independently made the discovery from laboratory measurements that there exists a component of the mass absorption coefficient which is dependent on water vapour pressure e. McCoy et al. (1969) discovered the same phenomenon with horizontal atmospheric transmittance measurements at the ground. The phenomenon has subsequently been confirmed by slant path measurements in the atmosphere (Houghton and Lee, 1972; Platt, 1972; Grassl, 1973a) and appears to go a long way towards explaining the large range in previously measured values of mass absorption coefficient. Bignell termed the effect "*e*-type absorption" and wrote the total mass absorption coefficient for the continuum, $\kappa_{\nu c}$, as:

$$\kappa_{\nu c} = \kappa_{\nu 1}(T)p + \kappa_{\nu 2}(T)e \qquad [7.43]$$

where $\kappa_{\nu 1}$ is the absorption coefficient for foreign broadening and $\kappa_{\nu 2}$ is the absorption coefficient for water vapour broadening — both are functions of temperature T and both are defined for standard pressure.

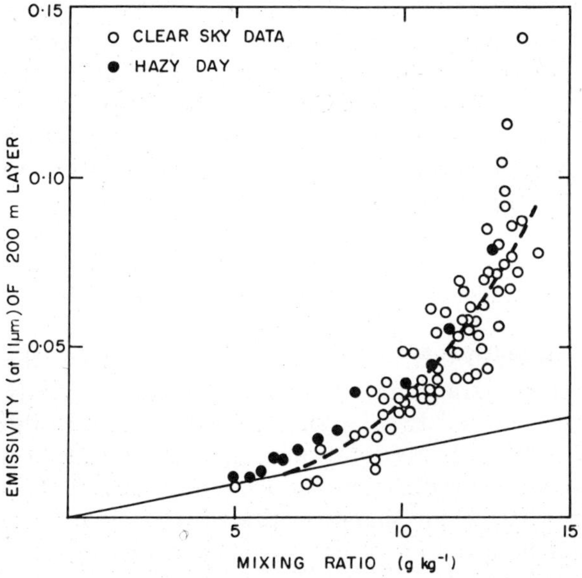

Fig. 7.10. Deduced values of layer zenith emissivity ϵ_{zen} (for layers 200 m thick) against humidity mixing ratio q. Straight line is for a constant value of absorption coefficient $\kappa_\nu = 0.1\ \mathrm{g}^{-1}\ \mathrm{cm}^2$. (From Paltridge and Platt, 1973.)

Although *e*-type absorption is now accepted as a very important factor in the continuum absorption, its physical origin remains rather obscure. The mechanism must involve attractive interaction between water vapour molecules. Perhaps the most consistent theory offered is that of absorption by the water vapour dimer $(H_2O)_2$. Absorption measured in the submillimetre wave region by Gebbie et al. (1969) was found to have the dependence on water vapour and temperature expected for dimeric absorption. The absorption increased with decreasing temperature as required by the dimer theory (this was found also by Bignell) and the absorption increased with the square of the water vapour pressure. This latter effect can be explained as follows. For an absorbing volume of gas at constant temperature, the optical

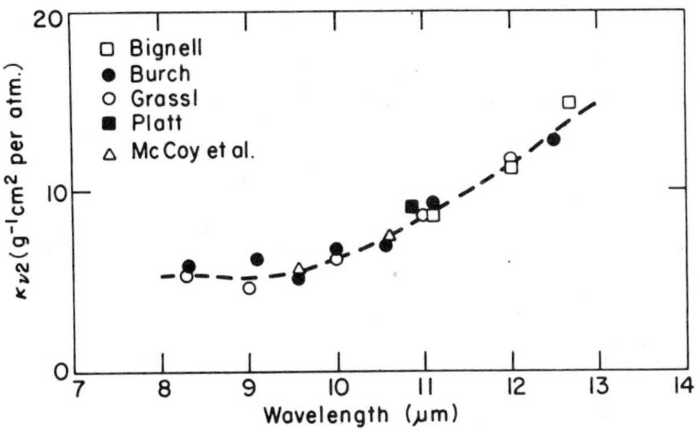

Fig. 7.11. Summary of values of the e-type mass absorption $\kappa_{\nu 2}$ obtained by the following: Bignell (1970), Grassl (1973a), Platt (1972), Burch (1970), McCoy et al. (1969).

path length u can be written as $B_1 e$, where B_1 is a constant. Then the integrated absorptance a for the e-type effect is:

$$a = \int_{\Delta\nu} [1 - \exp(-\kappa_{\nu 2} B_1 e^2)] \, d\nu \qquad [7.44]$$

which, for a small optical depth, reduces to:

$$a = \int_{\Delta\nu} B_1 \kappa_{\nu 2} e^2 \, d\nu \qquad [7.45]$$

The emissivity of thin layers of the atmosphere (i.e. $\bar{a}$ for the band 10–12 μm) at different water vapour pressures shows a similar effect (Fig. 7.10). These measurements were not corrected for any temperature effect. An interesting result found by both Burch (1970) and Grassl (1973a) was the small value of $\kappa_{\nu 1}$ particularly in the centre of the window. The results for $\kappa_{\nu 2}$ from all the above workers are summarized in Fig. 7.11. The agreement between the results is good. In the atmosphere, water vapour absorption is swamped by the CO_2 band between about 14 and 16 μm wavelength, but from 16 to 20 μm and beyond the water vapour spectrum contributes substantially to cooling. In this region, the e-type absorption should have a substantial effect, even in the middle atmosphere.

The theory of dimeric formation indicates a temperature dependence for $\kappa_{\nu 2}$ of the form $\exp(E/\hat{k}T)$. The factor $E/\hat{k} = 1745$ has been suggested by Lee (1973) as fitting Burch's data, a figure which agrees with that deduced from far infrared measurements by Burroughs et al. (1969).

The weak H_2O line spectrum present in the window can be treated with the statistical model. Roach and Goody (1958) give figures for the statistical band parameters between 10.55 and 12.85 μm, both from theoretical sources

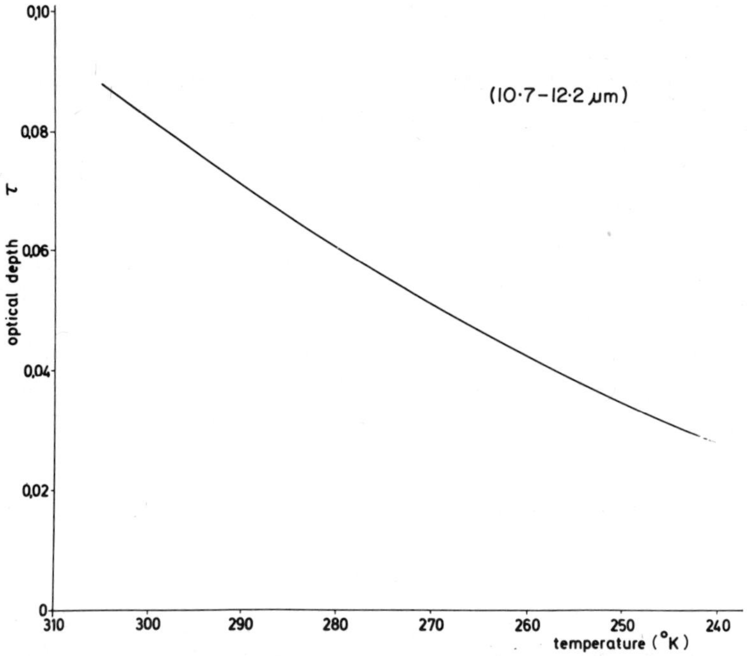

Fig. 7.12. The optical depth of the weak water vapour rotational line absorption (10.7–12.2 μm) as a function of temperature. (After Grassl, 1973a.)

and from experiment. Grassl (1973a) has calculated random mode parameters for the 10.2–11.7-μm and 10.7–12.2-μm regions, and for temperatures between 240 and 310 K. One of his curves is illustrated in Fig. 7.12, where the optical depth of water vapour line absorption for 1 g cm^{-2} water path is plotted against temperature. The temperature dependence is quite large.

Bolle et al. (1963) and Bolle (1965) have made substantial and detailed contributions to the subject through their interpretation of experimental line and continuum spectra and they list some useful data.

The total absorption coefficient of gases in the atmospheric window can be written as:

$$\kappa_\nu = \kappa_{\nu 1}(T)p + \kappa_{\nu 2}(T)e + \kappa_{\nu 3}(T)f(p) + \kappa_{\nu 4}(T)f(p) \qquad [7.46]$$

The last two terms represent selective absorption due to water vapour and to other gases respectively. The approximate contribution of the various terms to the atmospheric radiance is $\kappa_{\nu 1} \approx 15\%$, $\kappa_{\nu 2} \approx 70\%$, $\kappa_{\nu 3} \approx 11\%$, $\kappa_{\nu 4} \approx 4\%$.

6. RADIATIVE COOLING MODELS FOR THE TROPOSPHERE AND LOWER STRATOSPHERE

The last section has outlined the basic theory for deriving gaseous transmittance functions and the effects of temperature and pressure on these functions. It becomes quickly apparent that the problem of radiative transfer, although formally quite simple, is very complex in practice. This section is concerned with practical methods which have been used to compute atmospheric fluxes and cooling at different levels in the atmosphere.

Initial attempts at computing the cooling involved considerable simplification, often a grey-body treatment of the emissivity, and the use of generalized absorption coefficients. The success of these models, particularly in the troposphere, has been considerable. Later work has concentrated on various refinements with the aim of an improvement in accuracy. This section is restricted to models of cooling below 30 km altitude. Above that altitude special problems are encountered which are considered in section 7.

We start by considering a comprehensive and detailed model of radiative cooling. This will be followed by simpler models which have been devised. The consequent loss in accuracy with the various simplifications will be discussed.

6.1. *Multispectral, two-parameter model*

Rodgers and Walshaw (1966) (RW) devised a model which they claim is suitable for use at any level and which is probably the most reliable model available, at least for the troposphere. In such a multispectral model the division of the wavenumber spectrum into suitable intervals must be considered very carefully. In the cooling equation (equation [7.19]), it is apparent that the total computation time is directly proportional to the number of intervals into which the total spectrum is divided. The general technique is to choose a suitable number and assign a mean Planck function $\tilde{B}(T)$ to each interval $\Delta\nu$. One criterion for an accurate model is that $\tilde{B}(T)$ should not vary significantly over the spectral interval. RW started initially by dividing the H_2O rotation band into 9 spectral intervals and the H_2O 6.3-μm band into 9 intervals. They had one interval for the CO_2 band, one for the O_3 band and one for the continuum.

The techniques for performing the height integration are varied. The cooling at altitude z can be found either by calculating the local differentials $d\tilde{T}r/dz$ and $d\tilde{B}(T)/dz$ and using equation [7.19] or by taking finite differences of the net flux densities between the two levels. In the former case, equation [7.19] can be integrated with respect to height by standard methods of quadrature (Chandrasekhar, 1950; Margenau and Murphy, 1956). RW used a convenient matrix method developed by Curtis (1956). A useful discussion of this method is given by Stensrud et al. (1971). For the variable z, RW used the local scale height, $\hat{z} = -\ln(p/p_s)$.

RW used the random model (equation [7.37]) to calculate the H_2O transmittance in the water vapour bands. They made a detailed comparison of the random model parameters for the H_2O rotation band which were fitted to the theoretical computations of Benedict and Kaplan (1959) with

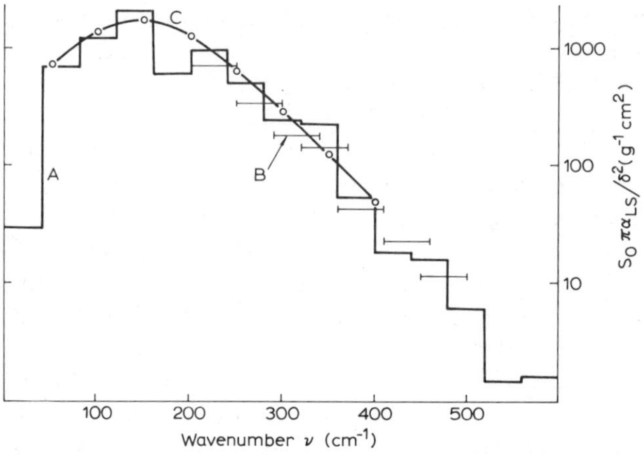

Fig. 7.13. Random model fit (strong region) to water vapour rotation band. (After Rodgers and Walshaw, 1966.) *A*: from Benedict and Kaplan's calculations (Goody, 1964). *B*: from measurements of Palmer (1960). *C*: from transmittance functions of Cowling (1950).

those fitted to the laboratory measurements of Palmer (1960) and those fitted to the universal transmission function obtained by Cowling (1950) from his earlier data. The comparison of the parameter $S\pi\alpha_{Ls}/\delta^2$ is shown in Fig. 7.13. This parameter gives the absorption in those spectral regions where the "strong-line" approximation is valid (see equation [7.38]). RW in fact chose the values of Benedict and Kaplan for their calculations (see Table 7.2.). Note that the product of the two parameters (1) and (2) in Table 7.2 gives the parameter in Fig. 7.13. The factors in the exponential of equation [7.37] are, respectively, parameter (1) and parameter (1)/parameter (2) of the Table. The spectral intervals ($\approx$100 cm^{-1}) are larger than those specified by Palmer as a necessary minimum for his universal curve. Values of the RW parameters of the 6.3-μm band of H_2O, fitted to experimental measurements of Burch et al. (1962) are given in Table 7.3. The H_2O continuum was allowed for crudely by adopting a constant absorption coefficient of 0.1 g^{-1} cm^{-2}.

The CO_2 bands present a difficult problem in that neither the regular nor random models are a good fit. For simplicity, RW used a random model and neglected overlap between H_2O, CO_2 and O_3 bands. They justified their method on the basis that, above about 30 km where the radiative cooling is dominated by CO_2 bands, the pressure is low enough for the lines to be virtually independent, so that the model used is not critical. In the troposphere the cooling is dominated by water vapour, so that errors in the treatment of CO_2 bands are relatively unimportant. However, as pointed out by RW, the correct band model is desirable and overlap between the different bands must be considered. RW also did not attempt detailed calculations for

TABLE 7.2

Water vapour rotation band parameters (after Rodgers and Walshaw, 1966)

	Interval (cm^{-1})	Parameter (1) S_0/δ (g^{-1} cm^2)	Parameter (2) $\pi\alpha_{LS}/\delta$	a (10^{-3} deg^{-1})	b (10^{-6} deg^{-2})	a' (10^{-3} deg^{-1})	b' (10^{-6} deg^{-4})
1	0–40	579.75	0.093	−6.75	8.55	−8.38	4.56
2	40–160	7210.3	0.182	−2.93	2.01	−2.68	1.57
3	160–280	6024.8	0.094	1.43	−13.0	2.03	−10.3
4	280–380	1614.1	0.081	9.59	−41.8	9.08	−38.1
5	380–500	139.03	0.080	14.3	−23.7	15.1	−54.1
6	500–600	21.64	0.068	15.2	−30.1	16.2	−38.1
7	600–720	2.919	0.060	19.0	−44.6	18.6	−62.6
8	720–800	0.3856	0.059	21.7	−53.2	23.1	−74.7
9	800–900	0.0715	0.067	24.1	−40.3	26.2	−74.1
10	900–1000	0.0209	0.051	12.9	165.0	28.5	−86.8

TABLE 7.3

Water vapour 6.3-μm band parameters (after Rodgers and Walshaw, 1966)

	Interval (cm^{-1})	S_0/δ (g^{-1} cm^2)	$\pi\alpha_{LS}/\delta$
1	1200–1350	12.65	0.089
2	1350–1450	134.4	0.230
3	1450–1550	632.9	0.320
4	1550–1650	331.2	0.296
5	1650–1750	434.1	0.452
6	1750–1850	136.0	0.359
7	1850–1950	35.65	0.165
8	1950–2050	9.015	0.104
9	2050–2200	1.529	0.116

O_3. They used a random model with data given by Walshaw and Rodgers (1963).

A useful trick known as the Curtis-Godson (CG) approximation is often employed to remove the effects of pressure on the transmission function. The CG approximation is that the correct half-width $\bar{\alpha}$ of a line for a transmittance along an inhomogeneous path is given by:

$$\bar{\alpha} = \int \alpha \mathrm{d}u \bigg/ \int \mathrm{d}u \qquad [7.47]$$

where α is the line width at the point on the path where the integrated optical path is u. As α at a pressure p is given by $\alpha_s p/p_s$ (neglecting the temperature dependence), then:

$$\bar{\alpha} = \int q(p)p\mathrm{d}p \bigg/ \int q(p)\mathrm{d}p \qquad [7.48]$$

where $q(p)$ is the absorber mass mixing ratio at pressure p and the integration is between the two pressure levels being considered.

The effects of pressure *and* temperature on the transmission function are considered by RW with an extension of the CG approximation developed by Godson. An effective optical path $\bar{u}$ is defined by:

$$\bar{u} = \int \phi(T)\mathrm{d}u \qquad [7.49]$$

and an effective pressure $\bar{p}$ along this path by:

$$\bar{u}\bar{p} = \int \psi(T)p\mathrm{d}u \qquad [7.50]$$

The optical path to insert in equation [7.37] is then $\bar{u}$ and α is adjusted for an effective (uniform) pressure $\bar{p}$. Empirical values of ϕ and ψ are:

$$\begin{aligned} \log_e \phi(T) &= a(T-260) + b(T-260)^2 \\ \log_e \psi(T) &= a'(T-260) + b'(T-260)^2 \end{aligned} \qquad [7.51]$$

where T is in degrees Kelvin. The constants a, b, a', b' for the H_2O rotation band are given in Table 7.2.

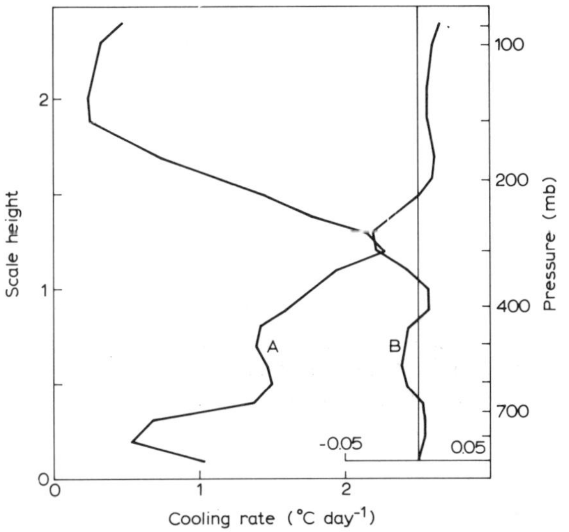

Fig. 7.14. Accuracy of the diffusivity factor (1.66 $\bar{u}$). A: "exact"; angular integration of transmission functions using the Curtis-Godson approximation. B: "exact" minus "approximate" ("approximate" refers to use of diffusivity factor). (After Rodgers and Walshaw, 1966.)

For an isotropic field, the flux density transmittance is given by the exponential integral of equation [7.9]. A way of expressing the transmittance $\tilde{T}r$ in this equation in terms of the vertical transmission is simply to increase the vertical optical path by a fixed amount so that:

$$\tilde{T}r(z, z') = Tr(1.66\bar{u}) \qquad [7.52]$$

where $\bar{u}$ is the CG-averaged *vertical* absorber amount between z and z'. The "diffusivity factor" of 1.66 is generally applied in all flux transmittance models.

The above appears to be a drastic approximation but RW considered a more detailed method for the angular integration using the Curtis-Godson approximation and concluded that errors involved in using the figure of 1.66 are surprisingly small. An example of a calculation of typical cooling rates in the atmosphere comparing the two methods is shown in Fig. 7.14.

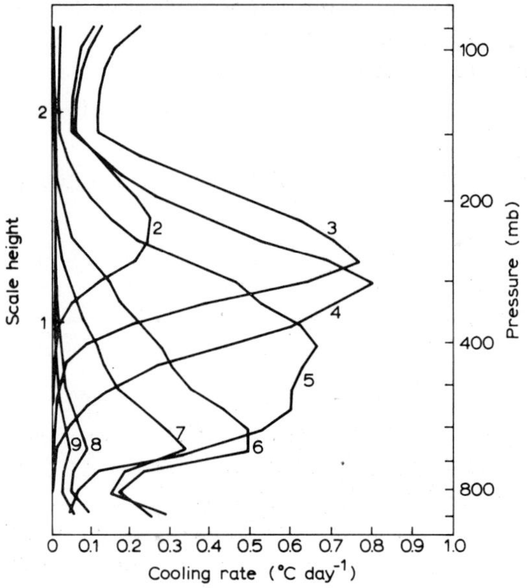

Fig. 7.15. Partial cooling rates for spectral intervals in the water vapour rotation band. The spectral intervals are numbered as in Table 7.3. Contributions from intervals *1* and *10* are negligible and have been omitted. Mid-latitude aerological data. (After Rodgers and Walshaw, 1966.)

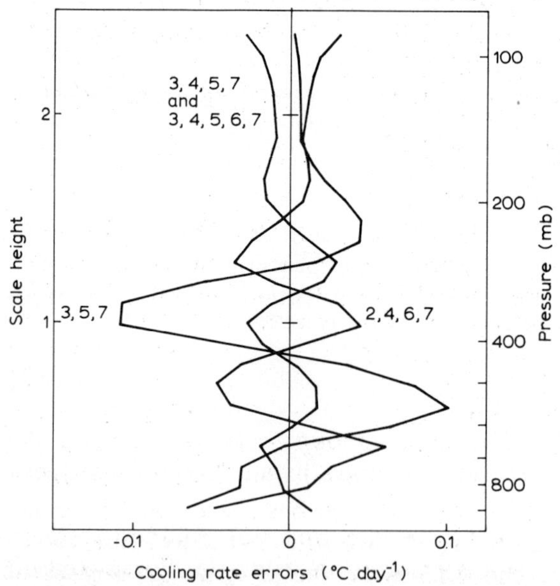

Fig. 7.16. Cooling-rate errors for various combinations of intervals in the water vapour rotation band. Same aerological data as for Fig. 7.15. (After Rodgers and Walshaw, 1966.)

TABLE 7.4

Weights for economized spectral intervals[1] (after Rodgers and Walshaw, 1966)

Band	Interval	
Rotation	3	1.657
	4	0.526
	5	1.435
	7	2.321
6.3 μm	1	1.317
	4	5.164

[1] Mean weights for two mid-latitude aerological profiles.

As mentioned previously, the time required to compute the cooling rate (or fluxes) numerically is directly proportional to the number of spectral intervals. The partial cooling rates for different spectral intervals are shown in Fig. 7.15. The closeness in shape between some of the spectral intervals suggests the use of a smaller number of spectral intervals with weighted band model parameters. Fig. 7.16 shows the cooling errors for various combinations of weighted intervals. Table 7.4 shows the weights applied to the random model (strong-line) parameter in the appropriate interval which give the lowest error in cooling. A similar analysis for the 6.3-μm band resulted in the weights and intervals also shown in Table 7.4. The error in using the economized intervals was found to be generally less than 0.03°C/day.

Rodgers and Walshaw (1966) investigated the effects of temperature and humidity errors. Their conclusions were as follows:

Systematic errors. An increase of 1°C in temperature or dewpoint over the whole profile gives on average about 1.4% increase in cooling rate.

Random errors. A random error of 1°C in temperature or dewpoint at a level gives on average a random error of about 5% in cooling rate at that level.

Thus according to RW "for meteorological studies of real atmospheres an accuracy of ±5% in the cooling rate calculation will suffice." This indicates that all the approximations in the computation described above are quite justified. However, it must be remembered that random errors from different causes are cumulative, so that a careful evaluation of computing technique against required accuracy should be made.

Another cause of error is the finite size of the vertical integration step. Fig. 7.17 shows the errors caused by using a step of $\Delta\hat{z} = 0.2$ instead of $\Delta\hat{z} = 0.1$, where $\Delta\hat{z}$ is the increment in scale height. The largest errors occur at points where temperature discontinuities have been smoothed by linear interpolation when using the larger step. If we add the errors due to the economy of spectral intervals, the diffusivity factor and the neglect of temperature dependence, the total error is about 10% which is greater than the random errors in meteorological data. However, errors due to incorrect

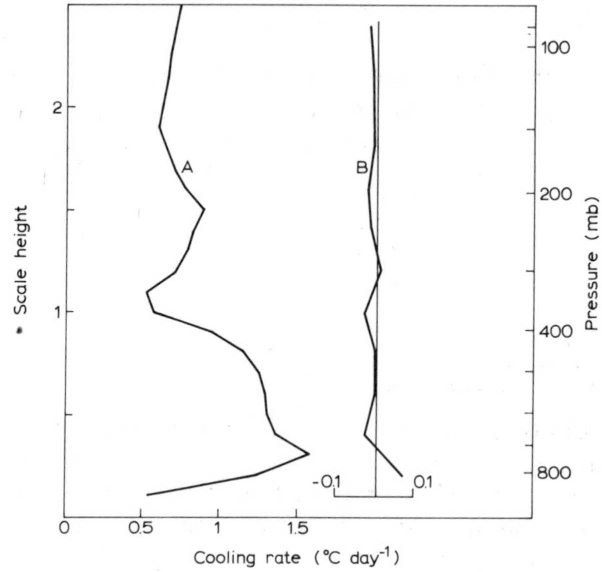

Fig. 7.17. Influence of finite thickness of layers ($\Delta\hat{z}$) in numerical computations. A: using a vertical step in scale height of $\Delta\hat{z} = 0.1$. B: difference when $\Delta\hat{z} = 0.1$ and $\Delta\hat{z} = 0.2$. (After Rodgers and Walshaw, 1966.)

band parameters were not considered by RW. In section 5 it was pointed out that such errors could themselves be 5–10%. Moreover, the errors due to incorrect modelling of the continuum in RW's model are considerable in tropical areas in the lower troposphere. However, modelling of the continuum can be treated separately from the other spectral intervals.

"Cutting off" their cooling calculation at a maximum height of 290 mb was found by RW to cause the top layer cooling to be underestimated by about 0.2°C per day, but the effect was not great at lower layers. Contributions from the top layer in the integral in equation [7.19] must therefore be small. Rodgers and Walshaw separated contributions to cooling at any level into (a) that from other levels, (b) that from the ground, (c) that from direct cooling from the level to space. Obviously the height at which the calculation is cut off affects only factor (a) and it turns out that the greatest contribution to cooling is from adjacent layers. RW suspected that the cooling to space term (b) might be dominant in the total cooling, and computations verified this for many situations. A simple "cooling to space" approximation gave good agreement with the complete cooling calculation in most tropospheric and stratospheric situations. It broke down when warming from the ground was significant (the bottom of the O_3 stratospheric layer) or when local transfer was as important as cooling to space (for instance, under the strong Antarctic inversion). A method of division of radiance beams into an internal transfer term and a cooling to space term is given by Green (1967).

6.2. *The emissivity approximation*

The emissivity or "grey-body" approximation treats the atmosphere as effectively having equal absorption at every frequency. It is thus a single-interval spectral model and the flux density from a layer of atmosphere at temperature T containing a (scaled) absorber optical path u is given by:

$$F(u,T) = \sigma T^4 \epsilon(u,T) \qquad [7.53]$$

where:

$$\epsilon(u,T) = 1/\sigma T^4 \int_0^\infty [1-\widetilde{Tr}_\nu(u,T)]\tilde{B}_\nu(T)\mathrm{d}\nu \qquad [7.54]$$

Here $\widetilde{Tr}_\nu(u, T)$ is the flux transmittance for an absorber amount u at temperature T (and some pressure p). The problem which the emissivity concept attempts to solve is that of numerical integration over the line structure of the atmospheric absorption spectrum. It uses approximations to perform the integration once and for all. In order to save computer time, some form of the emissivity approximation is often used in operational models.

We now consider equations which are suitable for calculating the flux density in terms of an emissivity. From equations [7.15] and [7.16], the equations for the total flux density may be written:

$$F(z)\uparrow = \int_0^\infty \tilde{B}_\nu(T_z)\mathrm{d}\nu - \int_{z_0}^{z}\int_0^\infty \widetilde{Tr}_\nu(z,z')[\mathrm{d}\tilde{B}_\nu(T_{z'})/\mathrm{d}z']\mathrm{d}\nu\mathrm{d}z' \qquad [7.55]$$

$$F(z)\downarrow = -\int_0^\infty \tilde{B}_\nu(T_z)\mathrm{d}\nu - \int_{z}^{Z}\int_0^\infty \widetilde{Tr}_\nu(z,z')[\mathrm{d}\tilde{B}_\nu(T_{z'})/\mathrm{d}z']\mathrm{d}\nu\mathrm{d}z' \qquad [7.56]$$

where in the original equations [7.15] and [7.16] it is assumed that $\tilde{B}_\nu(T_g) = \tilde{B}_\nu(T_{z_0})$ and $\tilde{B}_\nu(T_Z) = \tilde{B}_\nu(\mathrm{top})$. It is convenient to use a quantity $R(z, z', T_{z'})$ where:

$$R(z,z',T_{z'}) = \int_0^\infty [1-\widetilde{Tr}_\nu(z,z')][\mathrm{d}\tilde{B}(T_{z'})/\mathrm{d}z']\mathrm{d}\nu \qquad [7.57]$$

Note that R, like ϵ, is the result of an integration over all frequencies. Then equation [7.55] can be rewritten as:

$$F(z)\uparrow = \int_0^\infty \tilde{B}_\nu(T_z)\mathrm{d}\nu - \int_{z_0}^{z}\int_0^\infty [\mathrm{d}\tilde{B}_\nu(T_{z'})/\mathrm{d}z']\mathrm{d}z'\mathrm{d}\nu + \int_{z_0}^{z} R(z,z',T_{z'})\mathrm{d}z'$$

$$= \tilde{B}(T_{z_0}) + \int_{z_0}^{z} R(z,z',T_{z'})\mathrm{d}z' \qquad [7.58]$$

Similarly:

$$F(z)\downarrow \;=\; -B(T_Z) + \int_z^Z R(z, z', T_{z'})\mathrm{d}z' \qquad [7.59]$$

Now, as the water path $\mathrm{d}u = \rho \mathrm{d}z$, an integro-differential term such as the last term in equation [7.58] can be written:

$$\int_{z_0}^{z} Tr_\nu(u')[\mathrm{d}B(T_{u'})/\mathrm{d}u']\mathrm{d}u' \qquad [7.60]$$

where $u' = \int_{z_0}^{z'} \rho(z'')\mathrm{d}z''$.

Again as u is a function of T, we can write for expression [7.60]:

$$\int_{z_0}^{z} Tr_\nu(u')[\mathrm{d}B(T_{u'})/\mathrm{d}T']\mathrm{d}T' \qquad [7.61]$$

If we now assume that the downward flux at the top of the atmosphere is zero, the flux densities at a level z can be written:

$$F(z)\downarrow \;=\; \int_{T_z}^{T_Z} R\{[u(T_{z'}) - u(T_z)]\, T_{z'}\}\mathrm{d}T_{z'} \qquad [7.62]$$

and:

$$F(z)\uparrow \;=\; B(T_{z_0}) + \int_{T_{z_0}}^{T_z} R\{[u(T_z) - u(T_{z'})]\, T_{z'}\}\mathrm{d}T_{z'} \qquad [7.63]$$

In order to employ these equations, optical path u must be scaled in some manner to take account of the temperature and pressure dependence of $Tr(z, z')$ along the path z, z'. Similarly u must be scaled if the emissivity $\epsilon(u, T)$ is used, as equation [7.54] is defined for an optical path at a *constant* pressure and temperature. These aspects of the emissivity approximation will be discussed shortly. On the above assumptions, equations [7.62] and [7.63] can be integrated numerically to obtain $F\uparrow(z)$, $F\downarrow(z)$, net flux and cooling rate at any altitude. The simplifications over the multi-spectral model are the use of transmittance functions integrated over all frequencies and the neglect of any temperature dependence of transmittance functions.

At wavelengths where overlapping between absorption bands occurs, the transmittance $\widetilde{Tr}_\nu(u, T)$ is the product of the transmittances $\widetilde{Tr}_\nu(u_1, T)$ and $\widetilde{Tr}_\nu(u_2, T)$ of the two gaseous absorbers whose optical paths are u_1 and u_2 respectively. The equation for the emissivity is then:

$$\epsilon(u_1, u_2, T) = (1/\sigma T^4) \int_0^\infty [1 - \widetilde{Tr}_\nu(u_1, T)\widetilde{Tr}_\nu(u_2, T)]\widetilde{B}_\nu(T)\mathrm{d}\nu \quad [7.64]$$

$$= (1/\sigma T^4) \int_0^\infty [1 - \widetilde{Tr}_\nu(u_1, T)]\widetilde{B}_\nu(T)\mathrm{d}\nu$$

$$+ (1/\sigma T^4) \int_0^\infty [1 - \widetilde{Tr}_\nu(u_2, T)]\widetilde{B}_\nu(T)\mathrm{d}\nu$$

$$- (1/\sigma T^4) \int_0^\infty [1 - \widetilde{Tr}_\nu(u_1, T)][1 - \widetilde{Tr}_\nu(u_2, T)]\widetilde{B}_\nu(T)\mathrm{d}\nu \quad [7.65]$$

The first and second terms of equation [7.65] are identified as the separate emissivities $\epsilon(u_1, T)$, $\epsilon(u_2, T)$. The third term is a "correction" term $\Delta\epsilon$ for overlap (Staley and Jurica, 1970). For two gases whose absorption bands do not overlap, and are completely separated in wavenumber, the third term is zero. In this case the total emissivity $\epsilon(u, T)$ is simply the sum of the emissivities of the separate gases.

The overlapping of bands can similarly be incorporated into the quantity $R(u, T)$ of equation [7.57]. It should be remembered that overlapping occurs in only a limited spectral interval. Staley and Jurica (1970) give values of $\Delta\epsilon$ for overlap between the H_2O and CO_2 bands for various values of u_1 and u_2, and Elsasser and Culbertson (1960) give the correction factors for the quantity $R(u, T)$. Their methods give adequate accuracy (Braslau, 1972).

Sasamori (1968) numerically integrated equations [7.62] and [7.63] by a version of Simpson's rule. The absorber amount was scaled according to:

$$u^* = (1/g) \int_{p_1}^{p_2} [q(p)p/p_s]\mathrm{d}p \quad [7.66]$$

which is a form of the Curtis-Godson approximation. Values of $R(u, T)$ were interpolated from values tabulated by Elsasser and Culbertson (1960). In this way, Sasamori calculated $F\uparrow(z)$, $F\downarrow(z)$ and cooling rates. His cooling rates are compared with other methods in the next section.

Sasamori (1968) also used the normalized absorptivity $\hat{A}(u, T)$ or emissivity $\epsilon(u, T)$ defined as

$$\hat{A}(u, T) = R(u, T)/4\sigma T^3$$

$$= \int_0^\infty [1 - \widetilde{Tr}_\nu(u,T)][\mathrm{d}\widetilde{B}_\nu(T)/\mathrm{d}T]\mathrm{d}\nu \Big/ \int_0^\infty [\mathrm{d}\widetilde{B}_\nu(T)/\mathrm{d}T]\mathrm{d}\nu \quad [7.67]$$

Then the downward flux (for instance) is written:

$$F(z)\downarrow = 4\sigma \int_{T_z}^{T_Z} \hat{A}\{[u(T_{z'}) - u(T_z)]\ T_{z'}\}T_{z'}^3\mathrm{d}T_{z'} \quad [7.68]$$

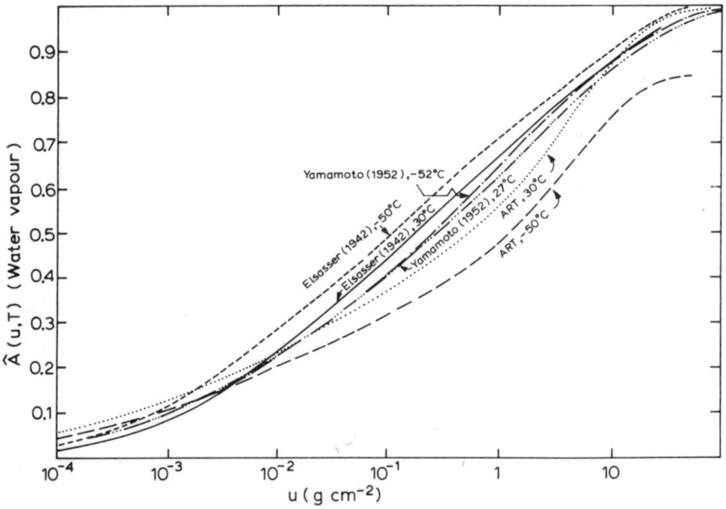

Fig. 7.18. Absorption (emissivity) curves for water vapour in terms of function $\hat{A}(u, T)$ and water vapour optical path u.

Sasamori compared values of $\hat{A}(u, T)$ for water vapour calculated by different authors. This comparison is shown in Fig. 7.18. The "ART" curves apply to Sasamori's values using the tables of Elsasser and Culbertson (1960). What is apparent from this figure immediately are (a) the low values of ART, and (b) the small temperature dependence of $\hat{A}(u, T)$ for Yamamoto's and Elsasser's results. The low value of ART was due to an error in the EC calculations of $R(u, T)$, as reported by Sasamori. Zdunkowski et al. (1966) tabulate corrected values. The point (b) is at first rather surprising. There are two factors which might make $R(u, T)$ dependent on temperature. First, the absorption strengths of the lines in a band are temperature dependent. Second, the intensity distribution with wavenumber of Planck's function changes with temperature. The former is a complex function of the lines in the bands. The latter causes the H_2O rotation band to gain in importance at the expense of the 6.3-μm band as the temperature decreases. This overall effect is illustrated in Figs. 7.19 and 7.20 which show the emissivity as computed by Staley and Jurica (1970). Rather fortuitously, the decrease in the contribution of the 6.3-μm band almost exactly compensates for the increase of the rotation band so that the *total emissivity is remarkably constant*. However, the *e*-type effect of the continuum was not included by Staley and Jurica.

Elsasser and Culbertson (1960) used a rather artificial method for obtaining the temperature dependence of the line strengths (see section 5.3). Yamamoto (1952) used a more detailed method and the temperature effect in Yamamoto's H_2O total emissivity is found to be small, in agreement with Staley and Jurica. Any emissivity technique cannot thoroughly account for

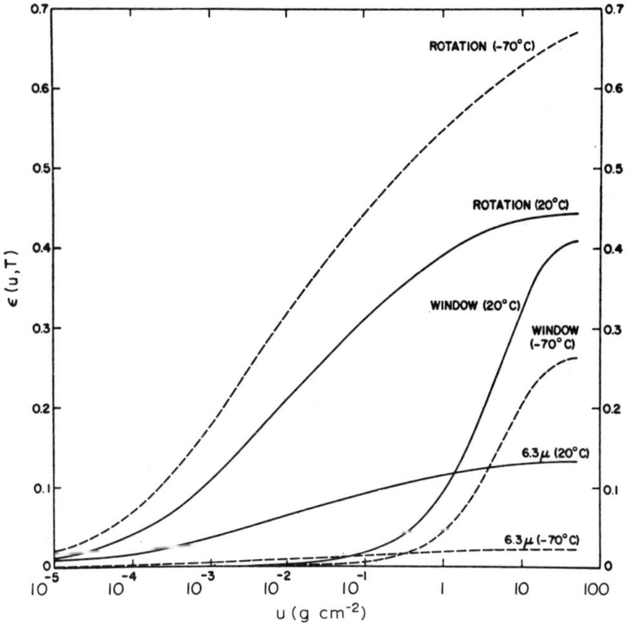

Fig. 7.19. Contributions of the water vapour rotation band, 6.3-μm band and atmospheric window to the water vapour emissivity at 20 and -70°C. (After Staley and Jurica, 1970.)

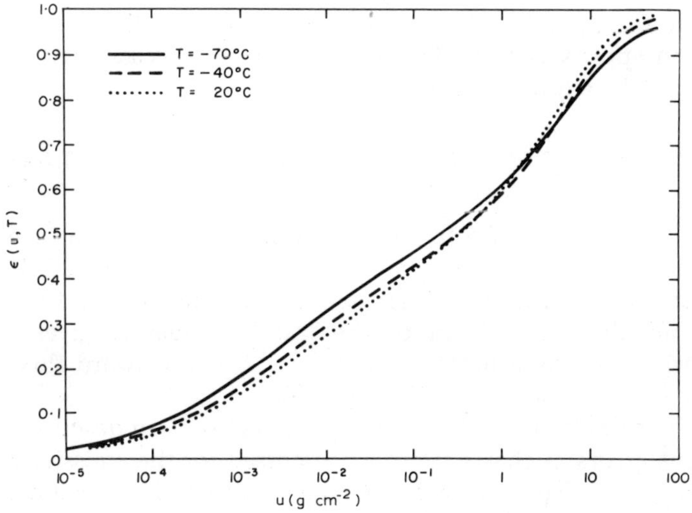

Fig. 7.20. Total water vapour emissivity at three temperatures. (After Staley and Jurica, 1970.)

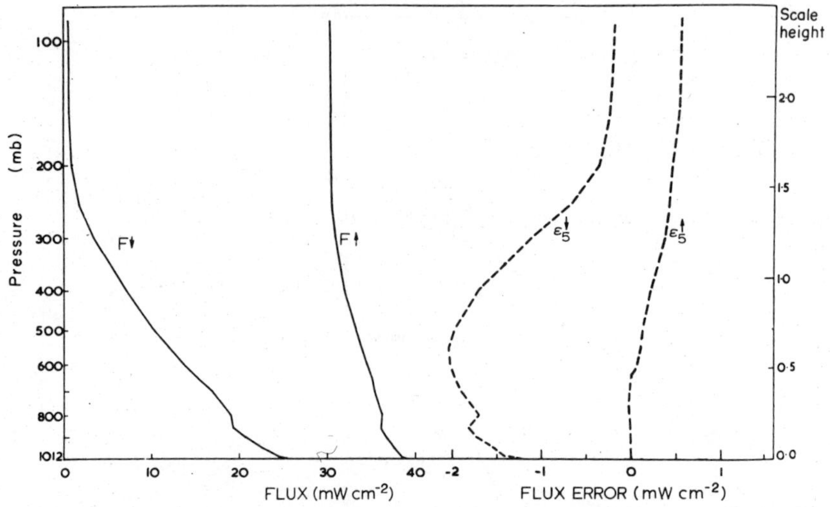

Fig. 7.21. Height dependence of the error in computed flux density using Rodgers' "emissivity" ϵ_5 (see text) compared to flux densities computed by the method of Rodgers and Walshaw (1966). (After Rodgers, 1967a.)

temperature effects, and this sets the limit on their accuracy. This conclusion is borne out by the work of Rodgers (1967a) who made a detailed investigation of emissivities in terms of the more accurate Rodgers and Walshaw model described earlier. One of his emissivity expressions (ϵ_5) is equivalent to $\hat{A}(u, T)$ defined in equation [7.67], but with the optical path u scaled with the Curtis-Godson approximation. The whole path $z - z'$ is assumed to be at a temperature $T_{z'}$. The expression for ϵ_5 is:

$$\epsilon_5(u, T) = \int_0^\infty [1 - \widetilde{Tr}_\nu(u^*, T_{z'})] [\mathrm{d}\widetilde{B}_\nu(T_{z'})/\mathrm{d}\widetilde{B}(T_{z'})] \mathrm{d}\nu \qquad [7.69]$$

The errors in upward and downward flux densities as attributed to the use of ϵ_5 rather than a full Rodgers-Walshaw treatment are shown in Fig. 7.21. It is seen that errors in the downward flux are greater than for the upward flux. Rodgers shows that an alternative form of the transfer equation gives a different emissivity which yields a better accuracy for the downward flux. This reduces the error from about 10–20% to about 5%.

The formalism used for radiation calculation by charts has been described above and the inherent errors in charts are those discussed for the emissivity approximation plus those involved in any graphical integration. The practical use of the radiation chart is described by Elsasser (1942), Yamamoto (1952) and Goody (1964). Suffice it to say here that the form of the transfer equation used in charts is similar to equations [7.55] and [7.56]. However, use is also made of the quantity Q given by (cf. equation [7.57]):

$$Q(z, z', T_{z'}) = \int_0^\infty \tilde{Tr}_\nu(z, z')[\mathrm{d}\tilde{B}_\nu(T_{z'})/\mathrm{d}z']\,\mathrm{d}\nu \qquad [7.70]$$

$$Q(u, T) = \int_0^\infty \tilde{Tr}_\nu(u, T)[\mathrm{d}\tilde{B}_\nu(T)/\mathrm{d}T]\,\mathrm{d}\nu \qquad [7.71]$$

The total flux density F can then be put in a form of a sum of integrals of the form $\int_0^T Q\mathrm{d}T$ along a closed path in a Q–T plane (Elsasser, 1942).

Returning to Sasamori's model, Sasamori obtained empirical relations between $\hat{A}(u, T)$ (equation [7.67]) and u using Yamamoto's transmittance curves. He then showed that a six-layer computation using these values gave agreement with Rodgers and Walshaw to within 10–20%. These results are consistent with the good agreement between Rodgers and Walshaw's and Yamamoto's radiation chart, and would appear to indicate that Yamamoto's radiation chart is perhaps the most reliable.

6.3. *Comparisons of model calculations*

We now consider a few comparisons of calculations of radiative cooling made with different radiation schemes. The question to be answered is "how do the emissivity-type schemes compare with more accurate methods?"

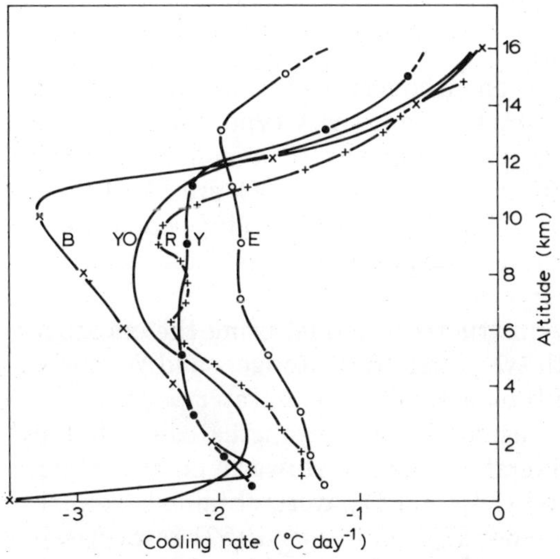

Fig. 7.22. Comparison of cooling rate using Rodgers and Walshaw's multispectral model with those calculated from various "emissivity approximation" models. (After Rodgers and Walshaw, 1966.) E = Elsasser (1942), B = Brooks (1950), Y = Yamamoto (1952), YO = Yamamoto and Onishi (1953), R = Rodgers and Walshaw (1966).

TABLE 7.5

Pressure, temperature and scaled water path from the top of the troposphere. Total water path for the stratosphere = 0.00017 g cm^{-2} at −65°C (after London, 1952)

Height (km)	p (mb)	T(°C)	u^* (g cm^{-2})[1]
0	1,011	27.7	4.325
1.0	900	22.8	2.575
1.5	856	20.6	2.070
1.8	826	19.2	1.775
2.0	802	18.2	1.560
2.3	780	16.9	1.378
3.0	715	13.3	0.932
4.0	636	7.8	0.541
4.4	602	5.5	0.416
5.0	562	2.0	0.297
6.0	496	−3.5	0.156
7.0	438	−10.0	0.0797
8.0	385	−16.6	0.0391
9.0	336	−23.2	0.0181
9.8	302	−28.5	0.00967
10.0	293	−30.0	0.00797
11.0	253	−38.6	0.00312
12.0	217	−47.2	0.00121
13.0	185	−55.7	0.00049
14.0	156	−62.3	0.00019
15.0	131	−69.0	0.000052
16.0	112	−75.5	

[1] Scaled according to $u^* = u(p/p_s)^{1/2}$.

Fig. 7.22 shows a comparison between Rodgers and Walshaw's detailed computation and several one-parameter emissivity-type models. The difference is indicative of the sort of errors inherent in an emissivity formulation. Systematic errors in the computation of $F\uparrow$ and $F\downarrow$ are likely to be magnified when the term $d(F\uparrow - F\downarrow)/dz$ in the cooling equation is considered. The aerological data used in these computations (London, 1952) are given in Table 7.5.

Stone and Manabe (1968) (MS) constructed a model using the emissivity approximation and compared it with two versions of Rodgers and Walshaw's model. For the MS model, transmittance functions of water vapour were calculated with the random model using the line strengths calculated by Yamamoto (1952) and the experimental data of Howard et al. (1955). Empirical transmittance functions of CO_2 and O_3 were obtained from the experimental data of Burch et al. (1961) and Walshaw (1957) respectively. Overlapping between bands was neglected by assuming zero absorption for water vapour in the overlapping region. Transfer in the stratosphere was handled by calculating an emissivity at 220 K. For the troposphere the

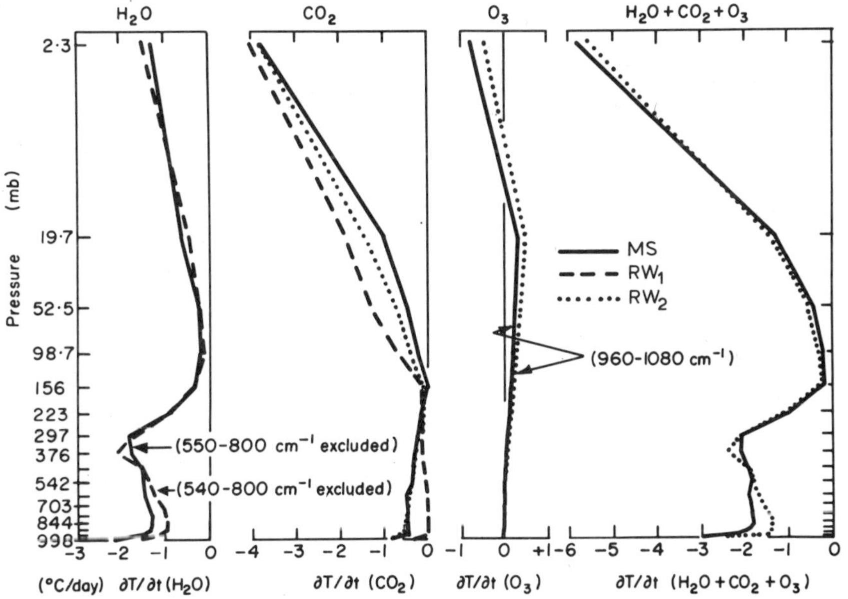

Fig. 7.23. Atmospheric cooling rates computed from the models of Stone and Manabe (*MS*), the Rodgers-Walshaw model (RW_1) and a modified version of their model (RW_2). (After Stone and Manabe, 1968.)

effective temperature of a layer between pressures p_1 and p_2 was taken as:

$$T_{\text{eff}} = \int_{p_1}^{p_2} T\mathrm{d}p \Big/ \int_{p_1}^{p_2} \mathrm{d}p \qquad [7.72]$$

The absorber amount u was scaled according to equation [7.41] with $i = 0.86$ for CO_2 and $i = 0.30$ for O_3.

The first version of Rodgers and Walshaw's model (RW_1) which MS used was identical with those authors' multispectral model described in section 6.1, but the cooling was calculated by taking flux differences rather than flux divergences. The second version (RW_2) incorporated Stone and Manabe's transmittance functions for CO_2 and O_3. Fig. 7.23 shows the cooling curves for the three models. The difference in H_2O cooling between MS and RW_1 is due to the difference in H_2O band parameters used and to the MS use of the emissivity approximation. The RW_2 and MS curves show much more CO_2 cooling in the troposphere than does the RW_1 model due to the different transmittance functions used. The aerological data used for the above comparisons is shown in Table 7.6.

Recent work on the atmospheric water vapour continuum has shown the important contribution of the so-called *e*-type absorption to the cooling in moist atmospheres (section 5.5). To date, only a few attempts have been

TABLE 7.6

Values of temperature, water vapour and ozone used for comparison of Stone and Manabe's and Rodgers and Walshaw's models — see text (after Stone and Manabe, 1968)

Pressure (mb)	Temperature (K)	Mixing ratio ($g\,kg^{-1}$)	
		water	ozone
0.0000	251.3879	0.0000	0.000000
2.2719	251.3879	0.0229	0.012436
19.6759	218.1012	0.0105	0.009136
52.5120	218.5738	0.0039	0.002766
98.7226	218.0662	0.0022	0.001148
156.2500	215.9686	0.0050	0.000676
223.0367	226.7878	0.0201	0.000372
297.0250	239.2173	0.0860	0.000290
376.1574	250.0018	0.2930	0.000202
458.3762	259.4072	0.5730	0.000139
541.6238	267.6333	1.0400	0.000139
623.8426	274.8176	1.7700	0.000080
702.9750	281.0544	2.6400	0.000085
776.9633	286.3196	3.4800	0.000062
843.7500	290.8185	4.3700	0.000071
901.2774	294.4844	5.2200	0.000044
947.4880	297.3169	5.9700	0.000058
980.3241	299.2888	6.5700	0.000033
997.7281	300.3594	7.0000	0.000094
1,000.0000	300.3594	7.0000	0.000094

made to incorporate modified continuum absorption functions into computing schemes. The simple method is to increase the originally accepted constant value of continuum mass absorption coefficient by a fixed amount across the whole atmospheric window. This method was used by Kuhn (1972) in order to bring computed fluxes into agreement with measured flux profiles. He used a value of $\kappa_{\nu c} = 0.2\ g^{-1}\ cm^2$ instead of the previously accepted value of $0.1\ g^{-1}\ cm^2$. Measurements of zenith sky radiance at the ground indicate that this kind of treatment is justified if the required accuracy is no more than about 30%. Fig. 7.24 shows measured zenith sky radiance (10–12 μm) compared with radiance computed from aerological data using a value of $\kappa_{\nu c}$ equal to $0.16\ g^{-1}\ cm^2$. The line of best fit indicates that a value of $\kappa_{\nu c}$ equal to $0.20\ g^{-1}\ cm^2$ would give best agreement between experiment and computation, in agreement with Kuhn's work.

Two recent detailed incorporations of *e*-type absorption into cooling models have been made by Grassl (1974) and Cox (1973). Grassl used Rodgers and Walshaw's scheme at wavelengths outside the window, and divided the window into seven intervals with the characteristics shown in Table 7.7 (see equation [7.43]). The cooling calculation was done by differencing between levels. Fig. 7.25 shows the resultant cooling in the

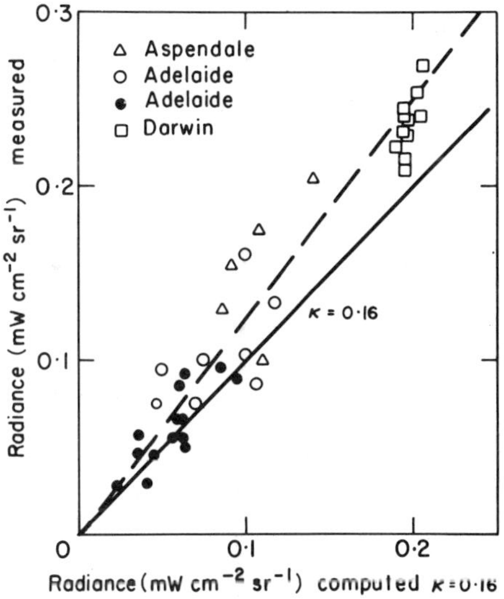

Fig. 7.24. Measured zenith sky radiance in the 10–12-μm wavelength interval plotted against sky radiance computed from aerological data using a continuum absorption coefficient $\kappa_{\nu c}$ equal to $0.16\,g^{-1}\,cm^2$. The dashed curve is the line of best fit corresponding approximately to $\kappa_{\nu c} = 0.2$. (C.M.R. Platt, unpublished.)

TABLE 7.7

Water vapour absorption coefficients in the atmospheric window — see equation [7.46] (after Grassl, 1973a, 1974)

Absorption coefficient	Wavelength (μm)						
	8.34	8.80	9.30	10.15	11.00	12.00	12.50
$\kappa_{\nu 1}$ [1]	0.005	0.005	0.004	0.003	0.004	0.010	0.012
$\kappa_{\nu 2}$ [1]	5.80	4.90	4.90	6.50	8.60	11.90	14.00
$\kappa_{\nu 3}$ [2]	0.155	0.085	0.048	0.028	0.034	0.089	0.144

[1] In $g^{-1}\,cm^2\,atm^{-1}$ at 273 K.
[2] In $g^{-1}\,cm^2$ at 273 K.

lower troposphere compared with Rodgers and Walshaw's model using the aerological data shown in Table 7.5 for a tropical atmosphere (see also Fig. 7.22). The enhancement in cooling of the lower atmosphere is clearly indicated. Grassl included the temperature and humidity dependence of the e-type and line absorptions in each layer. That the increased absorption is mainly significant in tropical atmospheres is shown in Fig. 7.26, where it is seen that cooling in high latitudes is much less affected by the additional absorption.

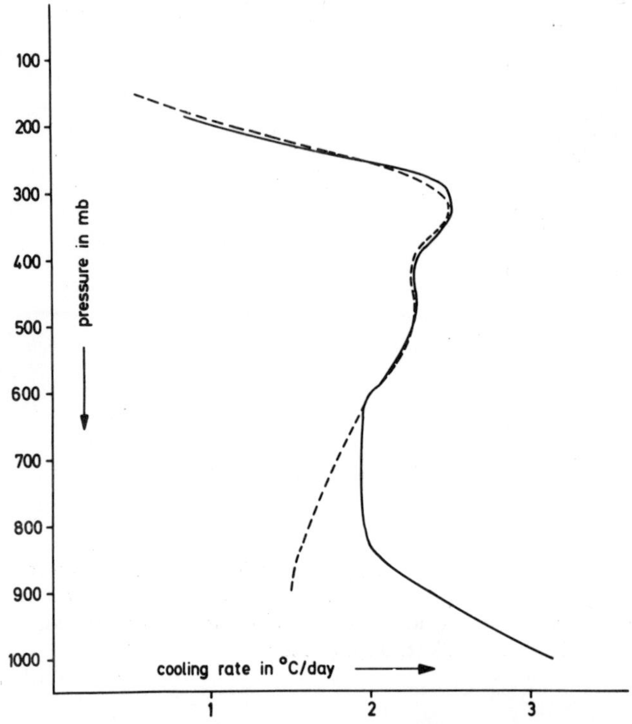

Fig. 7.25. The cooling rate in a tropical atmosphere with (solid curve) and without (dashed curve) inclusion of e-type absorption. (After Grassl, 1974.)

Ellingson and Gille (1972) have also included e-type absorption from Bignell's measurements in their model and find good agreement with experimental profiles in clear skies. Cox (1973) included the e-type absorption in an emissivity representation. In this case the vapour pressure e must be included as an additional variable. Cox considered the continuum as having only e-type absorption, with an exponential transmittance:

$$Tr(u, T, e) = \exp[-\kappa(\nu, T)eu] \quad [7.73]$$

He considered a linear dependence of $\kappa_\nu(T)$ and $B_\nu(T)$ on wavenumber. The temperature dependence of κ_ν was specified as proportional to the dimer mole fraction $\hat{C}$:

$$\hat{C} = [\hat{\gamma} - (\hat{\gamma}^2 - 4)^{1/2}]/2 \quad [7.74]$$

where $\hat{\gamma} = 2 + [p_s \exp(C_1 - C_2/T)e]$ and $C_1 = 13.0096$, $C_2 = 287.225$ K. The flux emissivity of water vapour at 20°C is compared with the values of Staley and Jurica (1970) in Fig. 7.27. Cox's results for a mid-latitude and a tropical profile show the same order of enhanced cooling in the tropical case as Grassl's results.

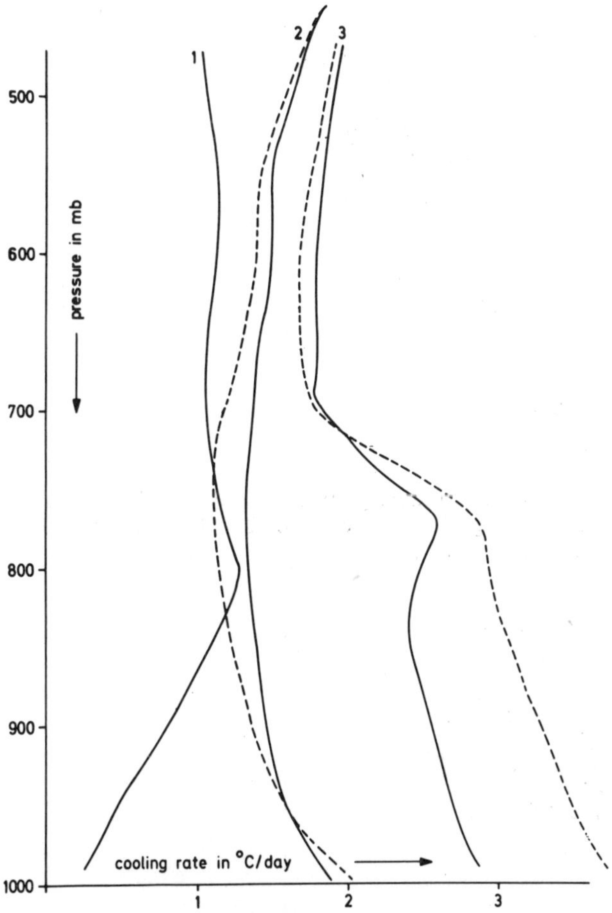

Fig. 7.26. Cooling rates for (*1*) Arctic atmosphere (75°N, January), (*2*) U.S. Standard Atmosphere (45°N), (*3*) tropical atmosphere (15°N) with (dashed curve) and without (solid curve) inclusion of *e*-type absorption. (After Grassl, 1974.)

It is evident that the *e*-type absorption is a very important factor in the cooling of the lower troposphere.

6.4. *Empirical relation between cooling and local temperature*

Radiative cooling in the troposphere in the absence of clouds depends mainly on the vertical distribution of temperature and water vapour and is usually dominated by direct "cooling to space". Thus the radiative cooling might have an approximate linear dependence with the local air temperature. London and Sasamori (1968) investigated this possibility by computing the cooling for winter and summer seasons at different latitudes of the northern

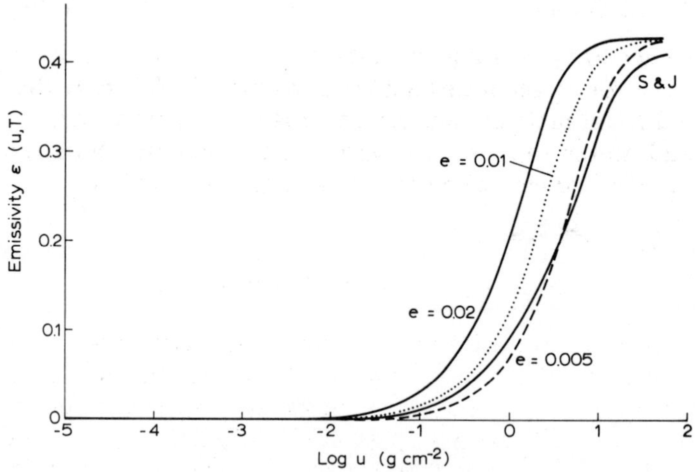

Fig. 7.27. Emissivity of the water vapour continuum given by Staley and Jurica (1970) (*S & J* curve) and that calculated from Bignell *e*-type data for $T = 20^{\circ}$C. *e* is the H_2O partial pressure in atmospheres. (After Cox, 1973.)

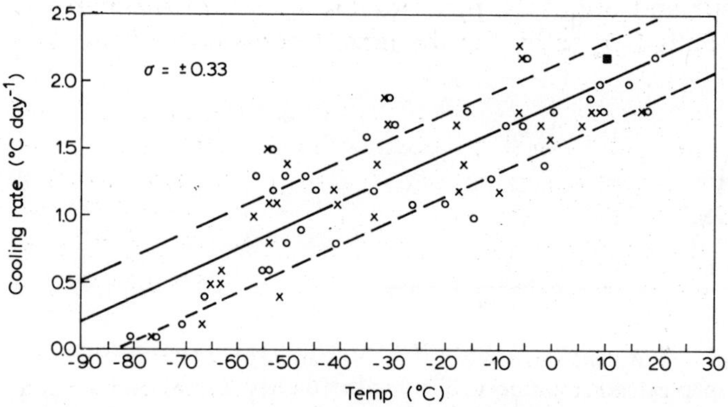

Fig. 7.28. Radiative cooling of individual layers given as a function of temperature of the layers for model atmospheres winter and summer for latitudes 0°, 30°, 60°. Crosses and circles are for the northern and southern hemispheres respectively. (After London and Sasamori, 1968.)

and southern hemispheres (0°, ±30°, ±60°). They used Sasamori's (1968) computational method. The models were based on observed mean temperature distributions up to 25 mb altitude and observed water vapour distributions up to 500 mb. Above this altitude the mixing ratio was extrapolated upwards to an assumed constant value of 5×10^{-6} g/g at 25 mb. A plot of cooling in degrees per day versus local air temperature is shown in Fig. 7.28. The crosses and circles are for the northern and southern hemisphere respectively. The cooling rate *CR* is represented by:

$$CR = 0.17T + 1.8(\pm 0.33°\text{C/day}) \quad [7.75]$$

This has about a 30% accuracy at a cooling rate of 1°C per day compared with an accuracy of 10% for an accurate (RW) cooling model. Naturally this formula cannot be used to estimate cooling near temperature discontinuities, where temperature and water vapour both vary rapidly and the radiative cooling consequently peaks, nor can it be used for a cloudy atmosphere.

7. RADIATIVE COOLING IN THE MIDDLE ATMOSPHERE (30–100 km)

The pattern of the general circulation in the upper atmosphere is determined by the positions and strengths of the local sources and sinks of energy. These are almost wholly radiative in origin, the exception being conversion of chemical to thermal energy (Hunt, 1972). As clouds are non-existent above the tropopause (apart from rare noctilucent clouds and a small aerosol component), all long-wave radiative absorption and emission is confined to gases. The relative contributions of CO_2, H_2O and O_3 to middle-atmospheric cooling are roughly proportional to their abundances. The mixing ratio of CO_2 is about 330 parts per million (ppm) compared to about 3 ppm for water vapour and about 12 ppm for O_3 at 30 km dropping to nearly zero at 90 km. Thus CO_2 is by far the largest contributor to middle atmospheric cooling.

Treatment of the transmittance functions is similar to that at lower altitudes, but with differences in the shape of the molecular absorption lines and in the types of band transmission employed. The region above 80 km is the hardest to treat because the air is rare enough for breakdown of local thermodynamic equilibrium (LTE). The source function in the equation of radiative transfer is then no longer equal to the Planck function but is itself a function of the net flux density.

The Lorentz line shape (see equation [7.20]) is more or less valid below about 30 km, but at higher altitudes the Lorentz line width becomes comparable with the Doppler line width (equation [7.27]). Between altitudes of 30 and 70 km a mixed line shape, or Voigt profile (Kuhn and London, 1969) is appropriate. Calculations using this line profile lead to larger cooling rates than those computed on the basis of a pure Lorentz profile. For CO_2 the increase is about 5% at 50 km, rising to more than 100% at 70 km. (Rodgers and Walshaw, 1966). Above 70 km, a pure Doppler profile is adequate.

Neither the Elsasser regular nor the Goody random band transmittance models are suitable for the accurate calculation of gaseous transmittance in the middle atmosphere. Furthermore, experimental transmission data at pressures appropriate to the middle atmosphere do not exist. Two methods have been evolved to handle the problem; a better band model and a direct integration model. The so-called "quasi-random" band model is a reasonably accurate representation of CO_2 transmittance. As the CO_2 molecular bands consist of a number of rather regular sub-bands which are randomly spaced with respect to each other, the main feature of the quasi-random model is the division of the whole band into a large number of spectral sub-intervals. Within each sub-interval the lines are regarded as randomly spaced, but each sub-interval contains a range of line strengths. The grouping of the line strengths within each sub-interval into five separate decades takes equal account of both very strong and very weak lines which are characteristic of CO_2 bands. The presence of, say, one or two strong lines in each spectral sub-interval gives the model a required amount of non-randomness. This band model is fully described by Wyatt et al. (1962) and Stull et al. (1964).

An alternative direct integration model in which the transmittance is obtained by integration over each individual line has been described by Drayson (1966) (see also section 5.4). Although this method is costly in computer time, the fact that the CO_2 mixing ratio is sensibly constant with altitude means that the transmittance between two pressure heights and for a given standard pressure and temperature profile can be computed for all time. These transmittances can then be used for different temperature profiles because atmospheric cooling is much more sensitive to temperature than is the actual transmittance.

Local thermodynamic equilibrium occurs when molecular collisions re-distribute the absorbed quantum energy before it can be re-radiated. The collisions preserve a Boltzmann distribution of energy between the molecules. Whether this occurs depends upon the molecular collision rate being much greater than the spontaneous radiative transition rate. Put another way, the collision relaxation time $\tilde{\theta}$ must be much less than the radiative lifetime $\tilde{\lambda}$. The value of $\tilde{\theta}$ is inversely proportional to pressure and becomes comparable with $\tilde{\lambda}$ at thermospheric heights.

When LTE breaks down, the source function $\tilde{J}$ for a molecular band depends on the Planck function $\tilde{B}$, $\tilde{\theta}$, $\tilde{\lambda}$ and the net flux density. Thus:

$$\tilde{J}(T_z) = [\bar{F}_{\Delta\nu}(T_z) + (\tilde{\theta}/\tilde{\lambda})\tilde{B}(T_z)]/[1 + (\tilde{\theta}/\tilde{\lambda})] \qquad [7.76]$$

where $\bar{F}_{\Delta\nu}(T_z)$ is the *absorbed* flux density per unit increment of optical path. Thus $\bar{F}_{\Delta\nu} = \int_{4\pi}\int_{\Delta\nu} N_\nu \kappa_\nu \rho \mathrm{d}s\mathrm{d}\nu\mathrm{d}\Omega / \int_{\Delta\nu} \kappa_\nu \rho \mathrm{d}s\mathrm{d}\nu$ where the $\Delta\nu$ represents the spectral extent of the molecular absorption band. $\tilde{B}$ and $\tilde{J}$ are representative mean values of the band. Consideration of equation [7.76] together with the equation of the net flux density divergence, and assuming that $\tilde{J}$ is isotropic, leads to the expression:

$$\tilde{J}(T_z) = \tilde{B}(T_z) + \{1/[4\pi S_\mathrm{b} w(z)\rho(\tilde{\theta}/\tilde{\lambda})]\}[\mathrm{d}F_\mathrm{N}(z)/\mathrm{d}z] \qquad [7.77]$$

where the band strength S_b is equal to $\int_{\Delta\nu} \kappa_\nu \mathrm{d}\nu$ and $w(z)$ is the absorber mixing ratio and ρ is the air density. The source function can be calculated from equation [7.77] if the net flux divergence is known. However, the net flux density is itself a function of the $\tilde{J}$'s at all other levels. Thus the equation of $\tilde{J}$ for an (homogeneous) atmospheric layer is coupled to the equations of all other layers through their source functions. In matrix-vector form:

$$\tilde{J}_k = \tilde{B}_k + \sum_j \alpha_{kj} \tilde{J}_j \qquad [7.78]$$

These equations must be solved numerically (Kuhn and London, 1969).

There is considerable uncertainty in the value of $\tilde{\theta}/\tilde{\lambda}$ due to uncertainty in the value of $\tilde{\theta}$. At standard pressure, the value of $\tilde{\theta}$ for carbon dioxide is believed to lie between 2×10^{-5} and 2×10^{-6} second. The radiative lifetime is 0.43 second.

In general, the actual cooling rate follows the temperature distribution so that maximum cooling is found at levels of positive curvature in the temperature profile and minimum cooling (or heating) at levels of negative curvature. The cooling per day due to CO_2 bands increases from a value of about 2° at 30 km to a maximum of 5° at 50 km, reducing to zero at 70 km and changing to a *heating* of about 4° per day at 80 km. Above about 88 km, cooling occurs again and thereafter reaches a value of about 18° per day at 100 km if $\tilde{\theta}$ at standard pressure is 10^{-6} second. The figure is only 3° per day if $\tilde{\theta}$ is 10^{-5} second. These latter results illustrate the importance of determining an accurate value of the collision relaxation time.

Radiative cooling due to the water vapour rotation bands is only significant in the upper stratosphere near the equator where (for a mixing ratio of $\approx 10^{-6}$ ppm), the cooling reaches about 1° per day. Ozone contributes a maximum cooling of 3° per day at 50 km, dropping to a negligible amount above 60 km.

CHAPTER 8

Clouds and Long-Wave Transfer

1. INTRODUCTION

The effect of clouds on *solar* radiation within the atmosphere is dominated by scatter, and the absorption of solar energy by liquid water drops and ice particles can often be neglected entirely. The situation is virtually the reverse for long-wave radiation. A dense low-level water cloud will absorb more than 90% of infrared radiation within a depth of only 50 m. The scatter by drops may be large but is usually swamped by absorption and can be neglected. Thus from the point of view of the atmosphere as a whole, reasonably thick clouds can be regarded most simply as black bodies to infrared radiation. In effect, a dense cloud layer forms an extra boundary in the atmosphere, the base forming an absorbing lid to the upwelling radiation from the surface and lower atmosphere, and the top forming another "surface" which radiates to space through the atmospheric window. The planetary long-wave emission to space is thus controlled by this surface rather than the ground. Unlike gaseous absorption, there are no spectral regions in the infrared where cloud absorption is small. The actual vertical profile of atmospheric cooling can be computed by exactly the same techniques as for clear skies, except that the transmittance at all wavelengths between any two levels is (usually) set to zero when there is a cloud somewhere between the levels. The source function at cloud level is set at the Planck function value $B_\nu(T)$ at all wavenumbers.

As far as operational models are concerned, the above procedure is virtually standard technique. Certain modifications may be introduced. For instance, it may be necessary to assign specific transmittances greater than zero to certain cloud types which are known to be optically thin — cirrus, for instance, usually has a long-wave transmittance greater than 0.5. In other words, one makes a grey-body rather than a black-body assumption. Further, in cases of less than 100% cloud cover in a particular layer, one can set the layer transmittance in direct proportion to the fraction of "clear sky" in the layer.

This black- or grey-body assumption has its limitations. On the one hand there is the problem of calculating just how grey a particular cloud layer may be, bearing in mind that (ultimately) cloud character must be predictable *a priori*. On the other, the process of cloud formation and dissipation can be influenced by the profile of radiative flux divergence (i.e. cooling rate) within the cloud itself. In either case it is necessary to look in some detail at the interaction of radiation and the cloud microphysics. Potentially one is faced with the problem of solving the complete equation of transfer in a Mie scattering medium. To date the subject has been examined only in terms of homogeneous horizontally stratified clouds. Most work has been done on ideal clouds whose liquid water content and drop size distributions are constant with vertical position (e.g. Yamamoto et al., 1970; Zdunkowski and Crandall, 1971; Hunt, 1973). Recently Zdunkowski and Davis (1974) have treated clouds whose liquid water content varies in a specific manner in the vertical. This sort of test-bed modelling is of importance in illustrating the possible magnitude of cooling rates within clouds and the modification by multiple scattering processes.

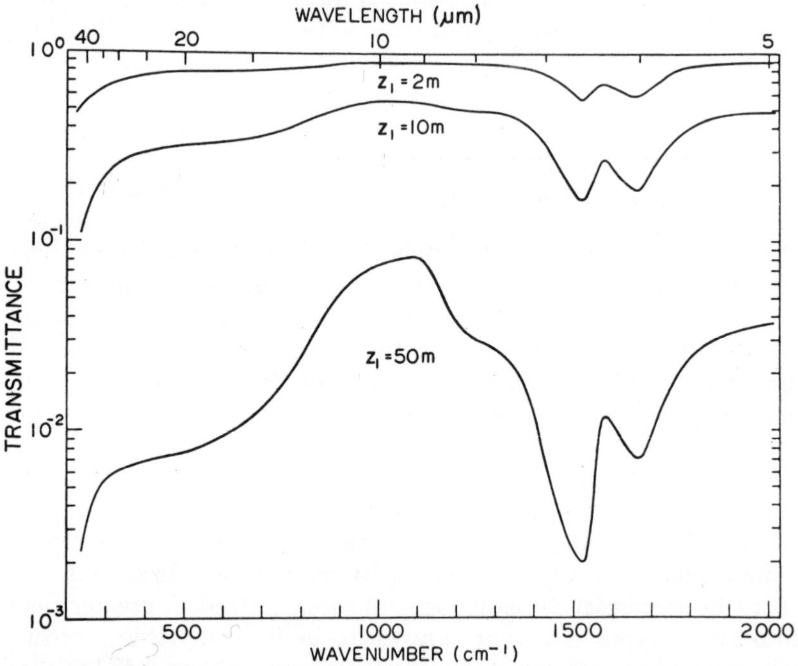

Fig. 8.1. Spectral transmittance of a cloud of various thicknesses. Altostratus cloud with a liquid water content of 0.28 g m^{-3}. (After Yamamoto et al., 1970.)

As an example of relevant magnitudes, the spectral flux transmittances, reflectances and emissivities of a water cloud of various thicknesses as given by Yamamoto et al., (1970) are shown in Figs. 8.1, 8.2 and 8.3. These curves represent a cloud of fixed liquid water content of 0.28 g m^{-3} and a drop size distribution appropriate to a typical deck of altostratus. The absorption by

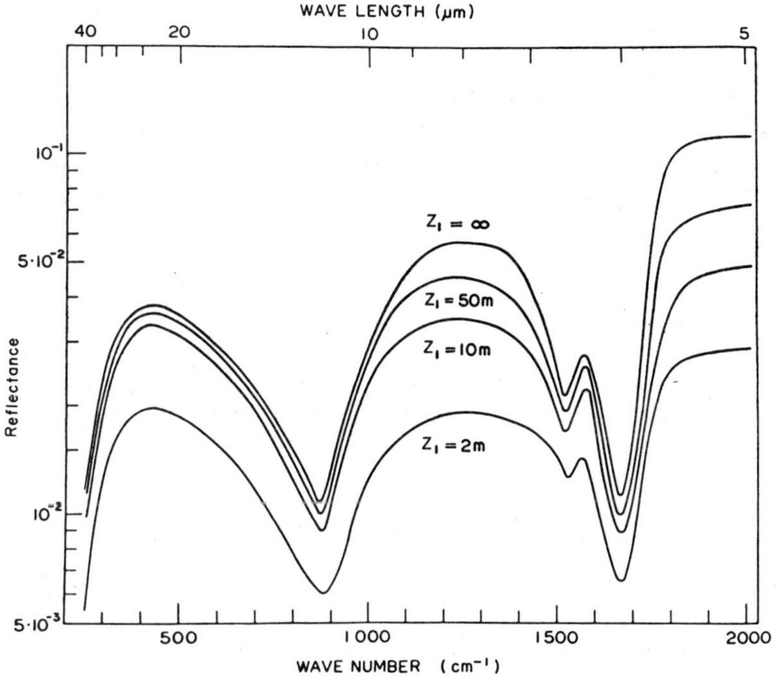

Fig. 8.2. Spectral reflectance of a cloud. Same characteristics as Fig. 8.1. (After Yamamoto et al., 1970.)

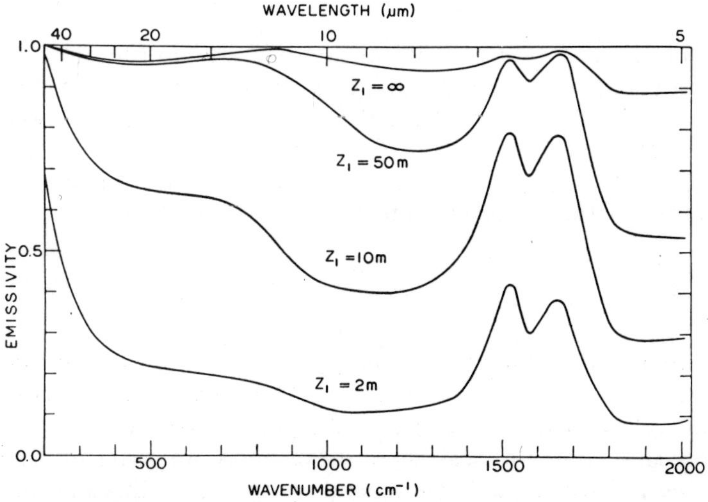

Fig. 8.3. Spectral emissivity of a cloud. Same characteristics as Fig. 8.1. (After Yamamoto et al., 1970.)

water *vapour* within the cloud is also included, but not the absorption due to carbon dioxide. As might be expected, there is a minimum of absorption in the region of the atmospheric window.

Cloud is usually a very inhomogeneous medium, so that defining the vertical flux density transmittance over an extensive area containing clouds always requires some averaging process. The meteorologist is forced to define "standard" cloud types and to use approximate models for the actual cloud transmittance.

2. WATER DROPLET AND ICE CRYSTAL INFRARED PARAMETERS

The total optical depth τ_ν of a cloud is the sum of the individual optical depths due to the drops or particles themselves and to the absorbing gases of water vapour and carbon dioxide. Here we shall discuss only the droplet or particle extinction. For the general case of a cloud whose volume extinction coefficient $\beta_{e\nu}$ varies in the vertical:

$$\tau_{e\nu} = \int \beta_{e\nu} \mathrm{d}z \qquad [8.1]$$

Note that this volume extinction coefficient comprises a scattering portion $\beta_{sc\nu}$ and an absorption portion $\beta_{a\nu}$ so that:

$$\beta_{e\nu} = \beta_{sc\nu} + \beta_{a\nu} \qquad [8.2]$$

and in terms of these parameters the single scattering albedo $\bar{\omega}_0$ is given by:

$$\bar{\omega}_0 = \beta_{sc\nu}/(\beta_{sc\nu} + \beta_{a\nu}) \qquad [8.3]$$

Unlike gases, the spectral extinction coefficient is not a unique function of the mass of the absorbing water molecules. For a given mass of water droplets (or ice particles), the volume extinction coefficient is in general dependent on the *size* of the droplets. Thus $\beta_{e\nu}$ must be expressed in terms of an extinction efficiency $Q_{e\nu}$ which is the ratio of the actual drop (or particle) extinction cross-section to its geometric cross-section. Thus:

$$\beta_{e\nu} = \pi \int_0^\infty n(r) r^2 Q_{e\nu} \mathrm{d}r \qquad [8.4]$$

where $n(r)$ is the number of drops of radius r and $Q_{e\nu}$ may also be a function of radius. Similar relations exist between the scattering and absorption coefficients ($\beta_{sc\nu}$ and $\beta_{a\nu}$) and the scattering and absorption efficiencies, $Q_{sc\nu}$ and $Q_{a\nu}$. For spheres (e.g. water drops) these efficiencies are given by the Mie equations discussed in Chapter 4. They depend on droplet radius and the real and imaginary parts of the refractive index of water. There is no exact theory which describes the absorption and scattering efficiencies of irregular

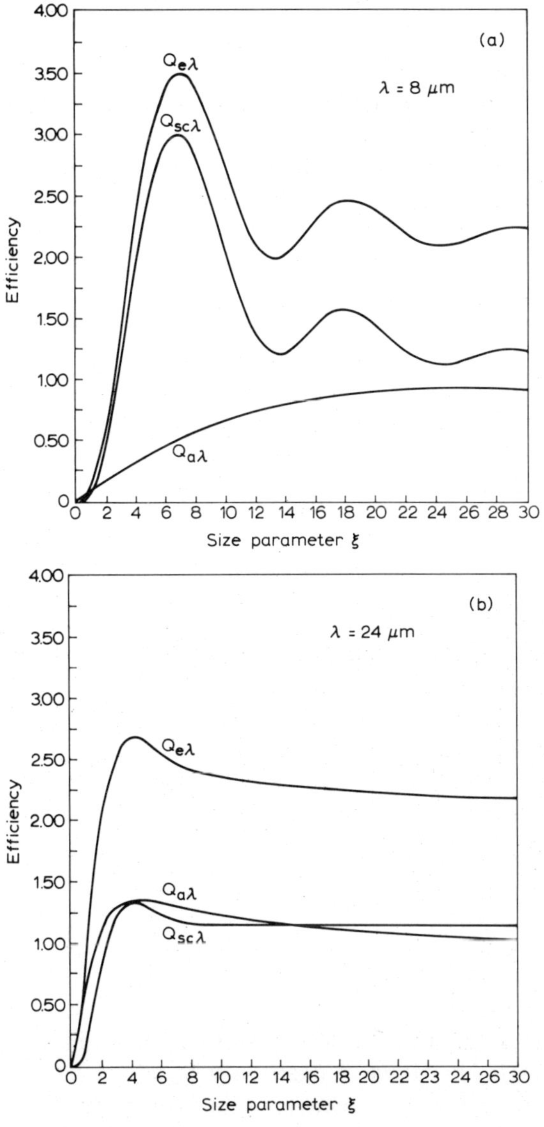

Fig. 8.4. (a) The extinction, scattering and absorption efficiencies of spherical water droplets for a refractive index $n_c = 1.293 - 0.0231i$ and a wavelength of 8 μm plotted against size parameter $\xi (= 2\pi r/\lambda)$. (b) Same as Fig. 8.4a but for a refractive index of 1.429–0.261i and a wavelength of 24 μm. (After Herman, 1962.)

particles such as ice crystals. Such particles are normally treated in terms of an equivalent sphere with a radius (providing the particle is convex) given as:

$$r = (A/4\pi)^{1/2} \qquad [8.5]$$

where A is the surface area of the particle (Van de Hulst, 1957).

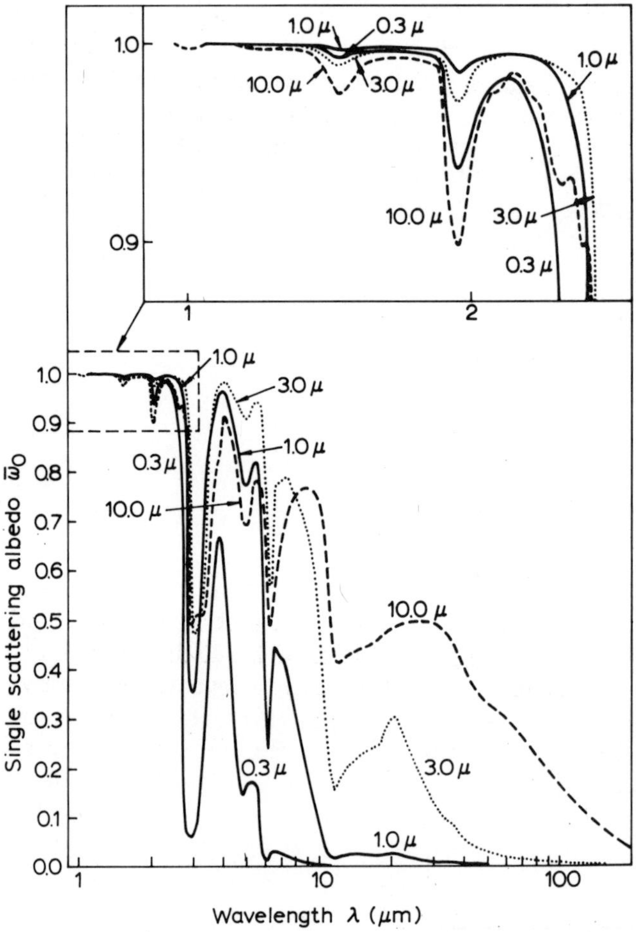

Fig. 8.5. Single scattering albedo $\bar{\omega}_0$ for liquid water droplets as a function of wavelength for several particle radii. (After Irvine and Pollack, 1968.)

The absorption coefficients of water and ice are extremely high in the infrared region of the spectrum. A *layer* of water need only be about 10 μm thick to absorb all radiation which enters it. However, Mie theory predicts that water spheres whose radii are less than the wavelength of the incident radiation have absorption efficiencies less than unity. For a given refractive index, $Q_{a\nu}$ is a function of the size parameter $\xi(=2\pi r/\lambda)$ and generally tends towards zero as ξ tends to zero. The behaviour of Q_a (and of Q_{sc} and Q_e) as a function of ξ is illustrated in Fig. 8.4 for two representative values of refractive index.

Single scattering albedo $\bar{\omega}_0$ also varies with wavelength and drop size as illustrated in Fig. 8.5. The equivalent curves for ice particles show similar oscillations, which differ only in detail from the case of water drops.

Within the range that it is less than unity, Q_a is approximately proportional to the imaginary part of the refractive index n_i and is relatively independent of the real part n_r. Thus n_i must be known accurately. Unfortunately, because of the high absorption, values of n_i for water are difficult to determine. Only recently have consistent values become available. Hale and Query (1973) have reviewed the available data and have given what they consider to be the most accurate values of both n_i and n_r (see the Appendix, Table A.4). The refractive index of water is quite insensitive to both temperature and pressure, excluding of course, the phase change from water to ice.

Cloud drop size distributions and some ice crystal distributions are described in the Appendix. Analytical formulae which have been devised as approximate fits to the measured distributions are given in Table A.1.

3. THE ABSORPTION APPROXIMATION

Considerable simplification occurs in the equation of radiative transfer if scattering by cloud droplets or ice particles is ignored. The scattering term then vanishes and the cloud radiative cooling is a function of the cloud absorption coefficient only. At first sight, considering the high values of $\bar{\omega}_0$ in Fig. 8.5, the neglect of scattering would appear to lead to prohibitive errors. There are two factors which reduce the effects of scattering. First, the phase function is peaked in the forward direction and much of the scattered radiation proceeds in a direction not too different to the forward direction. The radiance inside a cloud in some arbitrary direction is thus not affected very much by scattering. Second, the percentage of the incident radiation which is backscattered, particularly by an absorbing drop, is only small. It is usually less than 10%. The low reflectance of water clouds (Fig. 8.2) is a consequence of this fact.

Neglecting scatter, and dropping the formal parameter of wavenumber, the transmittance Tr and emissivity ϵ of a cloud of thickness z are defined by:

$$Tr = \exp(-\beta_a z) \qquad [8.6]$$

and:

$$\epsilon = 1 - Tr \qquad [8.7]$$

where in turn:

$$\beta_a = \pi \int_0^\infty n(r) r^2 Q_a(r) \mathrm{d}r \qquad [8.8]$$

The accuracy of the absorption approximation when applied to representative clouds can only be judged by comparison with a complete multiple scattering model of cloud absorption. Zdunkowski and Crandall (1971) (ZC) have made a comprehensive set of computations of cloud emissivity at several wavelengths in the "window" and at 6.5 μm within the water vapour absorption band. Their results for the *flux* emissivity ϵ of a cloud at several thicknesses and wavelengths are shown in Fig. 8.6. The flux emissivity $\epsilon(a)$ calculated from the absorption approximation is also shown, and has been calculated according to:

$$\epsilon(a) = \int_0^1 [1 - \exp(-\tau_a/\mu)]\,\mu d\mu \Big/ \int_0^1 \mu d\mu \qquad [8.9]$$

Here μ is the cosine of the zenith angle and $\tau_a = \beta_a z$ where z is the cloud thickness. The comparison is for a stratocumulus cloud with a liquid water content of 0.1 g m^{-3}.

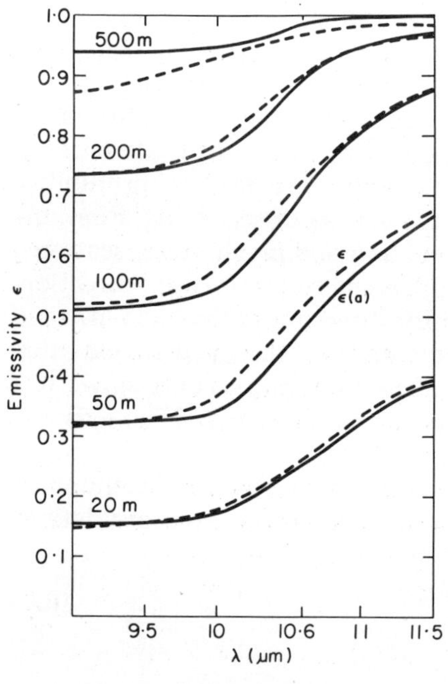

Fig. 8.6. The emissivity of a cloud as computed by Zdunkowski and Crandall (1971) (broken line) compared to the emissivity $\epsilon(a)$ computed from the absorption approximation (full line) — see text.

Referring to Fig. 8.6, for the 20-m (very thin) cloud, scattering has little effect on the emissivity. For clouds from 50 m to about 200 m in thickness, scattering causes ϵ to be greater than $\epsilon(a)$ because the effective path of radiation through the cloud is increased. When the cloud is 500 m thick, even the vertical path in the cloud is sufficient to absorb most radiation, at least beyond 10.6 μm wavelength. In this case, the finite reflectance *Re* of the real cloud causes the absorption approximation emissivity $\epsilon(a)$ to be greater than ϵ. The larger difference between $\epsilon(a)$ and ϵ at 9 μm compared to that at 11.5 μm is a consequence of the differing refractive indices at the two wavelengths. This variation is consistent with the values of *Re* which are given by Yamamoto et al. (1970) in Fig. 8.2.

The maximum error in using the absorption approximation to calculate flux emissivities in the atmospheric "window" is thus about 10%. Although the calculations have been made for a cloud of water droplets, the same arguments apply to an ice cloud since the refractive indices of ice differ in detail only from those of water.

We have so far considered only the window region. Appreciable cloud cooling occurs at wavelengths greater than 17 μm, particularly for thin clouds and high clouds. Both reflectance *Re* (Fig. 8.2) and single scattering albedo $\bar{\omega}_0$ of water spheres decrease at wavelengths greater than about 10 μm, so that the absorption approximation will, if anything, be more accurate beyond 17 μm.

4. "STANDARD" CLOUDS

Whatever procedure is used to calculate the long-wave flux densities within clouds one is still faced with the problem of "defining the medium" in the sense that no operational model is likely to predict the drop size distribution as a function of height within the cloud. The predictable parameters will probably be restricted to such things as liquid water content per unit volume w or the total liquid water W in the vertical column of the cloud.

The grossest approximation is simply to assign one volume absorption coefficient to each cloud type, irrespective of the variations in cloud water content and droplet size. The volume absorption coefficient then represents the "climate mean" for each cloud type, which can then be used in large-scale atmospheric modelling where clouds of certain depth and height are generated. The emissivity of a given cloud type is then a function only of the cloud depth. Kuhn et al. (1974) have used this type of formulation in a comparison of computed and theoretical fluxes in the atmosphere.

If possible it is obviously preferable to relate the volume absorption coefficient as directly as possible to liquid water content w. The ideal case is of a cloud which has a single unique drop size distribution, since then $\beta_{a\nu}$ is simply proportional to the total number of drops per unit volume and therefore to w. Unfortunately, the drop size distribution varies within one cloud and from one cloud to another, and at first glance it would seem that such an "ideal case" would exist only rarely. However, bearing in mind that drop size distribution and liquid water content are not independent, it is possible that some fairly simple relation does exist between $\beta_{a\nu}$ and w.

Consider a monodisperse cloud where all the drops have the same radius r and absorption efficiency $Q_{a\nu}$. Then:

$$\beta_{a\nu} = \pi n(r) r^2 Q_{a\nu} \qquad [8.10]$$

and:

$$w = 4\pi r^3 n(r)/3\rho \qquad [8.11]$$

where ρ is the density of water. Combining these two equations:

$$\beta_{a\nu} = 3wQ_{a\nu}/4r\rho \qquad [8.12]$$

and differentiating with respect to w:

$$\mathrm{d}\beta_{a\nu}/\mathrm{d}w = 3Q_{a\nu}/4r\rho \qquad [8.13]$$

If we return to Fig. 8.4a, which among other things, shows the variation of $Q_{a\nu}$ with radius, it can be seen that $Q_{a\nu}$ increases in a fairly linear manner with drop radius up to quite large drop sizes — and thereafter approaches its asymptotic value of 1.0. The point is that since $Q_{a\nu}$ increases with r, the *ratio* $Q_{a\nu}/r$ may be fairly constant, and in this case $\mathrm{d}\beta_{a\nu}/\mathrm{d}w$ would be constant. To the extent that this is true, one might therefore expect the volume absorption coefficient of a real drop size distribution to be simply proportional to w, and furthermore that this relation should hold for all points in a cloud provided the *shape* of the distribution does not alter drastically so as to give a preponderance of very large drops.

Whether such an approximation is valid can be tested by computations on measured drop size distributions. Thus Fig. 8.7 is a plot of the values of $\beta_{a\nu}$ at 11 μm wavelength calculated for the drop size distributions associated with various values of liquid water content measured at different positions in stratocumulus cloud decks over the sea. The linearity between $\beta_{a\nu}$ and w is quite remarkable. There is perhaps some tendency to low values of $\beta_{a\nu}$ at w greater than $0.2 \times 10^{-6}\,\mathrm{g\,cm^{-3}}$, which might be expected because large w generally is associated with large drop radii (see the Appendix). The slope of the curve drawn through the origin (i.e. $\mathrm{d}\beta_{a\nu}/\mathrm{d}w$) is 800 $\mathrm{cm^2\,g^{-1}}$, which is close to the values which can be computed on the assumption of Diermendjian model gamma distributions C1 and C2 of Table A.1 — that is, values of 840 and 1090 $\mathrm{cm^2\,g^{-1}}$ respectively. The C2 cloud has the smallest mean radius and narrowest distribution, explaining the larger value of $\mathrm{d}\beta_{a\nu}/\mathrm{d}w$.

As it turns out, the value of $\mathrm{d}\beta_{a\nu}/\mathrm{d}w$ for 11 μm is near the average value for the entire window region of the spectrum from 8 to 14 μm. The relation between the window and total spectral characteristics will be discussed shortly. The parameter $\mathrm{d}\beta_{a\nu}/\mathrm{d}w$ is in fact the *mass* absorption coefficient $\kappa_{a\nu}$ since by definition:

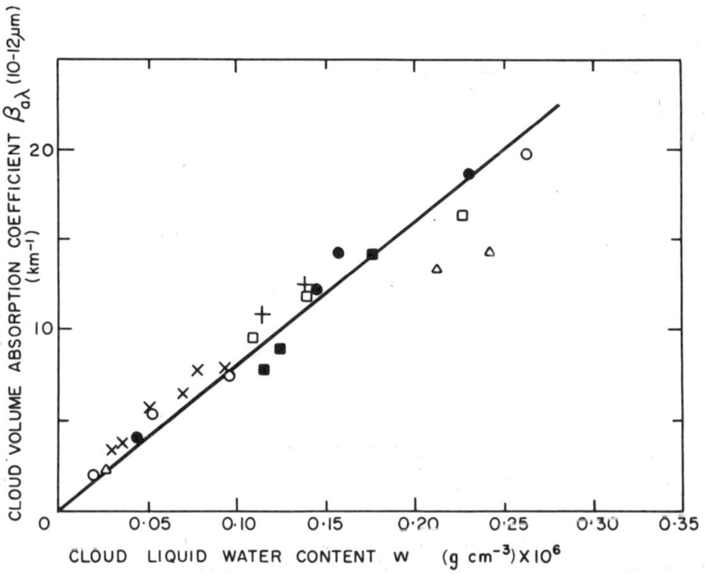

Fig. 8.7. The volume absorption coefficient at wavelength 10–12 μm of a stratocumulus cloud versus the cloud liquid water content for several different cloud decks. (C.M.R. Platt, unpublished.)

$$\beta_{a\nu} = \kappa_{a\nu} w \quad [8.14]$$

Thus if $\kappa_{a\nu}$ is constant with position in a cloud, the vertical emissivity at wavenumber ν can be expressed:

$$\epsilon_\nu = 1 - \exp(-\kappa_{a\nu} W) \quad [8.15]$$

where W is the total water path in the vertical column of the cloud.

The approximation that the mass absorption coefficient of a given cloud type is constant, although rough, is certainly better than the assumption of a constant volume absorption coefficient. For instance, C.M.R. Platt (unpublished) investigated six stratocumulus clouds and found experimentally that a plot of ϵ_ν against the integrated vertical water path (Fig. 8.8) gave much less scatter than an equivalent plot of ϵ_ν against actual path. Whether the particular curve of Fig. 8.8 is representative of water clouds in general has yet to be established.

For cirrus ice clouds, where particle radii are about 10 times as great as the typical drops of water clouds, $d\beta_{a\nu}/dw$ will be roughly proportional to $1/r$. The particle size distribution in cirrus clouds is not well known. Based on the work of Heymsfield and Knollenberg (1972) and Heymsfield (1975) the equivalent radius of cirrus particles is about 50 μm. Then the relation between $\beta_{a\nu}$ and w (assuming all particles to be this size and that $Q_{a\nu} = 1$) is:

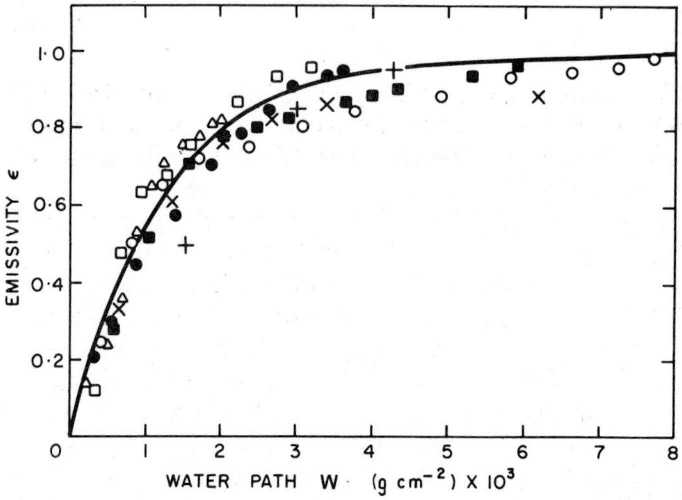

Fig. 8.8. The measured emissivity at 11 μm of layers of stratocumulus cloud decks versus liquid water path through the layers. Clouds sampled were the same as in Fig. 8.7. (C.M.R. Platt, unpublished.)

$$\beta_{a\nu} = 180w \quad \{\text{cm}^{-1}\}$$
$$\text{and } \kappa_{a\nu} = d\beta_{a\nu}/dw = 180 \quad \{\text{cm}^2\ \text{g}^{-1}\} \qquad [8.16]$$

This value of mass absorption coefficient should be used with caution as it is based on only a few measurements. However, it is the best figure available at the present time.

5. EMISSIVITIES OF CLOUDS

In the interests of simplicity, it is convenient in operational models to assume clouds to be grey bodies whose transmittance or emissivity is constant over the entire long-wave spectrum. This is identical in concept to the "emissivity approximation" for gases discussed in Chapter 7. The question arises as to the relation between the "window emissivity" ϵ_W discussed in the previous section and the total emissivity ϵ_T presented to the total long-wave field.

Imagine a roughly isothermal cloud of depth Z. The total downward flux out of the bottom of the cloud is given by:

$$F{\downarrow}(Z) = F{\downarrow}(0)[1 - \epsilon_T] + \epsilon_T \sigma T^4 \qquad [8.17]$$

where ϵ_T is the grey-body flux emissivity (approximately 1.6 times the emissivity of the vertical column), T is the temperature of the medium, σ is the Stefan-Boltzmann constant, and $F{\downarrow}(0)$ is the downward flux at the cloud top. Thus:

$$\epsilon_T = \frac{F\downarrow(Z) - F\downarrow(0)}{\sigma T^4 - F\downarrow(0)} \quad [8.18]$$

a relation which in effect defines the grey-body emissivity of the cloud. (It may not be a unique definition because the same cloud in a different radiation field might yield a slightly different value of ϵ_T — for instance, the same cloud referred to the upward radiation stream.)

Consider another extreme view of atmosphere–cloud radiative exchange, which involves the simplification of an idea originated by Simpson (1928). That is, regard the atmosphere as opaque to long-wave radiation at all wavelengths other than those in the atmospheric window ($\simeq$8–14 μm), where water vapour and CO_2 absorption is assumed to be negligible. On this model the downward flux $F\downarrow(0)$ at cloud top is confined to wavelengths outside the window region. Furthermore, at these optically black wavelengths the net radiative exchange is zero both outside and inside the cloud (provided the temperature profile is isothermal), and the downward flux is constant with position in the cloud. In the window region, the downward flux increases with depth from zero at cloud top to a total value depending on the window flux emissivity ϵ_W of the liquid water of the cloud. Mathematically, the total downward flux at the cloud base is given by:

$$F\downarrow(Z) = F\downarrow(0) + f\epsilon_W \sigma T^4 \quad [8.19]$$

where f is the fraction of total black-body radiation contained within the wavelength limits of the window region. That is $f = [\sigma T^4 - F\downarrow(0)]/\sigma T^4$. Thus:

$$\epsilon_W = \frac{F\downarrow(Z) - F\downarrow(0)}{\sigma T^4 - F\downarrow(0)} \quad [8.20]$$

which is an expression for window emissivity exactly the same as that derived for total emissivity on the basis of the grey-body assumption.

The above simplification of the process of radiative exchange is known to be reasonably accurate at levels near the surface, where water vapour mixing ratios are large (Swinbank, 1963). It is obviously much less appropriate in the upper levels, where mixing ratios are small, and partially black wavelengths are more significant. However, to the extent that the model is valid when it is applied to reasonably low-level cloud and its immediate surroundings, the above suggests that the total emissivity ϵ_T is dominated by the window emissivity ϵ_W. In any event the window region lies toward the peak of the black-body infrared spectrum at terrestrial temperatures, and ϵ_T is already heavily weighted by these wavelengths. Thus *because* of the considerable water vapour absorption at wavelengths outside 8–14 μm the two emissivities should at least be of the same order and have much the same form of dependence on liquid water path. In operational models therefore, it would seem legitimate to approximate low-level clouds as grey bodies of an emissivity equal to the window emissivity. It is not to be expected that the approximation would be valid for very high clouds such as cirrus.

The major lesson from the above exercise is that the effective emissivity of an atmospheric layer — the grey-body emissivity which *should* be used in an operational model — depends on the impinging spectrum. One can run into large problems when using the standard emissivity approximation in cloudy atmospheres. Referring back to equation [7.54] which is the basis of this approximation, the emissivity of the atmospheric layer between z_1 and z_2, say, is defined as:

$$\epsilon(u, T) = (1/\sigma T^4) \int_0^\infty [1 - \widetilde{Tr}_\nu(u, T)]\widetilde{B}_\nu(T)\mathrm{d}\nu$$

Mathematically this is only one definition of a mean emissivity appropriate to the whole spectrum, and is quite independent of the *impinging* spectrum. In cloudy atmospheres especially, the emissivity of a layer so defined may bear little relation to the emissivity which should be used in a grey-body model. This is true even when overlap between the absorption bands of water vapour and those of liquid water is taken into account in the integration of the above equation (that is, $\widetilde{Tr}_\nu(u, T) = \widetilde{Tr}_\nu(u, T)_{\mathrm{vapour}} \times \widetilde{Tr}_\nu(u, T)_{\mathrm{liquid}}$).

Thus as a general rule in cloudy atmospheres, fully fledged multispectral band models should be employed — if not on an operational basis, at least to the extent of calculating an *effective* grey-body emissivity which can then be used in the emissivity approximation.

In most operational models to date, it is assumed for simplicity that all clouds (except perhaps cirrus) are sufficiently thick and dense for them to be optically black — i.e. that their long-wave emissivity is 1.0. This assumption is questionable to say the least. Certain common types of stratocumulus clouds do not become black until they are about 500 m deep (Paltridge, 1974a; Platt, 1972). Middle-level and cirrus clouds can exist in depths of several kilometres without being black. Taking Fig. 8.8 as a guide, the vertical column of the cloud must contain at least 3 or 4 by 10^{-3} g cm^{-2} before its emissivity approaches unity. This condition is increasingly unlikely to exist the higher the cloud is in the atmosphere. As a very rough guide to the emissivities of clouds, the data given earlier in Table 6.3 are offered (see also Table 8.1).

Bearing in mind that the effective total emissivity of a cloud is dependent on the impinging spectrum, it may be appropriate in operational models to assign different values to the cloud emissivity when dealing with the upward and downward radiation streams. The technique bears some relation to that of Rodgers (1967a) who investigated the double-valued emissivity concept as it applies to clear-sky layers. The subject needs detailed examination.

The internal structures of some middle-level clouds as observed by laser

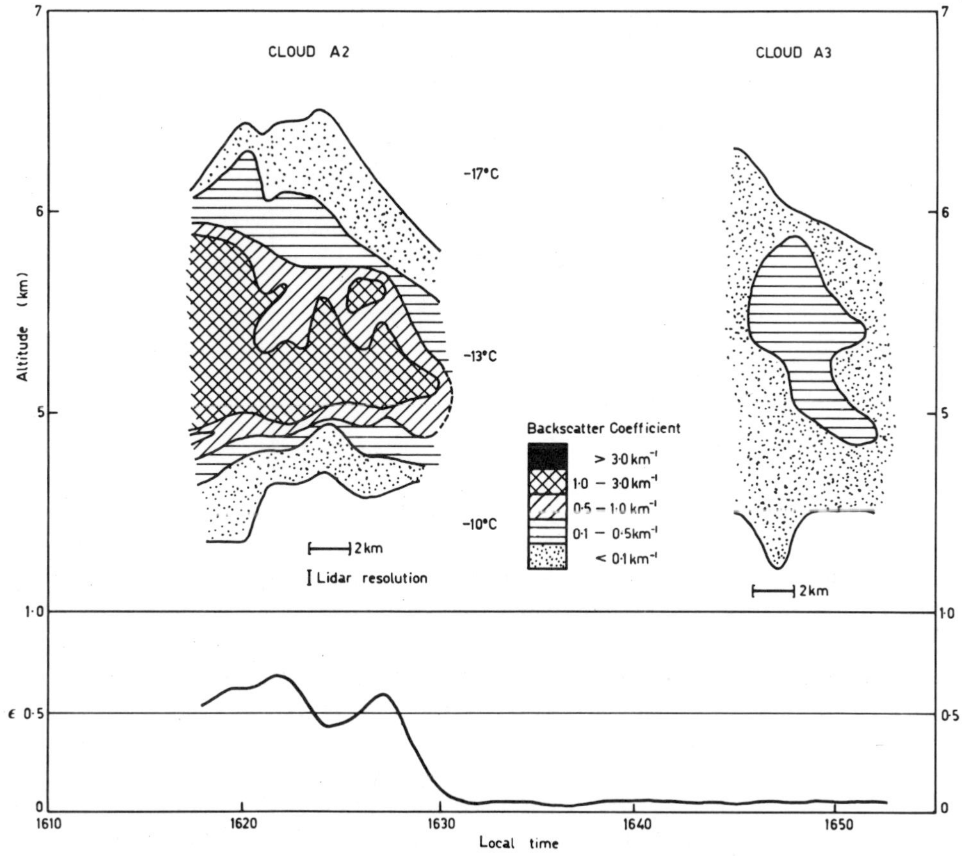

Fig. 8.9. The internal structures of middle-level altocumulus clouds as seen by laser back-scatter (lidar). The vertical emissivity (10–12 μm wavelength) of the clouds (ϵ) is shown below. (After Platt and Bartusek, 1974.)

back-scatter (lidar) measurements, together with infrared emissivities are shown in Fig. 8.9. They illustrate the difficulty of modelling the emissivities of clouds solely in terms of their thicknesses. Similar large variations in internal cloud structures of cirrus clouds have also been observed by Platt (1973) and Heymsfield and Knollenberg (1972). Direct measurements of cirrus cloud emissivities and absorption coefficients have been made by Kuhn and Weickmann (1969), Davis (1970), Allen (1971) and Cox (1971); and of aircraft contrail emissivities by Kuhn (1970). Their results, together with those obtained by one of the present authors, are shown in Table 8.1. The great range in emissivities of middle-level clouds is apparent. The figures obtained by Allen, whilst not containing cloud depth information, were nevertheless averaged over very many clouds and are therefore probably the most representative figures which are available. A frequency histogram for the *zenith* emissivity of jet-stream cirrus is shown in Fig. 8.10.

TABLE 8.1

Experimental absorption coefficients of clouds obtained assuming the absorption approximation and isotropic clouds. The "zenith" emissivity for clouds 1 km deep is also shown. The emissivity values of Allen (1971) are averages over many observations but contain no information on cloud depth

Cloud type	$\beta_{a\nu}$(km^{-1})	ϵ(1 km)	Spectral region (μm)	Researcher
Stratus and stratocumulus	10–50	≈ 1	5.0–35.0	Kuhn et al. (1974)
Stratus and stratocumulus	4–15	≈ 1	10–12	C.M.R. Platt (unpublished)
Altostratus and altocumulus	0.02–11	0.02–1	10–12	Platt and Bartusek (1974)
Cirrus and cirrostratus	0.19–0.66	0.17–0.48	10–12	Platt (1973)
Cirrus and cirrostratus	0.14	0.13	10–12	Kuhn and Weickmann (1969)
Cirrus and cirrostratus	0.54	0.42	10–12	Davis (1971)
Middle-level clouds		0.79[1]	8–13	Allen (1971)
Cirrus clouds		0.35[1]	8–13	Allen (1971)

[1] Mean values — no information on cloud depth.

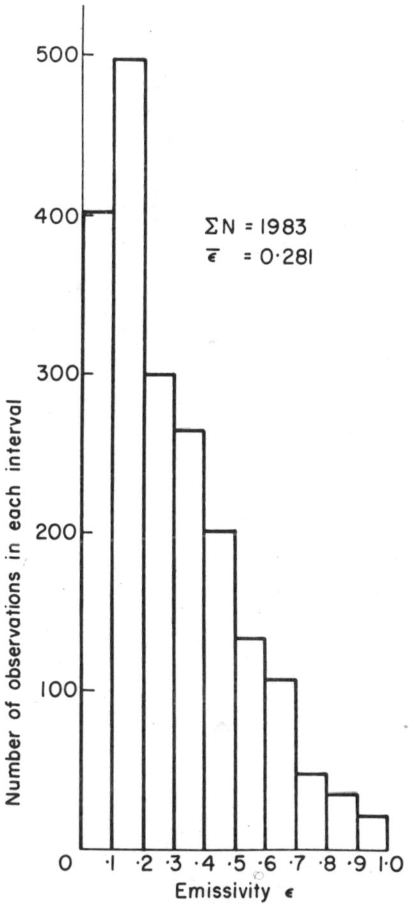

Fig. 8.10. Histogram of the frequency of emissivity observations in a cirrus deck observed by satellite Nimbus 4 versus the cirrus emissivity. (After Platt, 1975.)

6. EFFECTS OF CLOUDS ON ATMOSPHERIC COOLING

Low stratus and cumulus water clouds of typical thicknesses have emissivities which are close to unity. With respect to their effects on the atmospheric cooling rates above and below them they can be regarded as black objects radiating at cloud temperature. The radiation is thus "decoupled" above and below the cloud and the equations of radiative cooling can be integrated separately with appropriate boundary values of the Planck function $B(T)$. However, the actual magnitude of the cooling near the cloud boundaries, which may be important for the cloud dynamics (Chapter 10) is a function of the cloud absorption coefficient and thickness.

A useful procedure for non-black clouds is to consider the cloud to be

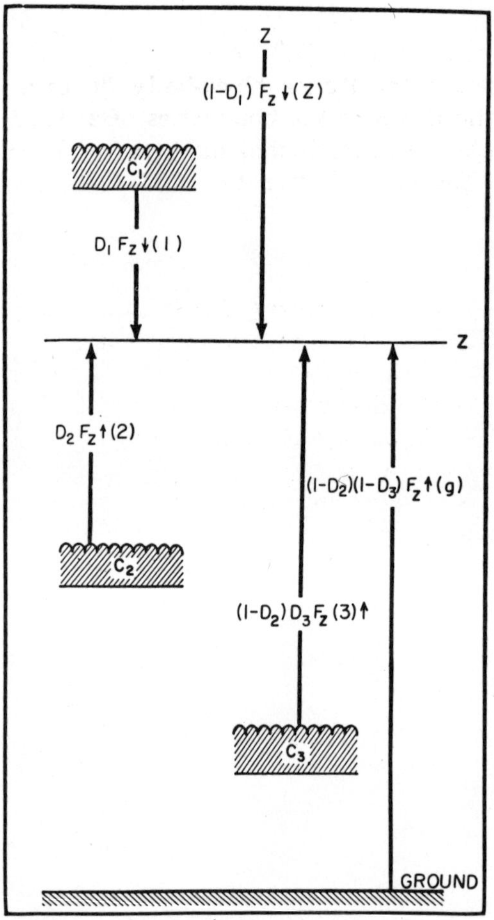

Fig. 8.11. Long-wave radiation flux densities at a height z in an atmosphere with clouds. (After Manabe and Strickler, 1964.)

black but to reduce the fractional cloud cover by the ratio ϵ. This interchange of fractional cloud cover and emissivity is possible only under the assumption of "grey" clouds of a single wavelength-independent emissivity. The method has been employed by Manabe and his co-workers (Manabe and Möller, 1961; Manabe and Strickler, 1964). Their scheme is shown in Fig. 8.11. The flux densities at a level z (using Manabe and Strickler's notation) are then:

$$F_z\downarrow = D_1 F_z\downarrow(1) + (1 - D_1) F_z\downarrow(Z) \qquad [8.21]$$

$$F_z\uparrow = D_2 F_z\uparrow(2) + D_3(1 - D_2) F_z\uparrow(3) + (1 - D_2)(1 - D_3) F_z(g) \qquad [8.22]$$

where terms such as $F_z\downarrow(1)$, etc., represent the flux density originating from the atmosphere between z and the cloud of amount D_1, etc., and from the cloud itself, and are calculated from equations [7.11] and [7.12]. The boundary flux densities in each case are the Planck black-body fluxes at cloud temperature. Fig. 8.12 shows the fluxes at the boundaries of a cloud layer of emissivity ϵ assuming that reflectance Re is zero and that the cloud is isothermal with a temperature T. The net radiation loss F_{NC} from the *cloud* is given by:

$$(F_{NC})_\nu = \epsilon_\nu [F_\nu\uparrow + F_\nu\downarrow - 2B_\nu(T)] \qquad [8.23]$$

The cooling of the whole cloud layer, which is proportional to F_{NC}, is thus a linear function of the cloud layer emissivity. The spectral emissivity of the cloud contains the emissivity of the water vapour and the CO_2 in the cloud layer. (Gaseous absorption is strongest in the H_2O rotation and 6.3-μm bands and CO_2 absorption is confined to the 15-μm band.) Thus when the cloud emissivity approaches zero, ϵ_ν approaches the gaseous value. For climatic models, the above treatment is presumably quite adequate. One restriction is that flux differencing across layers must be used to compute the cooling and that a cloud must completely fill one or more layers of a specified thickness.

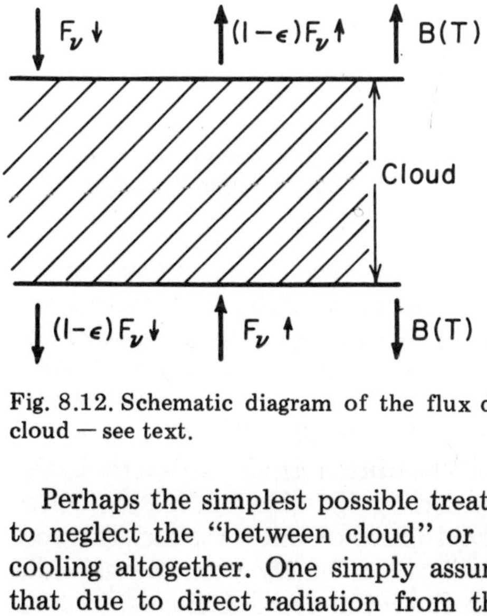

Fig. 8.12. Schematic diagram of the flux densities at the boundaries of a non-reflecting cloud — see text.

Perhaps the simplest possible treatment of clouds in operational models is to neglect the "between cloud" or "between cloud and ground" radiative cooling altogether. One simply assumes that the only significant cooling is that due to direct radiation from those cloud tops which are exposed to space. The approximation relies on the relatively small range of tropospheric temperatures (on an absolute scale), so that the air between cloud and ground or between cloud and cloud is close to being the classic black-body enclosure of no net radiative loss. The method would be compatible with an operational model which made use of the "cooling to space" approximation in clear skies.

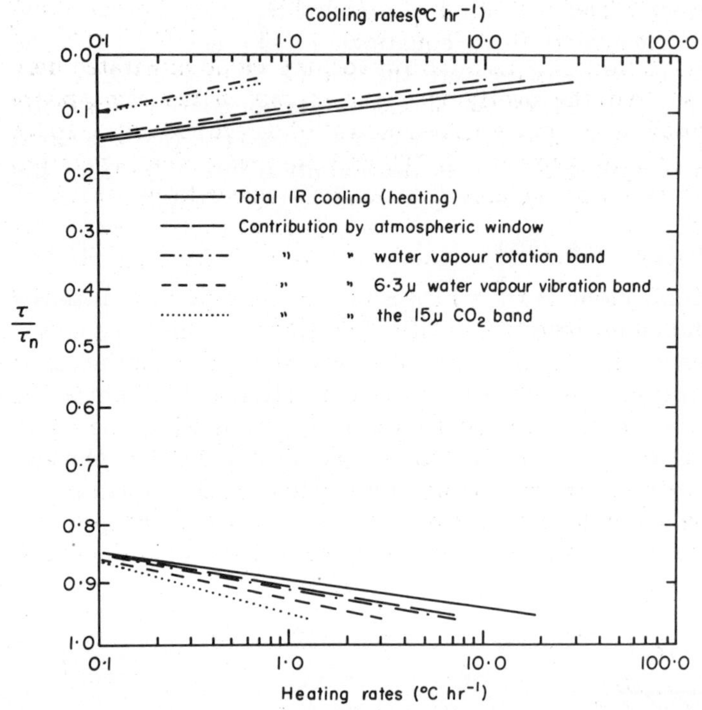

Fig. 8.13. Cooling rate profiles for various spectral regions through an isotropic cloud at approximately 4 km altitude for a standard atmosphere with surface temperature equal to 22°C. Cloud thickness = 1.6 km, extinction coefficient (10 μm) = 27 km^{-1}, $\bar{\omega}_0$(10 μm) = 0.57. (From G.L. Stephens, 1975, unpublished.)

When the interest is in the dynamics of cloud itself, it is necessary to calculate the detailed radiative cooling profile within the cloud. In principle one can do this sort of calculation with almost infinite precision provided the cloud microphysics is specified in sufficient detail. Yamamoto et al. (1970), for instance, have obtained detailed solutions of the transfer equation for three types of isotropic clouds placed at 4 km altitude in a standard atmosphere. They included water vapour absorption as an extra term in the single scattering albedo $\bar{\omega}_0 (= \beta_{sc}/\beta_e)$ by setting $\beta_e = \beta_{sc} + \beta_a + \beta_{wv}$, where β_{wv} is the effective volume absorption coefficient of water vapour. This sort of calculation is extremely valuable, but is still a far cry from the ultimate goal of parameterizing the cooling rate profile within realistic clouds — which are normally highly irregular and far from isotropic. Test-bed modelling very similar to Yamamoto's has been done by G.L. Stephens (1975, unpublished) who used the adding method to produce the cooling rate profiles of Fig. 8.13. Depth in the cloud is indicated in terms of normalized optical depth, and the various profiles indicate the cooling (or heating towards the cloud base) attributable to various spectral intervals. It is not surprising that cooling within the atmospheric window is a dominating term. It *is* somewhat surprising that cooling within the spectral region of the rotation band of water vapour is also a dominating term. However, the cloud illustrated here is at 4 km altitude so that the overlying water vapour in the atmosphere above the cloud is small, and a fairly large amount of direct cooling to space is possible. The cloud temperature is −10°C, and the surface temperature is 22°C. Both Kondratyev's (1972) and Yamamoto's calculations on cloud fluxes show the large contribution to cooling by the rotation band. Undoubtedly the rotational cooling by cirrus clouds is also highly significant, but it has not been investigated in detail.

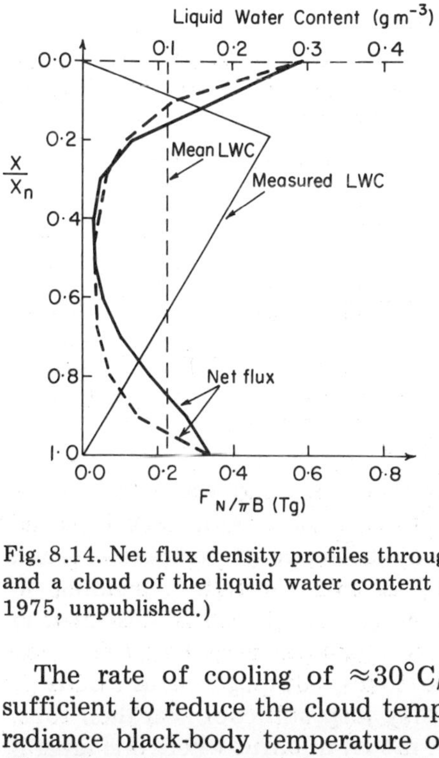

Fig. 8.14. Net flux density profiles through an homogeneous cloud (dashed curves refer) and a cloud of the liquid water content (LWC) profile as shown. (From G.L. Stephens, 1975, unpublished.)

The rate of cooling of ≈30°C/hour at the cloud top of Fig. 8.13 is sufficient to reduce the cloud temperature to that equivalent to downward radiance black-body temperature of 213 K within 100 minutes. Such large discontinuities at the cloud top never occur in practice and other heat transfer processes such as condensation, evaporation and turbulence come into play. It is obvious that radiation loss from a cloud must have a profound influence on the cloud dynamics (see Chapter 10.) While on the subject, it is interesting to note the difference between an homogeneous cloud and a cloud with a more realistic profile of liquid water content. Thus Fig. 8.14 (again from Stephens) shows the net flux profile calculated for a cloud with an experimentally measured liquid water profile, and also the net flux profile calculated for an homogeneous cloud of the same vertical dimension and the same total liquid water. Bearing in mind that cooling rate is proportional to the gradient in net flux density, the cooling rates in a realistic cloud are much less than those which might be calculated for an ideal homogeneous cloud.

No direct comparisons have yet been made of radiative cooling when scattering is neglected. However, the net loss of radiant energy from a layer of cloud of emissivity ϵ is roughly proportional to ϵ (cf. equation [8.23]) and thus the error in cooling from use of the absorption approximation will be roughly proportional to the error in ϵ. This was shown in section 3 to be about 10% in the worst case.

TABLE 8.2

Atmospheric cooling rates in the individual layers of a two-layer model of cirrus cloud and for two cloud flux emissivities. ϵ is here defined by $(1 - \tilde{Tr})$ where $\tilde{Tr}$ is the *cloud droplet* flux transmittance and is independent of wavenumber

Latitude	Altitude (km)	Cooling rate[1] (°C/hour)	
		$\epsilon = 0.44$	$\epsilon = 0.98$
0–10°	14.8–15.4	−0.98	−5.2
	15.4–16.0	−0.66	+1.5
30–40°	9.7–10.3	−0.15	−1.6
	10.3–10.9	−0.08	+1.0
60–70°	6.6–7.2	+0.11	−0.35
	7.2–7.9	+0.24	+1.05

[1] Negative sign indicates warming

Middle- and low-level clouds cool near their tops and warm to a lesser extent near their bases so that there is a net loss of heat. As a cloud is placed higher in the atmosphere, the discrepancy between its temperature and surface temperature becomes greater and at some altitude depending on latitude the cloud experiences a net warming. This change is illustrated in Fig. 10.9 of Chapter 10 where the net loss or gain of energy by a thin black cloud is shown for three different latitude zones. Although cirrus clouds (if high enough) experience a net warming, they may still cool near their tops. This can be demonstrated with a two-layer model of cirrus where one layer is adjacent to the next and the cooling is obtained by calculating the difference in net fluxes between adjacent layers. The results of such a computation using a Rodgers and Walshaw multispectral model are shown in Table 8.2. The clouds were treated as grey bodies with the same emissivity at each wavelength. Water vapour and CO_2 absorption were taken into account by multiplying the gas transmittance by the cloud transmittance in each spectral interval. When the total cloud emissivity is 0.44, both layers either warm or cool depending on cloud temperature. When ϵ approaches unity, the situation is similar to lower clouds but with the warming near the base usually being the dominant energy change. The cooling rates shown in Table 8.2 are averages over the layer depths, so that the actual temperature changes near to the cloud tops or bases will be greater than those shown.

7. COMPARISON OF ATMOSPHERIC COOLING IN CLEAR SKIES AND IN A CLOUDY ATMOSPHERE

The general form of clear sky atmospheric cooling in the troposphere was discussed in Chapter 7. The cooling is roughly proportional to the ambient temperature. Introduction of zonal average cloudiness does not much change the seasonal mean latitude-height distribution of infrared cooling (London, 1952). The main effects are a decrease in radiative cooling near the surface and a more sharply defined maximum in cooling in the middle troposphere. This additional cooling is at about the height of typical middle-level clouds and the overall cooling is then no longer fairly proportional to ambient temperature.

It is interesting to compare the effects of changes in both water vapour and cloud distributions with altitude on the *average* infrared cooling rates of the two layers of atmosphere above and below 500 mb. Fig. 8.15a illustrates the average cooling rates above and below 500 mb for a clear sky corresponding to the artificial water vapour distributions illustrated in Fig. 8.15b. The total water vapour content of the atmosphere is more or less the same from one distribution to another. The temperature profile chosen was a typical mid-latitude case. Fig. 8.16 shows what happens when 5/8 of cloud is placed in a layer about 25 mb in depth at the pressure levels indicated. This model uses the same temperature profile and an average mid-latitude water vapour distribution. The much greater effect on cooling of changing the

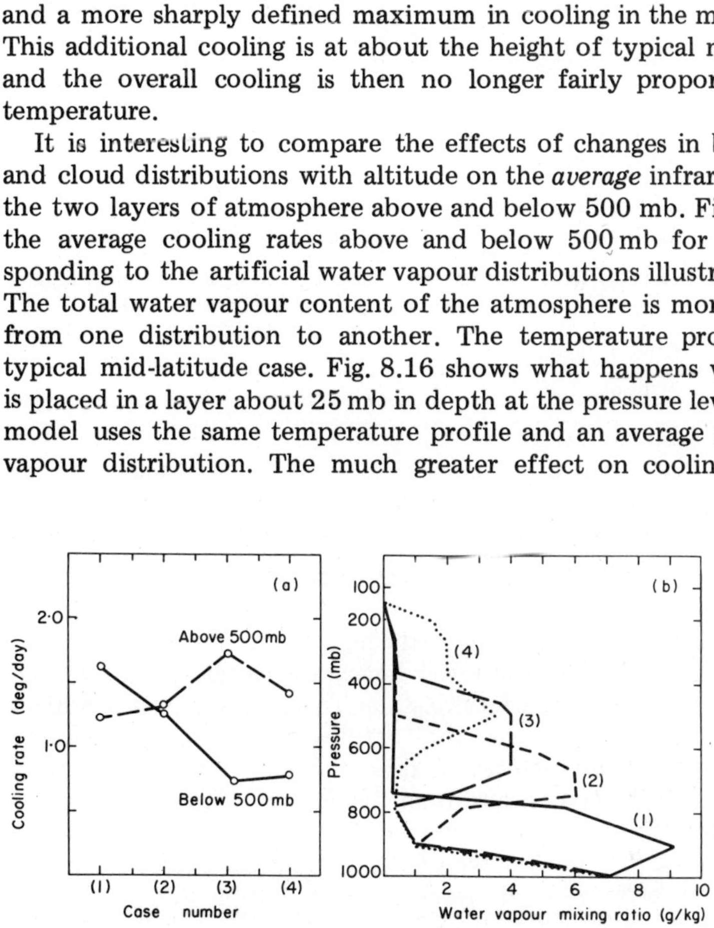

Fig. 8.15. Average atmospheric cooling per day above and below 500 mb pressure altitude for the four (artificial) water vapour mixing ratio profiles shown in Fig. 8.15b.

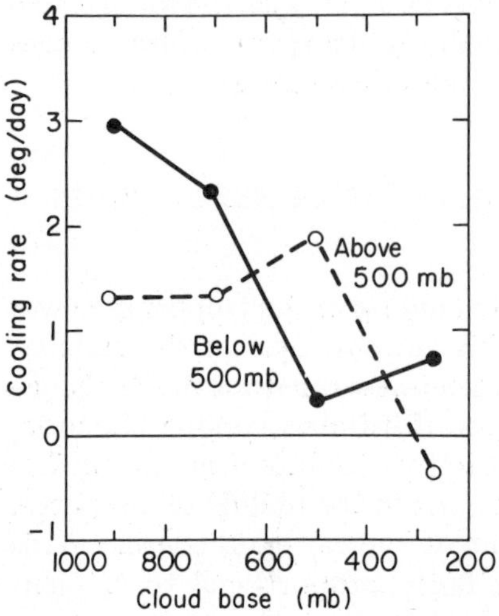

Fig. 8.16. Average atmospheric cooling per day above and below 500 mb pressure altitude for thin black clouds at the bases shown.

TABLE 8.3

Percentage of the total atmospheric cooling occuring in various atmospheric layers, assuming no cooling below the lowest cloud layer (Petrova and Feygel'son, 1969)

	Kind of layers		
	single-layered	double-layered	triple-layered
Between clouds	0	3	4
Cloud	43	62	80
Above clouds	57	35	16

cloud distribution compared to changing the water vapour distribution is very apparent. Cloudiness does not greatly affect the atmospheric cooling of the upper layer until the cloud is sufficiently high for its warming to affect the energy loss significantly. For the lower layer, low clouds almost double the cooling rate, but high clouds effectively suppress its cooling. It is noticeable that high clouds reduce the total atmospheric average cooling. Physically, this occurs because high clouds form a tropospheric "lid" cooling to space at a much lower temperature than the surface. The percentage of the total atmospheric cooling which occurs *within* clouds is a function of the number of cloud layers. The percentage of the cooling which occurs inside, between and above up to three cloud layers is shown in Table 8.3.

CHAPTER 9

Atmospheric Aerosols

1. INTRODUCTION

Aerosols are present throughout the atmosphere. As the source is the ground with a few exceptions such as man-induced aerosols from aircraft, their numbers (at least in the troposphere) decrease rapidly and almost exponentially from the surface upwards. This was illustrated in Fig. 6.6 where a "standard" aerosol height distribution was shown. Aerosols increase again in the stratosphere to form a layer which is largely decoupled from tropospheric turbulence by the tropopause. The nature, size distribution, refractive index and size-humidity relationships of atmospheric aerosols are summarized in the Appendix.

Aerosols are important in the meteorological context because they absorb and scatter both solar and infrared radiation. The magnitude of their infrared effect is generally much smaller than that of clouds because of their much smaller optical depths. This is due to their small particle sizes and cross-sections compared to water droplets, and to the fact that in the long-wave spectrum their size parameters, and thus extinction efficiencies, are much less than unity. Their extinction optical depth in the visible region is about 10 times the optical depth in the long-wave region but is still fairly small, lying in the range of 0.02–0.2. However, the effects of aerosols on the heat budget of the earth-atmosphere system can be very significant.

Because it is a simple measurement to make in principle, the reduction by aerosols of the direct solar beam irradiance at the ground has been studied extensively. The quantity of interest is the total aerosol optical depth in the vertical (at specific short wavelengths) or *turbidity*, and this is used generally as an index of the dust loading of the atmosphere. However, the turbidity gives only the total extinction of the solar beam and does not give the component absorbed. Nor can it give the distribution of extinction with height or even the ratio of solar radiation scattered to that which is scattered back to space. These are the quantities which are needed to determine the aerosol atmospheric effects. In the past, perhaps too much emphasis has been placed on the simple turbidity concept although it remains as a useful index of changes in the total atmospheric optical depth. Spectral turbidity measurements do go some way towards characterizing a given aerosol population. The problem which faces meteorologists is that aerosol absorption and aerosol backscatter to total scatter ratios are notoriously hard to measure in the real atmosphere. The turbidity measurement is therefore an attractive (if rather inaccurate) alternative for simple aerosol indexing. To quote one example, the five-year turbidity survey of the United States by Flowers et al. (1969) was invaluable in pointing out the broad distribution of aerosol extinction over that continent and highlighting the high turbidity in the eastern industrialized states. However, turbidities of the atmosphere above, say, remote oceanic locations are small and the solar extinction method has not at all a suitable accuracy at these low levels of turbidity. As it is just this "background" aerosol which is now considered to be an index of the general average rise of polluting materials in the atmosphere, more accurate methods seem appropriate. One such method, employing the scattered solar intensity at selected scattering angles, is mentioned briefly in section 4.

2. THE EXTINCTION OF SHORT-WAVE SOLAR RADIATION

At a given wavelength λ the direct beam of the sun is depleted on its path down through the atmosphere according to equation [6.4] of Chapter 6. The equation involves the total vertical extinction optical depth of the atmosphere due to aerosols $\tau_{D\lambda}$ which in turn is given by:

$$\tau_{D\lambda} = \int_0^\infty \beta_{e\lambda} \mathrm{d}h \qquad [9.1]$$

where $\beta_{e\lambda}$ is the volume extinction coefficient of the aerosols. $\beta_{e\lambda}$ depends markedly on the particle size *distribution* which may vary with height. However, this possibility is usually ignored in order that the wavelength dependence of the measured values of $\tau_{D\lambda}$ can be equated directly to that of the theoretically calculated $\beta_{e\lambda}$ of some single standard aerosol appropriate to the region.

We consider first a derivation of the spectral variation of $\beta_{e\lambda}$ for a Junge size distribution which indicates a simple dependence of $\beta_{e\lambda}$ on λ. For this purpose we follow Junge (1963) and Bullrich (1964). The size distribution of particles is described by the number density $n(r)$ of particles per unit radius interval at radius r. In terms of the single particle extinction efficiency (the extinction cross-section over the geometrical cross-section):

$$\beta_{e\lambda} = \int_0^\infty \pi r^2 Q_{e\lambda}(r)\, n(r)\, \mathrm{d}r$$

$$= \int_0^\infty \pi r^2 Q_{e\lambda}(r)\, \mathrm{d}n(r) \qquad [9.2]$$

The extinction coefficient $\beta_{e\lambda}(r_1, r_2)$ attributable to particles between radii

r_1 and r_2 is then:

$$\beta_{e\lambda}(r_1, r_2) = \int_{r_1}^{r_2} \pi r^2 Q_{e\lambda}(r)[\mathrm{d}n(r)/\mathrm{d}r]\mathrm{d}r \qquad [9.3]$$

Now, for a Junge size distribution (equation [A.2] of the Appendix):

$$\mathrm{d}n(r)/\mathrm{d}r = (c_1/2.3)\, r^{-(V+1)} \qquad [9.4]$$

Substituting equation [9.4] into [9.3]:

$$\beta_{e\lambda}(r_1, r_2) = (\pi c_1/2.3) \int_{r_1}^{r_2} Q_{e\lambda}(r)\, r^{1-V}\, \mathrm{d}r \qquad [9.5]$$

We now change the variable to the size parameter ξ $(= 2\pi r/\lambda)$ and with some rearrangement obtain:

$$\beta_{e\lambda}(r_1, r_2) = (\pi c_1/2.3)(\lambda/2\pi)^{2-V} \int_{\xi_1}^{\xi_2} Q_e(\xi)\xi^{1-V}\mathrm{d}\xi \qquad [9.6]$$

The Junge size distribution is an idealized case of a true atmospheric aerosol distribution and the linear slope of the $\mathrm{d}n(r)/\mathrm{d}r$ curve actually occurs over a restricted range of radius values (say, between r_1 and r_2). However, it was argued by Junge that the "straight" part of the distribution occurred over just that region which is most effective optically. For radii smaller than r_1, particles are too small, and for radii greater than r_2, particles are too few to contribute significantly to the extinction. Then the limits of the integrals in equations [9.5] and [9.6] can be changed to zero and infinity. Now the dependence of $Q_e(\xi)$ on size parameter ξ is a function only of the refractive index of the aerosol particles. If this refractive index is assumed to be constant over the spectral region of interest, then the integral in equation [9.6], when taken between the bounds of zero and infinity is some constant depending on particle refractive index, but *not* on wavelength. There is some restriction on the value of V for which the integral in equation [9.6] converges. The restriction is that V must be greater than 2.

As the refractive index of aerosol particles is known to be fairly constant with wavelength in the visible region then with the above restrictions:

$$\beta_{e\lambda} = \text{const.}\, \lambda^{2-V} \qquad [9.7]$$

A typical value for V is 3, corresponding to a $\beta_{e\lambda}$ proportional to λ^{-1}. This agrees fairly well with Ångström's (1964) finding from turbidity measurements which were averaged over many situations that the mean wavelength dependence was about equal to λ^{-1}. That is, the extinction decreases with increasing wavelength.

Equation [9.7] holds only to the extent that the aerosol number distribution $\mathrm{d}n(r)/\mathrm{d}r$ or $\mathrm{d}n(r)/\mathrm{d}(\log r)$ is "straight" over that range of values of r which contribute significantly to the extinction coefficient. Table 9.1

TABLE 9.1

Percentage of the total extinction occurring in different radius intervals at two wavelengths for a Junge non-absorbing aerosol distribution ($V = 3$)

$\lambda(\mu m)$	Radius interval (μm)		
	<0.1	0.1–1.0	>1.0
0.40	14.0	82.6	3.4
0.60	4.4	90.6	5.0

illustrates the percentage contribution to the total extinction coefficient of a non-absorbing aerosol, in different radius intervals, for two wavelengths and a value of 3 for the exponent V in equation [9.7]. Departures from a Junge distribution outside the range, say, from 0.1 to 1 μm radius are obviously not too significant.

The Junge aerosol distribution may not be appropriate in oceanic areas or even continental areas remote from local sources of "pollution". Distributions typical of such areas are shown in Figs. A.1 to A.3 of the Appendix. In Fig. A.1 the aerosol particles with radii less than 0.01 μm do not contribute significantly to the extinction and the size distribution for optical effects can be represented by a single log-normal distribution or a gamma distribution (Table A.1 of the Appendix). We choose the standard gamma distribution to illustrate its extinction characteristics. Fig. 4.3 of Chapter 4 shows the volume scattering efficiencies as defined by equation [4.41], plotted against effective size parameter ξ. The parameter b is a measure of the variance of the size distribution about the effective radius. Fig. 4.3 thus shows the spectral extinction (absorption is zero) for different size distributions, but with $\lambda = (2\pi r_e/\xi)$ increasing from right to left. It indicates that if the effective radius r_e of the gamma distribution is greater than about 0.4 μm, then the extinction coefficient slowly *increases* with wavelength over the visible spectrum. If r_e is somewhat less than about 0.1 μm the extinction will show a peak at some wavelength. An increase of the extinction coefficient with wavelength is shown very well by the calculations of De Luisi et al. (1972) who calculated values of $\beta_{e\lambda}$ for the measured size distributions shown in Fig. A.4. These authors generalized their distributions somewhat by fitting them with Diermendjian-type gamma distributions. Their computed values are shown in Fig. 9.1. In all cases, $\beta_{e\lambda}$ increases slowly with λ.

This type of spectral extinction is termed "anomalous" because if equation [9.7] is applied to spectral measurements made at two wavelengths the exponent in the equation is negative. In fact equation [9.7] should not be applied at all. Porch et al. (1973) made a very useful investigation of the conditions under which anomalous extinction can occur. They defined an

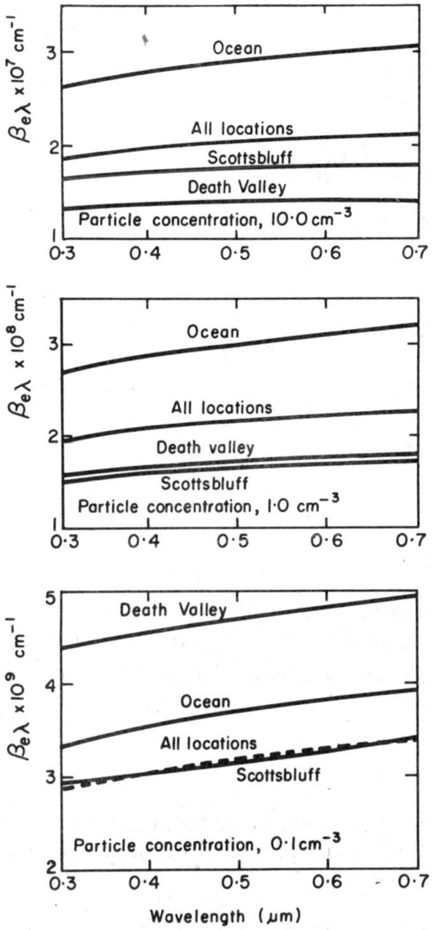

Fig. 9.1. Mie extinction coefficients ($\beta_{e\lambda}$) as a function of wavelength for the size distribution curves of Fig. A.4. Refractive index of 1.4–0.07i. (After De Luisi et al., 1972.)

exponent $\hat{\alpha}_{BV}$ by:

$$\hat{\alpha}_{BV} = \log(\beta_{eV} - \beta_{eB})/\log(\lambda_B - \lambda_V) \quad [9.8]$$

where β_{eV} and β_{eB} are aerosol extinction coefficients at wavelengths of 0.440 and 0.537 μm respectively. They considered lognormal size distributions and their computed values of $\hat{\alpha}_{BV}$ plotted against geometric number mean radius r_g are shown in Fig. 9.2. σ_g is the geometric standard deviation analogous to b in Fig. 4.3. These curves show that for "anomalous" extinction to occur, r_g must be about 0.3 μm or greater and σ_g must be less than about 1.5, which is consistent with the curves in Fig. 4.3. If the imaginary part of the refractive index n_i is greater than zero, then anomalous extinction becomes less likely. This is because the peak in scattering efficiency

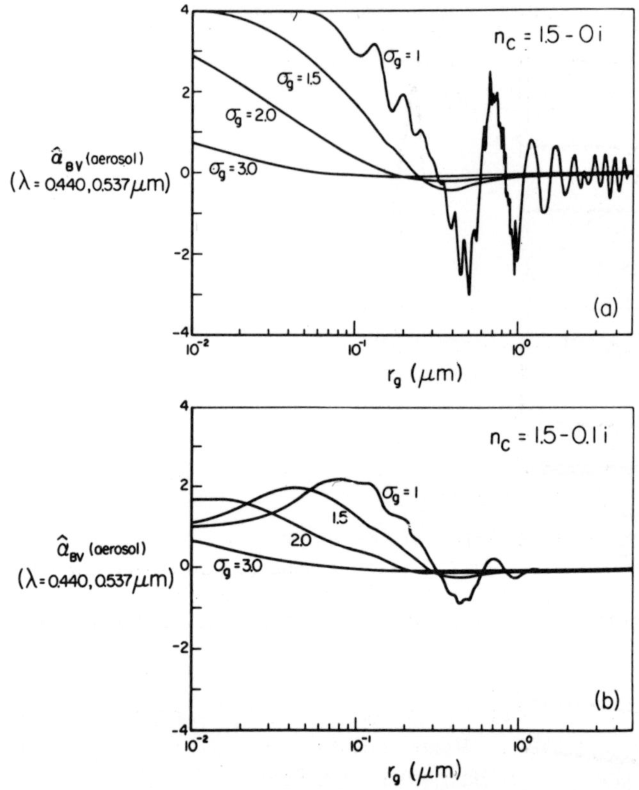

Fig. 9.2. Wavelength dependence of Junge exponent $\hat{\alpha}_{BV}$ versus the mean radius of aerosol particles (r_g) for various geometric standard deviations in size (σ_g) for a lognormal size distribution of absorbing particles. (After Porch et al., 1973.)

at a certain size parameter becomes progressively more suppressed as n_i increases.

Measurements of atmospheric extinction of the solar beam by aerosols have been made for many years. As early as 1939 three broad types of spectral extinction were recognized by Ångström (1964): (1) roughly neutral extinction (i.e. more or less independent of wavelength), (2) extinction maximal at 0.5 μm ("anomalous"), and (3) extinction gradually decreasing in the 0.4–0.6-μm wavelength interval (i.e. appropriate to a Junge distribution).

Recent spectral extinction measurements showing a Junge behaviour have been reported by Shaw et al. (1973). Measurements made by Quenzel (1970) in the tropical Atlantic showed the anomalous effect and are illustrated in Fig. 9.3. Quenzel fitted his measurements with curves of extinction coefficients calculated from selected lognormal distributions which are shown on the right-hand side of Fig. 9.3. Solar extinction measurements

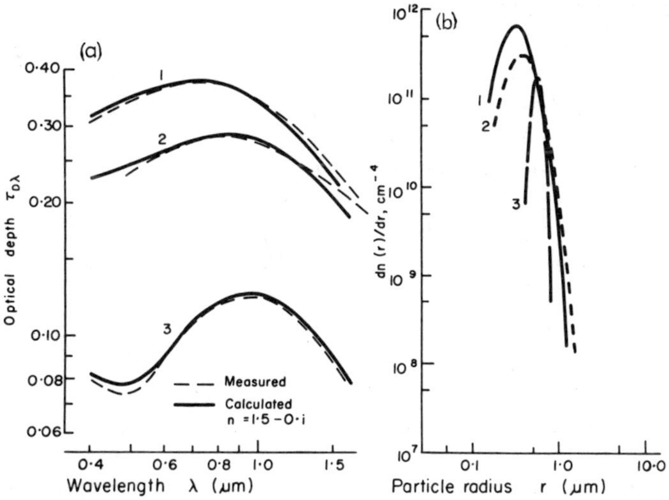

Fig. 9.3. (a) Vertical optical depths (turbidities) of aerosol particles versus wavelength. From on board R.V. "Meteor" in the tropical Atlantic Ocean. (b) Corresponding size distributions of aerosol particles versus particle radius. Results obtained on three different days. Calculated optical depths are from corresponding size distributions. (After Quenzel, 1970.)

made by Volz (1970) in the Caribbean gave a very similar spectral dependence to those of Quenzel. Intermediate behaviour in spectral extinction is illustrated in the data of Barnhardt and Streete (1970).

The effect of humidity on the size of aerosol particles is discussed in the Appendix. In an aerosol distribution of mixed composition the increase in radius with humidity is best represented by an empirical formula. Typically, particles double their radii with increase of humidity from about 10% to 95%. Their cross-sections are thus quadrupled. In addition, the aerosol refractive index changes from about 1.50 for a dry aerosol to 1.33 for a wet aerosol. This change can also be related to humidity by an empirical formula (equation [A.8]). If the size distribution is of the lognormal type with a peak in $n(r)$ at some radius r_p, the effect of droplet growth is to shift r_p to longer wavelengths. Then it could happen that a size distribution which gives a positive value of $\hat{\alpha}_{BV}$ at a low humidity may give a negative value at high humidity. Several authors have investigated the increase in extinction with relative humidity. The experimental measurements of aerosol extinction near the ground made by Fillipov and Mirumyants (1972) and Georgievskiy and Rozenberg (1973) verify broadly the above type of behaviour. Theoretical curves of Gambling (1972) covering the region from 0.5 to 15 μm are shown later in Fig. 9.9. Similar theoretical work has been done by Barnhardt and Streete (1970).

3. ABSORPTION OF SHORT-WAVE RADIATION

Absorption by aerosol particles occurs when the imaginary part of the particle refractive index is non-zero. Normally a certain percentage of aerosols in a given distribution are absorbing and they can increase the total extinction and even change its spectral dependence. The evidence from measurements to date is that the refractive index of absorbing particles is quite spectrally independent in the short-wave region. It is thus possible that the spectral dependence of $\beta_{e\lambda}$ for Junge particle distributions could be given by an equation such as [9.7]. However, it has been shown theoretically that Aitken particles ($r < 0.1\,\mu$m) with a non-zero value of n_i absorb quite effectively, so that it is not obvious that equation [9.7] is a good representation of absorbing aerosols. Furthermore, the absorption properties of aerosol layers are very significant regarding their effects on the global heat budget (see section 6). It is therefore desirable to look more closely at the relation between the aerosol scattering and absorption coefficients and the particle radius.

A useful way to investigate this relation is to consider a hypothetical monodisperse (all particles have the same radius) aerosol layer with a given mass $w(a)$ per unit volume of air and a given density ρ of particle material. Then (see equation [8.13] of Chapter 8, and also Bergstrom, 1973):

$$\beta_{a\lambda}(r) = 3w(a)Q_{a\lambda}(\mathrm{r})/4r\rho \qquad [9.9]$$

A similar expression holds for $\beta_{e\lambda}$ in terms of $Q_{e\lambda}$. It follows that:

$$\mathrm{d}\beta_{a\lambda}(r)/\mathrm{d}w(a) = 3Q_{a\lambda}(r)/4r\rho \qquad [9.10]$$

$$\mathrm{d}\beta_{e\lambda}(r)/\mathrm{d}w(a) = 3Q_{e\lambda}(r)/4r\rho \qquad [9.11]$$

When the size parameter ξ (i.e. $2\pi r/\lambda$) is less than 0.1:

$$Q_{sc\lambda} = (8/3)(2\pi r/\lambda)^4\ \mathrm{Real}[(n_c^2-1)/(n_c^2+2)]^2 \qquad [9.12]$$

$$Q_{a\lambda} = -4(2\pi r/\lambda)\ \mathrm{Imag}[(n_c^2-1)/(n_c^2+2)] \qquad [9.13]$$

Compared with $Q_{a\lambda}$ which falls off only in proportion to ξ, $Q_{sc\lambda}$ decreases very rapidly with decreasing ξ.

In the small-particle limit equation [9.13] indicates that $Q_{a\lambda}(r)$ is proportional to r, $Q_{a\lambda}(r)/r$ is a constant and that therefore $\beta_{a\lambda}$ is independent of the size distribution. On the other hand, $Q_{sc\lambda}$ is proportional to r^4 (the Rayleigh limit for small particles) so that $\beta_{sc\lambda}$ increases as r^3. Fig. 9.4 shows the quantities $\mathrm{d}\beta_{a\lambda}(r)/\mathrm{d}w(a)$ and $\mathrm{d}\beta_{e\lambda}(r)/\mathrm{d}w(a)$ plotted against r for a wavelength of 0.55 μm. This figure demonstrates the range of radii for which $\mathrm{d}\beta_{a\lambda}/\mathrm{d}w(a)$ is a constant. It shows that the value of $\mathrm{d}\beta_{a\lambda}/\mathrm{d}w(a)$ is approximately proportional to n_i. It also shows that $\beta_{e\lambda}$ becomes independent of n_i for values of r greater than about 0.2 μm. Small absorbing aerosol particles with radii less than 0.1 μm obviously contribute substantially to the total

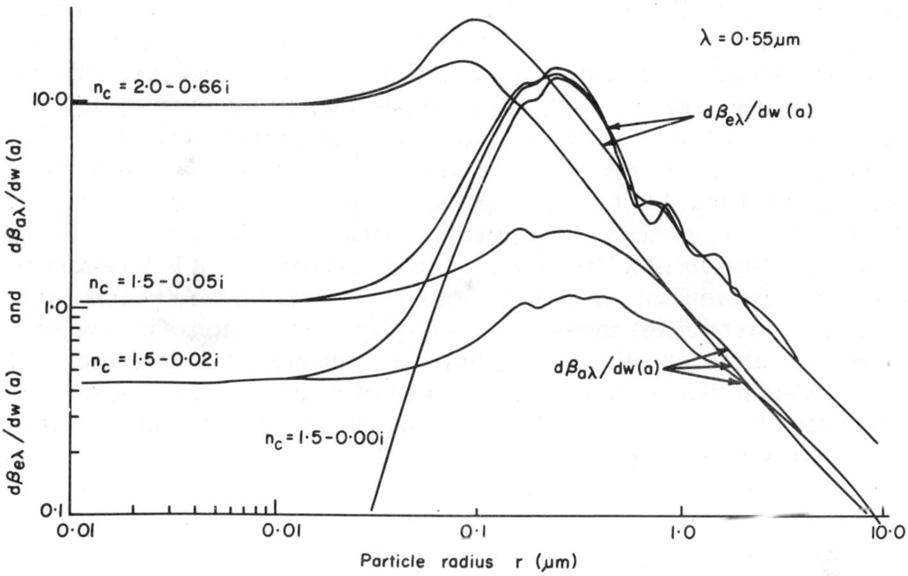

Fig. 9.4. Extinction and absorption coefficients per unit mass of spherical particles as a function of particle radius for four different refractive indices at a solar wavelength of 0.55 μm. The values of the ordinate scale must be multiplied by 1.5×10^4 to obtain units of $g^{-1}\,cm^2$ (for $\rho = 1$; see text). (After Bergström, 1973.)

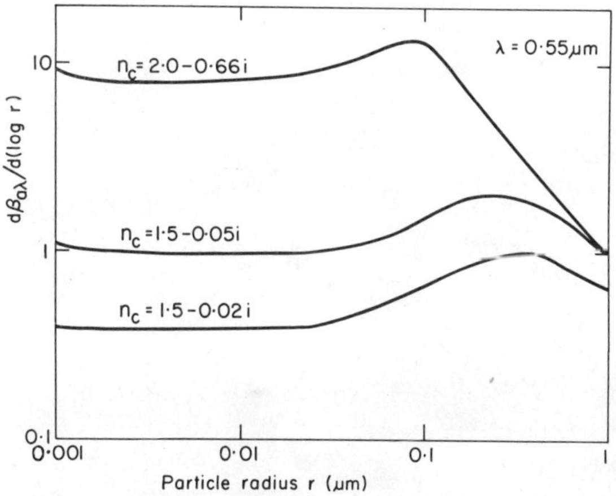

Fig. 9.5. Volume absorption coefficients per decade of radius as a function of particle radius for three different refractive indices at a solar wavelength of 0.55 μm (calculated from Bergström, 1973).

aerosol absorption. The actual amount is dependent on the fraction of the total mass of aerosols contained in the small particles, which is in turn a function of the aerosol size distribution. The volume absorption coefficient per decade of radii $[d\beta_{a\lambda}/d(\log r)]$ for a Junge distribution ($V = 3$) is shown in Fig. 9.5. For high values of n_i typical of carbonaceous aerosols ($n_i = 0.66$) all decades of radii below 0.1 μm contribute almost equally, but above 0.1 μm the contribution falls off. As n_i decreases, the radius interval of 0.1–1.0 μm gradually becomes dominant but smaller particles still contribute significantly. For precise calculations of aerosol absorption it is necessary to have detailed information on the aerosol size distribution and the imaginary part of the refractive index. Aerosol size distributions in polluted industrialized areas or other semipolluted continental areas (see Figs. A.4 and A.5) contain many small particles, some of which are highly absorbing. It is obviously a challenging task to determine the solar absorption properties of such aerosol populations.

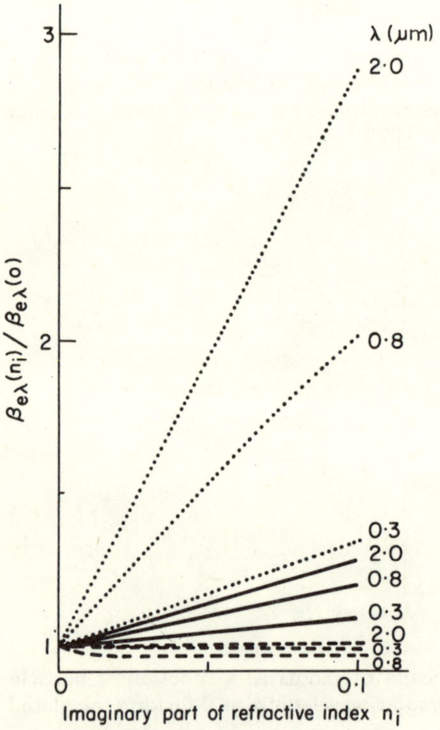

Fig. 9.6. Ratio $\beta_{e\lambda}(n_i)/\beta_{e\lambda}(0)$ of the extinction coefficient versus imaginary part n_i of the refractive index for different wavelengths and Junge exponents V. The boundary radii to the calculation were $r_1 = 0.032$ μm; $r_2 = 10$ μm; $n_r = 1.50$; $V = 4$ (dotted lines); $V = 3$ (solid lines); $V = 2$ (dashed lines). (After Hänel and Bullrich, 1970.)

The effect of n_i on aerosol extinction is shown in Fig. 9.6 for three Junge distributions. As the distribution with $V = 4$ has the greatest number of small particles relative to large ones, the effect of n_i on the extinction coefficient $\beta_{e\lambda}$ is also maximum. As different wavelengths will be affected differently by absorption, the spectral extinction alters when absorption is present and the power law for a Junge distribution is no longer strictly true. Although the spectral dependence is still approximately proportional to some negative power of λ (i.e. $\lambda^{-\hat{\alpha}}$ say) the exponent $\hat{\alpha}$ is decreased by absorption. For $n_i = 0.1$ and a Junge distribution with $V = 4$ the work of Hänel and Bullrich (1970) indicates that $\hat{\alpha}$ is reduced from 2 to about 1.5. A typical average value of n_i for atmospheric aerosols is 0.01 (but see below). Thus the effect of n_i on the spectral change in *extinction* is generally not large.

Calculation of absorption by an aerosol layer is harder than for a water cloud if only because the aerosol case includes another "degree of freedom". At one wavelength, water drops all have the same value of n_i, but in a typical aerosol layer a small percentage of the particles (e.g. soot, ash, etc.) may be highly absorbing while the rest may be slightly absorbing or even transparent. It is the *absorbing* particles which give the aerosol layer its "mean complex refractive index". Such mean figures must be used with caution in any computation of aerosol absorption.

4. ATMOSPHERIC TURBIDITY

Atmospheric turbidity is defined as the optical depth $\tau_{D\lambda}$ due to aerosols in the entire vertical column of the atmosphere. Turbidity measurements have been made for a long time and are usually accomplished by measuring the extinction of the solar beam at wavelengths unaffected by gaseous absorption and subtracting the known optical depth due to Rayleigh scatter. A practical approach to analysis adopted by Ångström (1964) is obtained by writing:

$$\tau_{D\lambda} = \hat{\beta}\lambda^{-\hat{\alpha}} \qquad [9.14]$$

where λ is in microns. $\hat{\beta}$ is known as the Ångström turbidity coefficient and is equal to $\tau_{D\lambda}$ at a wavelength of 1 μm. The actual measurement is usually made at a smaller wavelength and in the absence of other information $\hat{\beta}$ is calculated by assuming $\hat{\alpha}$ to have a "standard" value of 1.3. If it is assumed that the aerosol size distribution is constant with height then $\tau_{D\lambda}$ is proportional to the volume absorption coefficient $\beta_{e\lambda}$. Thus $\hat{\alpha}$ can be identified with the wavelength exponent $2 - V$ in equation [9.7], and Ångström's turbidity equation should hold for Junge distributions. A value of 1.3 for $\hat{\alpha}$ corresponds to a Junge exponent of $V = 3.3$.

If the turbidity has been measured at two wavelengths, the exponent $\hat{\alpha}$

can be calculated and used directly — again on the assumption of a Junge distribution. However, in order to classify the extinction properties more precisely (i.e. to remove the assumption about distribution shape), measurements are required in a number of narrow spectral intervals. Further, in order to measure low "background" turbidity accurately, a very precise measurement is needed. Photometric instruments which measure the direct solar extinction require prior calibration in terms of the solar irradiance outside the atmosphere. This calibration needs to be very precise. A 1% error in instrument calibration is equivalent to an error of 50% in a turbidity measurement of 0.02. Such low turbidities are not uncommon in very clear atmospheres. Photometric instruments such as the Volz sun photometer are known to be subject to this sort of error (Laulainen and Taylor, 1974). An alternative method of measuring low turbidities is to measure the solar spectral scatter at angles close to the direction of the solar beam (Newkirk, 1956; Green et al., 1971). This scatter is very intense so that even for low turbidities the aureole radiance is easily measured. Such measurements can always be normalized against the radiance of the solar disc, which provides a calibration for each measurement. Ångström (1974) has analyzed the solar aureole radiance between angles of $3\frac{1}{2}°$ and $13\frac{1}{2}°$ away from the direct solar beam from measurements made over many years and found a good relation between turbidity and solar aureole radiance. More work is required to examine the sensitivity of the measured spectral angular radiance to varying size distribution (Kuriyan and Sekera, 1974).

The necessity for long-term monitoring of the characteristics of the "background" aerosol population has been stressed in recent international meetings (SMIC, 1971). The optical turbidity method is obviously attractive for reasons of simplicity. However, the large number of parameters involved in aerosol optical characteristics forces us to interpret turbidity measurements with extreme caution. For instance, background aerosols are thought to be composed largely of sulphates. These hygroscopic materials respond in their own unique fashion to atmospheric humidity (Fig. A.7). Thus comparison of turbidities in the long-term must include the variable humidity factor. Turbidity measurements should be supplemented by radiosonde data and aerosol height profiles. The latter can be obtained by lidar measurements (although their interpretation must also be treated with caution). No reliable method has yet been developed of obtaining the aerosol absorption directly and such a technique is needed urgently. Finally the necessity of multispectral turbidity measurements cannot be too highly stressed.

5. EXTINCTION OF INFRARED LONG-WAVE RADIATION

5.1. *Calculation of extinction coefficients*

The main meteorological interest in the infrared extinction of long-wave radiation is the extent to which the heating of aerosol layers by solar short-wave radiation is offset by infrared cooling in the same layers. The extinction of infrared radiation by aerosols is limited by the small values of the size parameter ξ at such wavelengths. Therefore infrared cooling by aerosols is in general rather small except perhaps when a large number of aerosols become trapped below a temperature inversion. The difficulties involved in accurate calculation of the infrared extinction are considerable. The size distribution, composition and refractive indices of the particles must be known. However, in most circumstances only a rough estimate of the infrared *absorption* coefficient is required and the following very simple treatment may be quite adequate. For much the same reasons as discussed in Chapter 8 with reference to clouds, the scattering can often be neglected in radiative cooling calculations.

Referring back to Fig. 8.4, the absorption efficiency $Q_{a\lambda}$ increases from zero with increasing particle radius r but asymptotes to the value 1.0 at large r. Perhaps the simplest approximation to this curve is of two straight lines — $Q_{a\lambda} = c_2 r$ for radii less than some value r_2, say, and $Q_{a\lambda} = 1.0$ for radii greater than r_2. The slope c_2 can be calculated from the small size-parameter approximation of equation [9.13] for any given wavelength and refractive index.

If one assumes that the particle size distribution has cut-off radii of r_1 and r_3 at the small and large ends respectively, and that the distribution has the Junge shape, then the volume absorption coefficient $\beta_{a\lambda}$ is given by equation [9.5]. This can be broken into the separate contributions [(1) and (2)] from particles of radius less than r_2 and those of radius greater than r_2. That is:

$$\beta_{a\lambda}(1) = (\pi c_1 c_2/2.3)\int_{r_1}^{r_2} r^{2-V}\,\mathrm{d}r; \qquad Q_{a\lambda} < 1 \qquad [9.15a]$$

and:

$$\beta_{a\lambda}(2) = (\pi c_1/2.3)\int_{r_2}^{r_3} r^{1-V}\,\mathrm{d}r; \qquad Q_{a\lambda} = 1 \qquad [9.15b]$$

Substituting $V = 3$ in these equations and integrating:

$$\beta_{a\lambda}(1) = (\pi c_1 c_2/2.3)[\log_e(r_2/r_1)] \qquad [9.16a]$$

$$\beta_{a\lambda}(2) = (\pi c_1/2.3)[(1/r_2) - (1/r_3)] \qquad [9.16b]$$

The constant c_1 in the above equations is a measure of the absolute number of particles per unit radius interval at some radius of the distribution. For city polluted conditions $\mathrm{d}n(r)/\mathrm{d}r$ is typically about 10^4 cm^{-4} at a radius of 0.1 μm. Substituting this value into equation [A.2] yields a value of $c_1 = 2.3 \times 10^{-11}$. The constant c_2 in the above equations depends on both wavelength and the particle refractive index via equation [9.13], which can be re-written as:

$$Q_{a\lambda} = (2\pi r/\lambda)24 n_i n_r/[(n_r^2 - n_i^2 + 2)^2 + (2n_i n_r)^2] \qquad [9.17]$$

and simplified for cases where $n_i \ll n_r$ to:

$$Q_{a\lambda} = (2\pi r/\lambda)24 n_i n_r/(n_r^2 + 2)^2 \qquad [9.18]$$

A representative value of n_c for atmospheric aerosol material in the infrared is not available. However, on the basis of values quoted in the Appendix we assume here a figure $n_c = 1.55 - 0.1i$. Thus for a wavelength of 9 μm, the value of $c_2 (= dQ_{a\lambda}/dr)$ is about 1.34×10^3 cm^{-1}.

It now remains to determine the values of r_1, r_2 and r_3. As Junge distributions usually peak at about 0.01 μm, we make $r_1 = 0.01$ μm. For the above refractive index and wavelength, $Q_{a\lambda}$ attains the value 1.0 when $r = 7$ μm. This then is chosen as the value of r_2. Finally we make r_3 equal to $10r_2$. According to equation [9.16b], it is not important if r_3 is in fact greater than $10r_2$. These conditions lead then to the following numerical values for $\beta_{a\lambda}$ at 9 μm:

$$\beta_{a\lambda}(1)(0.01, 7\,\mu m) = 2.8 \times 10^{-7}\ cm^{-1} = 2.8 \times 10^{-2}\ km^{-1} \qquad [9.19]$$

$$\beta_{a\lambda}(2)(7 \rightarrow 70\,\mu m) = 4.4 \times 10^{-8}\ cm^{-1} = 4.4 \times 10^{-3}\ km^{-1} \qquad [9.20]$$

The total value of $\beta_{a\lambda}[= \beta_{a\lambda}(1) + \beta_{a\lambda}(2)]$ is 3.24×10^{-2} km^{-1}. As c_2 is proportional to n_i for small values of n_i then $\beta_{a\lambda}$ is obviously very dependent on this quantity. For instance, for a carbonaceous aerosol n_i has a value of about 0.6 and for such an aerosol with the above number distribution $\beta_{a\lambda} = 1.9 \times 10^{-1}$ km^{-1}.

In problems involving the infrared properties of aerosols, the above formula can be used as an *estimate* of the absorption coefficient over the most significant part of the infrared spectrum. For other power law distributions, it is only a matter of changing the value of V. For instance, if $V = 2$ then:

$$\beta_{a\lambda} = (\pi c_1' c_2/2.3)(r_2 - r_1) \qquad \text{for radii when } Q_a < 1 \qquad [9.21a]$$

$$\beta_{a\lambda} = (\pi c_1'/2.3)[\log_e(r_3/r_2) \qquad \text{for radii when } Q_a = 1 \qquad [9.21b]$$

where the constant c_1' has a different value and dimension to c_1.

Some interesting points emerge from the forms of equations [9.16] and [9.21]. The dependence of $\beta_{a\lambda}$ on $\log_e r$ in equation [9.16a] means that each decade of aerosol radius is equally important to the absorption, whereas in equation [9.21a] each successive decade of radius below r_2 contributes an order of magnitude less as r is decreased. Then, however, the value of r_2 is important. Again, for equation [9.16b], r_2 is relatively unimportant, whereas in equation [9.21b] each decade beyond r_2 is equally important. When $V = 4$ in the Junge distribution it is important to determine a suitable value of r_1 at the lowest radius of the size distribution because the number density $n(r)$ increases rapidly with r. When $V = 2$, it is important to determine the distribution $n(r)$ in the large-particle range of the spectrum.

Given a knowledge of the aerosol size distribution and refractive index,

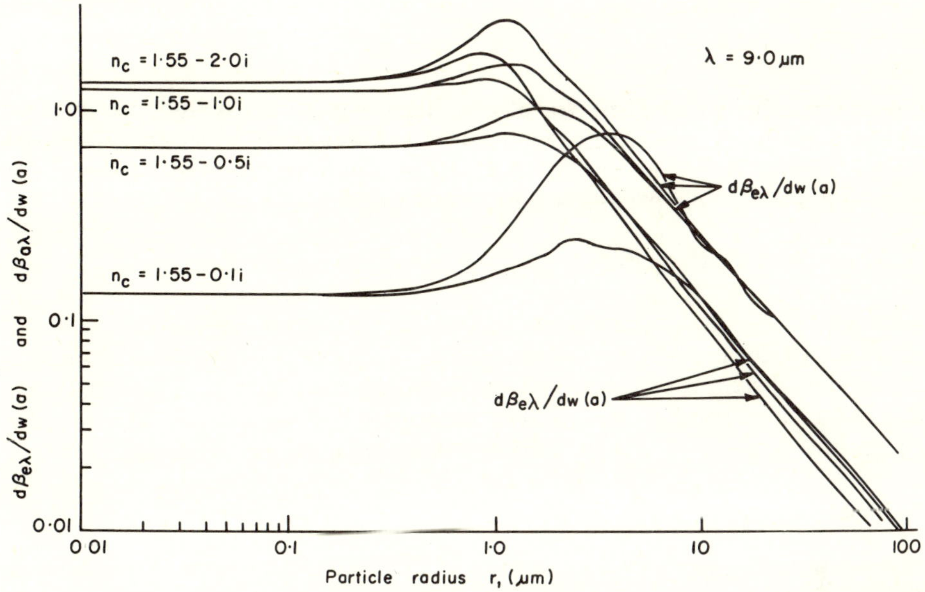

Fig. 9.7. Same as Fig. 9.4 but for an infrared wavelength of 9 μm. (After Bergström, 1973.)

accurate computations of $\beta_{a\lambda}$, $\beta_{e\lambda}$, $\beta_{sc\lambda}$ can be made using the Mie theory (Chapter 4) and an integration of the corresponding efficiencies over the size distribution. Approximations for $Q_{sc\lambda}$ and $Q_{a\lambda}$ such as equations [9.12] and [9.13] can be used within the allowable values of ξ. The work of Bergstrom (1973) gives a good up-to-date account of the state of the art. Fig. 9.7 shows values of $d\beta_{a\lambda}/dw(a)$ and $d\beta_{e\lambda}/dw(a)$ (equations [9.10] and [9.11]) for a wavelength of $\lambda = 9\ \mu m$. The region where $d\beta_{a\lambda}/dw(a)$ is constant corresponds to where the absorption efficiency $Q_{a\lambda}$ is proportional to r and it can be seen that there is a fairly sharp transition between that region and where $d\beta_{a\lambda}/dw(a)$ falls off as $1/r$ ($Q_{a\lambda}$ = const.). The figure also illustrates the negligible effects of scattering in the horizontal regions of the curves. The fact that $d\beta_{a\lambda}/dw(a)$ and $d\beta_{e\lambda}/dw(a)$ are constant with r over quite a wide range of significant radii suggests that $\beta_{a\lambda}$ and $\beta_{e\lambda}$ are substantially independent of the aerosol size distribution. An alternative approximation to that described above is therefore to compute the extinction coefficients in terms of $w(a)$, the mass of haze per unit volume of air. The ranges over which $d\beta_{e\lambda}/dw(a)$ and $d\beta_{a\lambda}/dw(a)$ are roughly constant decrease somewhat with increasing values of n_i so that the approximation that extinction is proportional to aerosol mass must be used with care.

The swelling of aerosol radii with atmospheric humidity creates even more of a problem in the infrared region than it does in the short-wave region. This is because absorbing aerosols, which show a wide spectral variation in

n_i when dry, must gradually assume the values of n_i characteristic of water as the humidity increases. Only an empirical form of equation such as [A.10] of the Appendix is at present available to describe this change. The aerosols also increase in *radius* with increasing humidity. Any attempt to treat the humidity problem must be very approximate and geared towards a specific aerosol distribution. The significance of the humidity can best be judged by the few computations which have been made and which are described below.

5.2. *Examples of extinction coefficients for specific distributions*

Values of $\beta_{a\lambda}$ for typical aerosol size distributions and calculated from equations [9.16a] and [9.16b] are shown in Table 9.2. The characteristics of the size distributions and cut-off radii are also shown. Table 9.2 also includes a value of $\beta_{a\lambda}$ which was computed by Grassl (1973b) for the background aerosol of Fig. A.1 of the Appendix for different values of n_i.

TABLE 9.2

Volume absorption coefficients of aerosols at 9 μm wavelength using an approximate model (see text)

n_c	c_2 (cm^{-1}) $= 1/r_2$	Volume absorption coefficient (km^{-1})		
		continental urban	continental country	Grassl (1973b)
1.55–0.10i	0.134×10^4	1.1×10^{-2}	0.6×10^{-3}	
1.88–0.66i	0.693×10^4	4.2×10^{-2}	2.1×10^{-3}	
1.33–0.04i	0.063×10^4	6.0×10^{-3}	0.3×10^{-3}	0.4×10^{-3}
1.55–0.02i	0.068×10^4	2.0×10^{-3}	0.1×10^{-3}	

Aerosol characteristics

Continental urban: Junge distribution $V = 3$. $r_1 = 0.05\ \mu$m. $r_3 = \infty$. $dn(r)/d(\log r) = 10^4$ cm^{-3} at 0.1 μm. $c_1 = 10^{-11}$.

Continental country: Junge distribution $V = 3$. $r_1 = 0.05\ \mu$m. $r_3 = \infty$. $dn(r)/d(\log r) = 5 \times 10$ cm^{-3} at 0.1 μm. $c_1 = 5 \times 10^{-13}$.

The "continental country" model curve approximates in both $n(r)$ and V Junge's curve of Fig. A.1 for radii greater than 0.1 μm. The approximate method of computing $\beta_{a\lambda}$ gives good order of magnitude agreement with Grassl's computation. Fig. 9.8 shows computations of absorption optical depth $\tau_{a\lambda}$ of the atmosphere made by Grassl (1973a) at different relative humidities for a Junge continental aerosol. The increase in relative humidity actually decreases the absorption coefficient at wavelengths where n_i for the dry aerosol is greater than n_i for water at the same wavelength. Figs. 9.9a and 9.9b illustrate the theoretical extinction and scattering coefficients for a model continental aerosol and maritime aerosol respectively at different relative humidities. The expressions for the model size distributions were

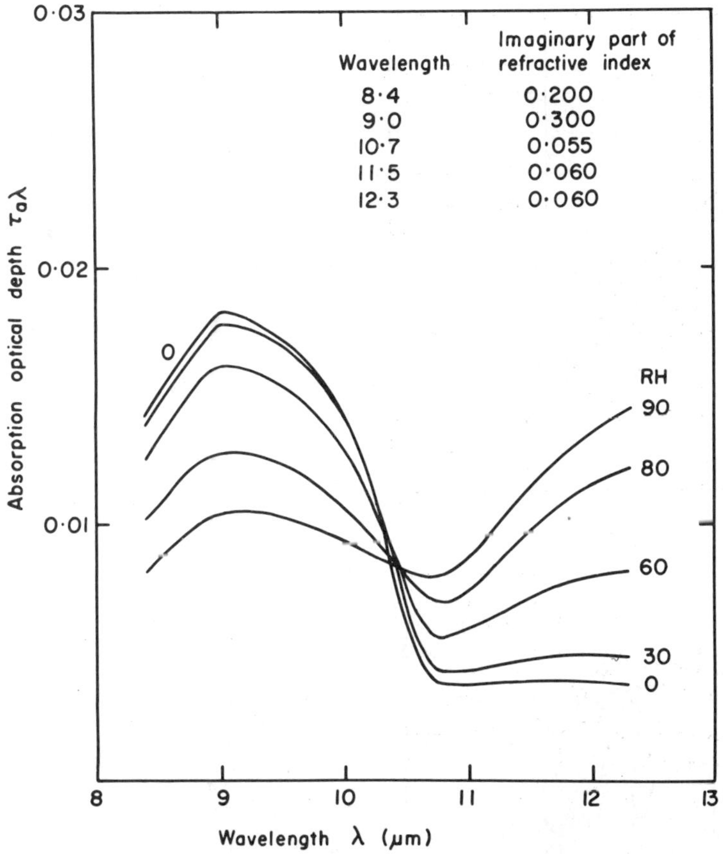

Fig. 9.8. Absorption optical depth of atmospheric aerosol (when the short-wave optical depth is 0.2 at 0.58 μm wavelength) versus wavelength and for different relative humidities (*RH*). (After Grassl, 1973a.)

$dn(r)/dr = (2.02/r)\exp(-0.66r)$ for the maritime distribution and $dn(r)/dr = 0.27r^{-4}$ for the continental distribution. Gambling (1972) used Barnhardt and Streete's (1970) proposed curve for the growth of particle radius (equation [A.6]), equation [A.8] for the change in n_r and an equation similar to [A.8] for the change in n_i.

The above curves illustrate the importance of the aerosol size distribution to the *spectral* distribution of extinction and the importance of the relative humidity, when greater than about 70%, to the value of extinction at any wavelength. The value of $\beta_{a\lambda}$ (that is $\beta_{e\lambda}$ minus $\beta_{sc\lambda}$) for the maritime distribution, a wavelength of 10 μm and a relative humidity of 50% is about 10^{-2} km^{-1}, increasing to about 10^{-1} km^{-1} when the relative humidity is 95%. For the continental aerosol Gambling has included a 70% insoluble component. The size distribution is also different from the maritime case

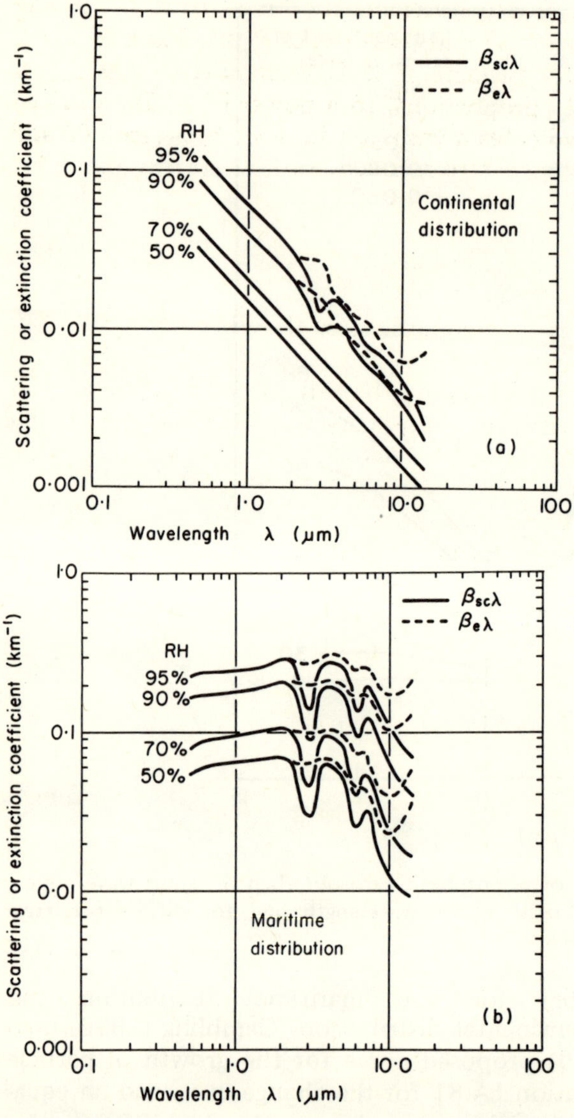

Fig. 9.9. Calculated spectral scattering and extinction coefficients of aerosol distributions for different values of relative humidity (*RH*). (a) Continental aerosol (see text); (b) maritime aerosol (see text). (After Gambling, 1972.)

and $\beta_{a\lambda}$ is much smaller than for the maritime aerosol and becomes significant only at the higher humidities. Spectral extinction coefficients of model aerosols of specific compositions have been computed by Twitty and Weinman (1971) for two models of carbonaceous aerosols and by Conel (1969) for models of silicate hazes.

Spectral variations of absorption coefficients for Junge distributions can be calculated from equations [9.16a] and [9.16b], or [9.21a] and [9.21b]. The variation is quite easy to estimate if it is remembered that $\beta_{a\lambda}$ is proportional to n_i and inversely proportional to a power of λ. The work of Volz (1970) (see Appendix) indicates a variation in n_i of between 0.05 and 0.3 in the window region of the spectrum and n_r varies between about 1.4 and 1.8.

5.3. *Infrared cooling by atmospheric aerosols*

Infrared cooling by aerosol is largely confined to the 8–14-μm wavelength atmospheric window both because the aerosol absorption coefficient is small compared to those of water vapour and carbon dioxide and because most aerosol is confined to the lower atmosphere. Grassl (1973b) has shown that the contribution of aerosol cooling in the molecular absorption bands is less than 10%.

Because of the low infrared absorption coefficients, cooling by atmospheric aerosol becomes appreciable only in "pollution" situations

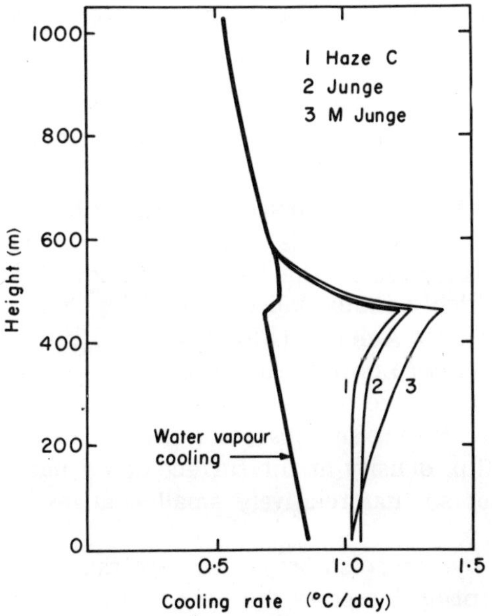

Fig. 9.10. Height dependence of the cooling rate for different aerosol types and imaginary parts of the refractive index for a constant short-wave optical depth of 0.2 at 0.55 μm. The spectral region is 8–13 μm. All aerosol particles are below 500 m altitude with relative humidity increasing from 60% to 90% from ground to 500 m. Size distributions are as follows: *Haze C* — "Haze L" in Table A.1; *Junge* — Junge distribution with $V = 3$; *M Junge* — similar to Fig. A.1. The change in refractive index with relative humidity is given by Fig. A.12. (After Grassl, 1973b.)

when dense aerosols may be trapped below a temperature inversion. The aerosol cooling rate below an inversion at 500 m altitude is shown in Fig. 9.10 for different aerosol distributions. The optical depth of 0.2 at 0.55 μm specified by Grassl corresponds to a rather dense haze. The specified increase in humidity with altitude creates a sharp cooling near the inversion top particularly for the maritime distribution M which has a low value of n_i at a low humidity. In this sort of situation the aerosol cooling is appreciable. In a more typical situation where the aerosol optical depth at 0.55 μm is about 0.1, and the aerosol is spread throughout 2 or 3 km altitude, then the associated cooling rates are usually less than 0.1°C per day which is only a fraction of the total cooling due to water vapour.

Few measurements have been made in which aerosol infrared emissivity or cooling has been detected unambiguously. Grassl (1973a) was able to identify optical depths of ≈0.01–0.03 per 0.2 optical depth at 0.58 μm, and with the right spectral dependence. Platt and Gambling (1971) measured an aerosol optical depth of 0.02 at 10–12 μm outside the visible edges of small cumulus clouds. Corresponding increases in aerosol content near the clouds were detected by lidar. The infrared extinction coefficient when flying through a haze layer was found by Paltridge and Platt (1973) to be less than 0.02 km^{-1} (Fig. 7.10) whereas the solar absorption and scatter was appreciable and easily measurable.

6. THE EFFECT OF AN INCREASE IN ATMOSPHERIC AEROSOLS ON CLIMATE

The attention of meteorologists has been focussed recently on the potential modifying influences of atmospheric aerosols on the climate of the earth. There is some evidence for an increase in the background level of aerosols in remote areas adjacent to highly industrialized countries and a recent survey of atmospheric turbidity over America (Flowers et al., 1969) shows the high average values over the eastern populated areas. Atmospheric aerosols generally do not have a great influence on the long-wave fluxes but they influence the solar flux by both scattering and absorbing the radiation. Only a 1 or 2% change in the solar flux density at the surface *could* have important global climatic consequences so that relatively small changes in turbidity could be significant.

Although scattering and absorption by aerosols always decrease the solar input to the earth's surface, the absorption increases the heat input to the layer itself. Then the net effect on the *planet* may be either a heating or a cooling. Further, if the aerosol layer is in convective contact with the surface, the dynamics may ensure that the surface, as well as the planet, is warmed by an absorbing aerosol. There are many factors involved. The optical properties of the layer depend on the composition and size distribution of the aerosols. The thermal and dynamic effects of the layer depend on the distribution of the aerosols in the vertical. The solar radiation absorbed depends in part on the albedo of the underlying surface because the aerosol layer also absorbs reflected radiation. Nevertheless, a simple non-dynamical model of a homogeneous aerosol confined to a single layer above the earth's surface serves to point out the salient features of the problem.

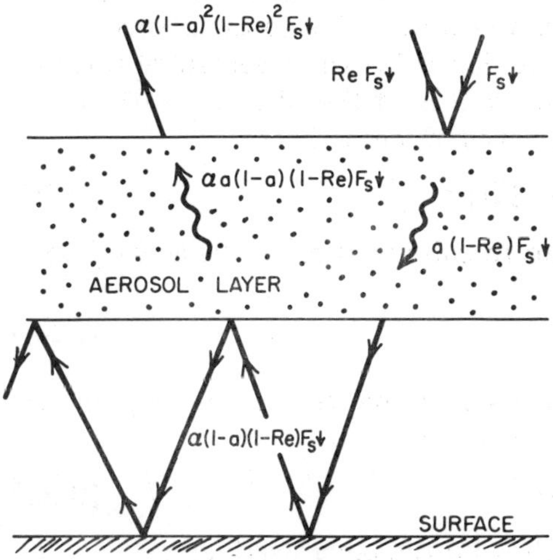

Fig. 9.11. Schematic diagram of solar flux densities in a reflecting and absorbing aerosol layer above a surface of albedo α.

The situation is illustrated in the simple model of Fig. 9.11. The parameters involved are the downward global flux density at the top of the layer $F_S\downarrow$; the reflectance Re of the layer itself; the absorptance a of the layer itself; and the earth's surface albedo α. The layer reflectance is assumed to be the same for both the up and downcoming radiation streams.

Considering first the case of zero aerosol absorption, it can be seen that the total outgoing flux density $F_S\uparrow$ is the infinite sum:

$$F_S\uparrow \ = \ ReF_S\downarrow + \alpha(1-Re)^2 F_S\downarrow + \alpha^2(1-Re)^2 ReF_S\downarrow + \ldots \qquad [9.22]$$

Assuming that Re is of order 0.01 or less so that the terms in Re^2 can be neglected:

$$F_S\uparrow \ = \ F_S\downarrow[Re + \alpha(1+\alpha Re - 2Re)] \qquad [9.23]$$

and the "planetary albedo" (that is $F_S\uparrow/F_S\downarrow$) is given by the term in square brackets. Thus the insertion of the aerosol layer changes the planetary albedo by an amount $\Delta\alpha$ such that:

$$\Delta\alpha \ = \ Re - 2\alpha Re + \alpha^2 Re \qquad [9.24]$$

Note that the model ignores the effect of the intervening atmosphere so that the planetary albedo in the absence of the aerosol is the same as the surface albedo. The global short-wave flux at the surface $F'_S\downarrow$ is again given by an infinite sum which, when higher terms in Re are neglected, reduces to:

$$F'_S\downarrow = F_S\downarrow(1 - Re + \alpha Re) \qquad [9.25]$$

This equation formally makes the obvious point that a pure reflecting aerosol always reduces the solar input to the surface and to the planet.

Consider now the more general case of Fig. 9.11 where aerosol absorption is included. Assuming that the absorptance a is of the order of 0.01 or less, the expression for the outgoing flux density reduces to:

$$F_S\uparrow = F_S\downarrow[Re + \alpha(1 + \alpha Re - 2Re - 2a)] \qquad [9.26]$$

and the change in planetary albedo due to the insertion of the aerosol layer is given by:

$$\Delta\alpha = Re - 2\alpha Re + \alpha^2 Re - 2\alpha a \qquad [9.27]$$

The downward flux density at the surface is given by $F_S\downarrow(1 - Re + \alpha Re - a)$ so that the net input there is:

$$F_{NS} = F_S\downarrow(1-\alpha)(1 - Re + \alpha Re - a) \qquad [9.28]$$

The flux density F_S (lay) actually absorbed by the layer itself is:

$$F_S\,(\text{lay}) = F_S\downarrow[a + (1 - Re)(1 - a)\alpha a/(1 - 2Re)]$$
$$\approx F_S\downarrow(a + \alpha a) \qquad [9.29]$$

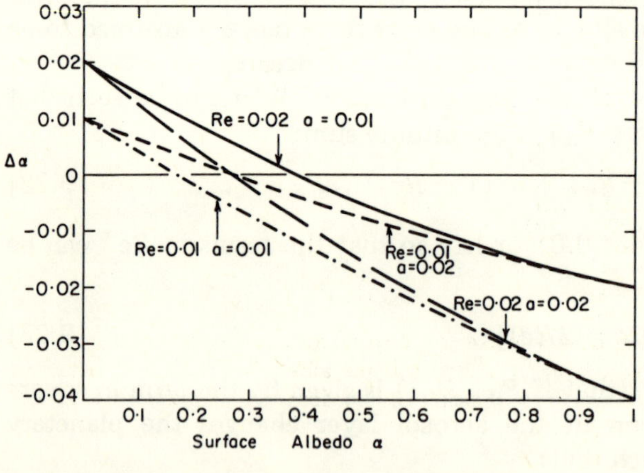

Fig. 9.12. The apparent change in surface albedo α when a layer of aerosol with the given properties is interposed above the surface.

Any change in the planetary albedo alters the radiation balance and causes a warming or a cooling of the system as a whole. Other feedback processes such as an alteration of cloud amount may modify the "original" change in albedo. It must be emphasized that the following discussion of the above equations makes the improbable assumption that all other factors remain fixed. Thus the predicted coolings and warmings are predictions only of initial tendencies.

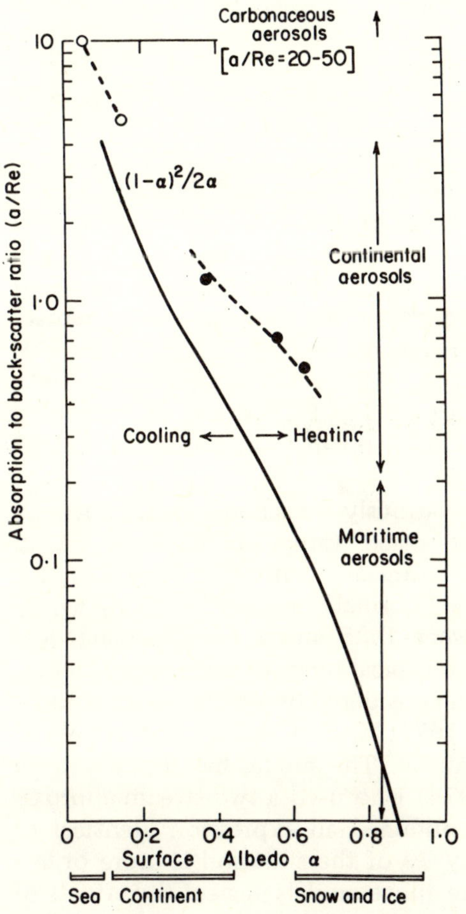

Fig. 9.13. Value of the absorption to backscatter ratio of an aerosol layer for which there is zero net change in the planetary albedo versus surface albedo α. Full line: equation [9.30]; dashed line with solid circles from Reck (1974); dashed line with open circles from Yamamoto and Tanaka (1972).

Fig. 9.12 is a plot of equation [9.27] which shows the change $\Delta\alpha$ of planetary albedo as a function of surface albedo α when an aerosol layer of given *Re* and *a* is inserted in the atmosphere. The significant point is the change in *sign* of $\Delta\alpha$ as the surface albedo increases. The reversal comes about because reflected radiation is also absorbed by the aerosol layer and this absorption increases with increasing α.

Using equation [9.27] one can calculate the surface albedo for which the insertion of an aerosol layer of given *Re* and *a* makes no difference to the planetary albedo. That is, $\Delta\alpha$ is zero when:

$$a/Re = (1-\alpha)^2/2\alpha \qquad [9.30]$$

This function is plotted in Fig. 9.13. To the left of the curve, insertion of an aerosol layer gives an initial tendency to planetary cooling. To the right of the curve, the initial tendency is to a warming.

Now for a layer of small optical thickness and a vertical depth z, both a and Re can be simply expressed in terms of the volume scattering and absorption coefficients appropriate to the aerosol. Thus $a = \beta_a z$ and $Re = \beta_{sc} \cdot bf \cdot z$ where the factor bf is the fraction of solar radiation scattered into the backward hemisphere. Thus the ratio a/Re depends on the aerosol size distribution and the real (n_r) and imaginary (n_i) parts of the refractive index of the aerosol material. Ensor et al. (1971) worked out a/Re for a number of Junge aerosol size distributions. Fig. 9.14 shows the value of a/Re for $n_r = 1.5$ plotted against values of n_i taken from Ensor's tabulated values.

As discussed in the Appendix, n_i is variously considered to lie between 0.005 and 0.02. (However, n_i might reach values as high as 0.1 in industrialized areas.) According to the figure and with $V = 3$, $(a/Re)_{\Delta\alpha=0}$ lies between 0.6 and 2.3 for the above "normal" range of n_i. This would indicate a critical surface albedo between 0.35 and 0.18 for changeover from heating to cooling, which is within a typical range of terrestrial albedos.

The above simple model has also been considered by several other authors (Charlson and Pilat, 1969; Ensor et al., 1971; Mitchell, 1971) who arrived at substantially the same conclusions as above. The model has recently been extended by Chýleck and Coackley (1974) who used a two-stream approximation for a deep aerosol layer and obtained an expression identical to equation [9.30], indicating that the physics of the situation is more or less independent of aerosol thickness. All the above models neglect the effects of clouds, which are known to have a large modifying effect on solar flux densities. The models also neglect effects of varying solar angle. Recently Reck (1974, 1975) incorporated aerosol layers into a detailed radiative-convective equilibrium model (see Chapter 10). Reck considered a fixed ratio of a/Re equal to 0.62 and placed aerosol layers successively at heights of 958, 418 and 41.5 mb. Water cloud layers were fixed at altitudes of 925, 881 and 336 mb, with optical characteristics which are tabled in her article.

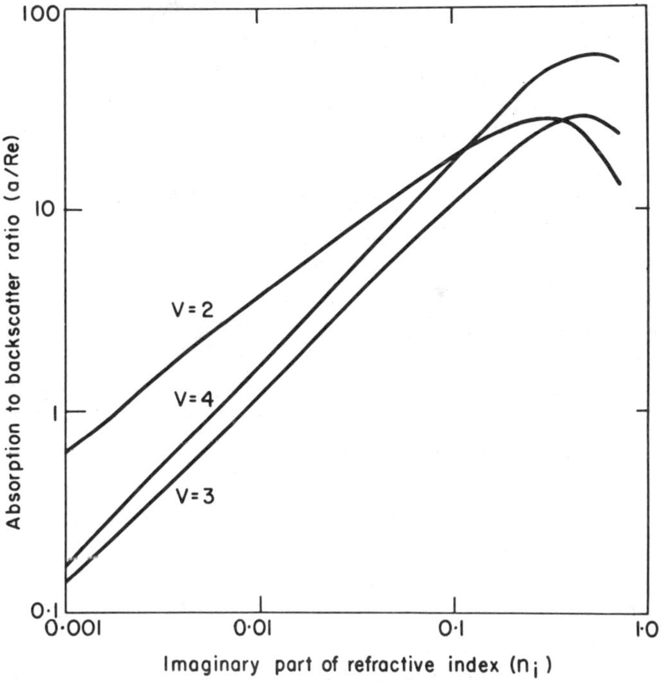

Fig. 9.14. Absorption to backscatter ratio (a/Re) from calculations by Ensor et al. (1971) for three atmospheric aerosol distributions with the given Junge exponents V versus the imaginary part of the aerosol refractive index n_i.

She also considered a small long-wave absorptance of 0.001 per 0.065 visible optical depth.

For $a/Re = 0.62$ one would expect from the simple model that the planetary albedo would increase with the insertion of an aerosol layer for surface albedos below 0.35. Reck's detailed model also showed this behaviour. For $\alpha = 0.07$ and 0.3 the planetary albedo increased and for $\alpha = 0.6$ the planetary albedo decreased. The dependence of the change in albedo with aerosol height was quite complex. Generally, the change in planetary albedo decreased with the height of the inserted aerosol layer. The model showed also that although the surface cooled at the low surface albedos in response to the increase in planetary albedo, the integrated long-wave cooling rate of the atmosphere decreased and thus tended to cause an atmospheric warming. However, the magnitude of this effect was very small.

Reck (1974) was able to calculate the ratio $(a/Re)_{\Delta\alpha=0}$ for the three aerosol layers at different heights and the actual earth-atmosphere albedos below the layer. Values are given in Fig. 9.13. The albedo "beneath" the highest layer is the largest because it includes the most cloud below the layer as well as Rayleigh scattering between the layer and the surface. As mentioned above, Reck's results give qualitatively the same picture as the simple model, but with a greater range of albedo α for which surface cooling is predicted. Yamamoto and Tanaka (1972) predicted the increase in planetary albedo for various polluted atmospheres using a multiple-scattering model. Two figures of $(a/Re)_{\Delta\alpha=0}$ deduced from their work are shown in Fig. 9.13. Their points are again shifted slightly to the right of the simple model curve. Their "atmosphere" did not contain clouds.

It is interesting to look at the order of magnitude of surface temperature change predicted for a given change in optical depth. According to Reck's work, a doubling in mean visible optical depth from its present average value ($\approx$0.065) would give about 2°C cooling at the surface over water, about 1°C cooling over land and about 1°C heating over snow. However, this is for one assumed value of a/Re. As can be seen from Fig. 9.14, considerable uncertainty about the value of a/Re exists. All that can be said at present is that for aerosols over sea, cooling will probably result. The aerosol over the sea consists of sulphates and chlorides and hydrated particles, which are all very transparent. On the other hand, as the "background" aerosol optical depth over the sea is quite low any cooling effect is likely to be small. For highly absorbing sooty substances over snow, a warming is likely. Between those extremes, the situation is still ambiguous.

The ratio a/Re is not well known because the value of n_i has not been determined to any high accuracy. Until it is known better, the effects of aerosols on the radiation budget will remain problematical. A direct measurement of atmospheric solar absorption and upward scattered (backscattered) solar flux density from an aircraft (Paltridge and Platt, 1973) in both a visually clear and a very hazy atmosphere indicated that the ratio of the excess backscatter to the excess absorption due to the haze changed dramatically with altitude. In the first 600 m above the sea surface the ratio of a/Re was about 5, indicating a strongly absorbing haze layer. This lower layer originated from coastal bush fires. The ratio a/Re reduced to an average of 0.8 between 0.8 and 2 km altitude, and to 0.25 between 2 and 3 km. The latter figure is more representative of a "normal" mixed continental and maritime layer. Values of a/Re deduced from the measurements of G.D. Robinson (1966) from surface and aircraft flights in Britain were in the range of about 4–8.

Large increases in turbidity over populated areas such as the eastern United States could have appreciable local effects. The turbidity in these regions at present is already about 3 times the ocean background level, although how much of this is a result of man's activities is uncertain.

It must be stressed again that the above analysis ignores many feedback processes which affect the radiation balance. Other factors such as cloud amount will be affected by a change in the earth's temperature and may adjust themselves accordingly to reduce or amplify that change.

Another point which should be considered is the spectral variation of the short-wave surface albedo α, which is particularly pronounced for land surfaces (see Fig. 6.10). The albedo tends to be low in the visible and high in the near-infrared. If the ratio a/Re of the aerosol layer is also spectrally dependent, then a wavelength integration is necessary to obtain an effective value of the resultant change in planetary albedo. Very little is known at present about the spectral dependence of a/Re. G.D. Robinson (1966) noted that the absorption in the near-infrared appeared to be larger than in the visible.

CHAPTER 10

Radiation and General Dynamics

1. THE MEAN TEMPERATURE PROFILE OF THE ATMOSPHERE

1.1. *Introduction*

Fig. 10.1 gives standard atmosphere temperature profiles at various latitudes in the northern hemisphere for both summer and winter. The gross features of these observed profiles emerge more or less automatically from modern general circulation models which contain among other things detailed "radiative convective equilibrium" schemes based on the classic series of papers by Manabe and his colleagues (Manabe and Möller, 1961; Manabe and Strickler, 1964; Manabe and Wetherald, 1967). However, these schemes in turn are based on numerical development of earlier semi-analytic models which, while necessarily too simple, provide a much better medium for understanding the basic physics of the situation.

The problem is to explain the observed profiles and in particular (1) the presence of the stratosphere and the fact that — at least in the mean — the lower stratosphere is fairly close to isothermal; (2) the fairly uniform lapse rate of temperature in the troposphere; (3) the existence of the tropopause; and (4) the latitudinal variation of temperature in the stratosphere which is such that high surface temperatures are correlated with low stratospheric temperatures — cf. the tropics and the polar regions. These features are intimately associated with the physics of atmospheric radiation and the historical approach provides excellent insight into the overall problem of global energy balance. It illustrates also the physical consequences of band absorption as opposed to grey-body absorption. For these reasons the beautiful review by Pekeris (1932) and the discussion by Goody (1964) are required reading.

Imagine a gaseous absorbing medium of infinite horizontal extent sandwiched between two surfaces of equal temperature. Such a system is the classic black body wherein the medium would adopt a state of both radiative and thermal equilibrium. The net radiative flux, the net radiative flux divergence and the gradient of temperature would be zero everywhere. If now the upper surface is removed the medium will tend to adjust itself to a new state of radiative equilibrium where the net flux divergence is zero (by definition of radiative equilibrium) but which now supports a net and constant upward flux and a temperature profile determined by this flux and

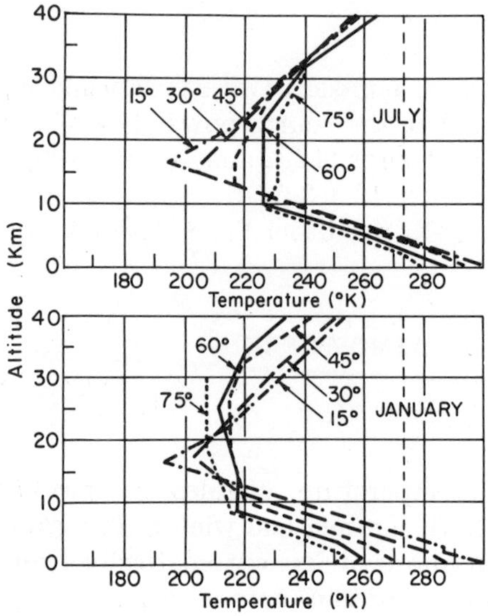

Fig. 10.1. Temperature profiles to 40 km altitude of the January and July supplemental atmospheres for 15°. 30°, 45°, 60° and 75°N. (After Cole et al., 1965.)

the boundary conditions. Provided that it is physically possible to achieve such a state and that the equilibrium is stable to small disturbances, this would be the expected state. A close analogy is the conduction of heat down a metal rod.

To a first approximation one can envisage the stratosphere as such a medium — as did Humphreys (1909). Its absorption of downcoming solar radiation F_0 is very small so that to preserve planetary energy balance its lower "boundary" emits long-wave radiation upwards equal to $(1 - \bar{\alpha}_0)\bar{F}_0$ where $\bar{\alpha}_0$ is the mean planetary short-wave albedo. The boundary consists of both the troposphere and the surface and has an effective mean radiative temperature $\bar{T}_0 \approx 254$ K (from equation [1.1] with $\bar{\alpha}_0 = 0.28$). Suppose now that the stratosphere is a grey body as far as long-wave radiation is concerned with an absorptance or emissivity ϵ constant with wavelength. It will absorb an amount $\epsilon\sigma\bar{T}_0^4$ of the upwelling long-wave radiation from the boundary. Energy balance dictates that its temperature must adopt a value $\bar{T}_2$ such that:

$$2\epsilon\sigma\bar{T}_2^4 = \epsilon\sigma\bar{T}_0^4 \qquad [10.1]$$

and therefore that:

$$\bar{T}_2 = \bar{T}_0/2^{1/4} \qquad [10.2]$$

This simple model predicts a stratospheric temperature $\bar{T}_2$ of about 213 K which is not far from the observed mean temperature of the lower stratosphere.

However, the agreement is largely fortuitous since the grey-body assumption ignores two phenomena associated with band absorption. On the one hand the spectral nature of atmospheric emission (with practically no emission in the 8–14-μm region) is such that the fourth-power law of the Stefan-Boltzmann equation does not apply. Pekeris (1932) made some rough calculations to show that the power is more like 3.5 and that therefore equation [10.2] becomes $\overline{T}_2 = \overline{T}_0/2^{(1/3.5)}$. On the other hand the radiation which is actually absorbed by the stratosphere will be at wavelengths outside the atmospheric window. This radiation derives from the water vapour and carbon dioxide emission of the troposphere rather than from the ground. In other words the absorption bands of the stratosphere will be looking only at corresponding emission bands in the upper troposphere whose radiative temperature is very much lower than that of the ground and lower than $\overline{T}_0$. Both these effects tend to reduce the predicted stratospheric temperature well below that which is observed.

Grey-body absorption by the atmosphere is physically untenable as a model and, as will appear shortly, becomes obviously so when attempts are made to predict the latitudinal variation of stratospheric temperature. Nevertheless, it is a useful basis on which to discuss the concept of "radiative convective equilibrium" which was introduced somewhat awkwardly by Gold (1909) and developed largely by Emden (1913). Emden showed that it is impossible for an atmosphere in radiative equilibrium to be simultaneously isothermal (cf. Humphreys and Gold). Further he developed the heuristic argument of Gold that radiative equilibrium for the whole atmosphere demands lower-atmosphere lapse rates which are hydrostatically unstable and which cannot therefore be maintained. On this basis the troposphere corresponds to that region of the atmosphere where equilibrium is maintained convectively and which develops *as a result* of the tendency towards radiative equilibrium which is achieved only in the stratosphere. The stratosphere is close to being isothermal only because of its small optical depth relative to its large actual depth. The lapse rates are correspondingly small.

1.2. *Radiative equilibrium and the stratosphere*

We follow here Goody's (1964) discussion of the Emden (1913) model since he develops the argument in the modern idiom of the transfer equation. It is necessary first to derive the profile of temperature in an atmosphere of given distribution of absorbing substance on the assumption that radiative equilibrium can be maintained. Such a profile is derived most conveniently by using a form of the Eddington approximation whereby the upward and downward flux streams are assumed to be isotropic within their respective hemispheres. The flux densities $F\uparrow$ and $F\downarrow$ can then be expressed in terms of single-valued radiances N^+ and N^- each independent of direction within their hemisphere. The associated error of a few percent is insignificant in the present qualitative context.

In terms of optical depth $\tau_\nu(=\kappa_{a\nu}\rho dz)$ at wavenumber ν the transfer equation:

$$\cos\theta \frac{dN_\nu(\theta)}{d\tau_\nu} = N_\nu(\theta) - \tilde{J}_\nu \qquad [10.3]$$

involves the directionally specified radiance $N_\nu(\theta)$ and the source function $\tilde{J}_\nu$ which, for a non-scattering medium in thermodynamic equilibrium, is equal to the Planck function B_ν. If both sides are multiplied by $d\omega(=\sin\theta\, d\theta d\phi)$ and integrated over the complete sphere we obtain:

$$\int_0^{2\pi} d\phi \int_0^{\pi} \frac{dN_\nu(\theta)}{d\tau_\nu} \cos\theta \sin\theta\, d\theta = \int_0^{2\pi} d\phi \int_0^{\pi} N_\nu(\theta) \sin\theta\, d\theta - \int_0^{2\pi} d\phi \int_0^{\pi} B_\nu \sin\theta\, d\theta$$

$$= 2\pi \int_0^{\pi/2} N_\nu^+ \sin\theta\, d\theta + 2\pi \int_{\pi/2}^{\pi} N_\nu^- \sin\theta\, d\theta - 4\pi B_\nu$$

$$= 2\pi(N_\nu^+ + N_\nu^-) - 4\pi B_\nu \qquad [10.4]$$

Comparing the left-hand side with the definition of net flux density wherein $F_{N\nu} = \int_0^{2\pi} d\phi \int_0^{\pi} N_\nu(\theta) \cos\theta \sin\theta\, d\theta$, we have:

$$dF_{N\nu}/d\tau_\nu = 2\pi(N_\nu^+ + N_\nu^-) - 4\pi B_\nu \qquad [10.5]$$

Again, by multiplying the transfer equation [10.3] by $\cos\theta\, d\omega$ and integrating over the sphere we have:

$$\int_0^{2\pi} d\phi \int_0^{\pi} \frac{dN_\nu(\theta)}{d\tau_\nu} \cos^2\theta \sin\theta\, d\theta = \int_0^{2\pi} d\phi \int_0^{\pi} N_\nu(\theta) \cos\theta \sin\theta\, d\theta$$

$$- \int_0^{2\pi} d\phi \int_0^{\pi} B_\nu \cos\theta \sin\theta\, d\theta \qquad [10.6]$$

That is:

$$\int_0^{2\pi} d\phi \int_0^{\pi/2} \frac{dN_\nu^+}{d\tau_\nu} \cos^2\theta \sin\theta\, d\theta + \int_0^{2\pi} d\phi \int_{\pi/2}^{\pi} \frac{dN_\nu^-}{d\tau_\nu} \cos^2\theta \sin\theta\, d\theta = F_{N\nu} - 0 \qquad [10.7]$$

so that:

$$\frac{2\pi}{3} \frac{d}{d\tau_\nu} (N_\nu^+ + N_\nu^-) = F_{N\nu} \qquad [10.8]$$

Eliminating $N_\nu^+ + N_\nu^-$ from equations [10.5] and [10.8] we have:

$$4\pi \frac{dB_\nu}{d\tau_\nu} = 3 F_{N\nu} - \frac{d^2 F_{N\nu}}{d\tau_\nu} \qquad [10.9]$$

which is a second-order differential equation describing the distribution of B_ν (i.e. of temperature) as a function of τ_ν (i.e. of depth in the medium) for a given net flux density $F_{N\nu}$. This equation requires two boundary conditions for solution but in the special case of monochromatic equilibrium where the net flux divergence is zero (i.e. $dF_{N\nu}/d\tau_\nu$ and $d^2F_{N\nu}/d\tau_\nu^2 = 0$) the equation reduces to one of the first order:

$$dB_\nu/d\tau_\nu = 3F_{N\nu}/4\pi \qquad [10.10]$$

requiring only a single boundary condition. Such a condition can be obtained from either of the net flux density continuity requirements at the top and bottom of the atmosphere, that is $F_{N\nu}/2\pi = B_\nu(T \text{ at } \tau_\nu = 0)$ at the top of the atmosphere since the downcoming long-wave flux is zero there; or $F_{N\nu}/2\pi = B_\nu(T^*) - B_\nu(T_g)$ at the bottom of the atmosphere where T^* is the temperature at the bottom of the atmosphere and $B_\nu(T_g)$ is the Planck function corresponding to actual surface temperature T_g. Note that for conditions of radiative equilibrium there must be a temperature discontinuity at both the top and bottom of the atmosphere. Further, since only one boundary condition is required, the situation allows calculation of the lower boundary temperature [i.e. of $B_\nu(T_g)$]. Thus solution of equation [10.10] leads to:

$$B_\nu(T \text{ at } \tau_\nu) = B_\nu(\tau_\nu) = F_{N\nu}(1 + 3\tau_\nu/2)/2\pi \qquad [10.11]$$

$$\text{and } B_\nu(T_g) = F_{N\nu}(2 + 3\tau_\nu/2)/2\pi \qquad [10.12]$$

If we assume a grey absorbing medium the subscript ν can be dropped and:

$$T(\tau) = [\pi B(\tau)/\sigma]^{1/4} \qquad [10.13]$$

$$T_g = [\pi B(T_g)/\sigma]^{1/4} \qquad [10.14]$$

Goody (1964) examined the properties of an atmosphere in radiative equilibrium and of various total optical depths τ^* and a distribution of absorbing substance with height h such that τ from h to the top of the atmosphere is given by

$$\tau = \tau^* \exp[-h/(2 \times 10^5)] \qquad [10.15]$$

This distribution corresponds to an emissivity varying in direct proportion to atmospheric density. For such a distribution and for a net infrared flux F_N (positive in the upward direction) equal to the net solar energy absorbed at the ground $\bar{F}_0(1 - \bar{\alpha}_0)$ the temperature profiles corresponding to various values of τ^* are given in Fig. 10.2. Note that for the present description solar absorption in the atmosphere is assumed zero.

As the transparency increases the atmospheric profile tends towards the

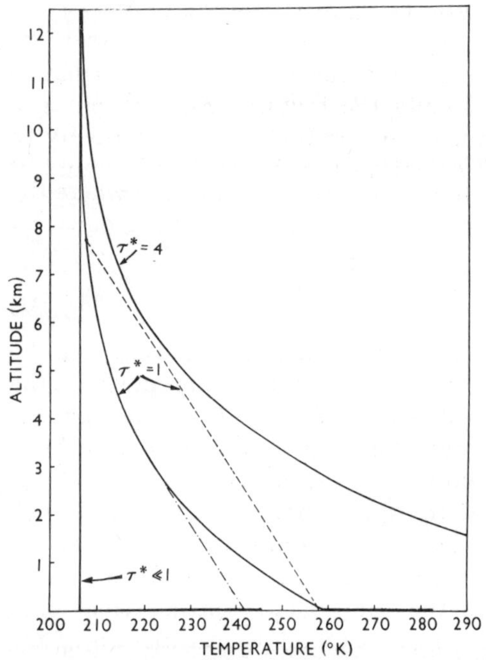

Fig. 10.2. Temperature profiles under radiative equilibrium (full lines) as modified by tropospheric convection (broken lines). (After Goody, 1964.)

isothermal case envisaged by Humphreys (1909) and Gold (1909). The negative temperature discontinuity at the surface ranges from about 40° for a very transparent atmosphere towards zero for a very opaque atmosphere. Such a discontinuity is inherently unstable, as is any lapse rate greater than about 6.5 K km^{-1} which is the typically observed lapse rate limit in the real convective troposphere. Supposing for example that the $\tau^* = 1$ curve is roughly representative of the grey atmosphere equivalent to the real atmosphere, we would therefore expect radiative equilibrium to be destroyed at altitudes below about 3.2 km and to be replaced by convective equilibrium with a linear profile corresponding to the left-hand dashed curve in the figure.

1.3. *Radiative-convective equilibrium and the troposphere*

It is important to realize that the state of radiative equilibrium in the upper atmosphere is completely unaffected by the physical processes occurring in the lower atmosphere provided that these processes still maintain the supply of net upward radiative energy F_N. This requirement specifies the upper boundary condition of the convective layer. It then becomes apparent that the temperature profile of the left-hand dashed curve of Fig. 10.2 cannot supply the required flux to maintain the $\tau^* = 1$ radiative equilibrium profile above 3.2 km, because the temperature at all points along the dashed curve below 3.2 km (including the ground) is always less than the temperature of the profile below 3.2 km demanded by pure radiative equilibrium. This latter profile does supply the required F_N. On physical grounds therefore, the convective equilibrium profile must move bodily to higher temperatures — i.e. towards the right in the figure. Furthermore, it must move sufficiently far to the right so as not to introduce a negative temperature discontinuity in the overall profile. In this example, it must move so that surface air temperature is at least equal to 258 K, with a corresponding rise in the lower boundary of the region of radiative equilibrium to 7.8 km. There is no mathematical requirement for continuity of temperature gradient at this boundary. Obviously it corresponds to the tropopause of the real atmosphere. Goody (1964) goes on to show that the tropopause height so defined must lie between 6.04 km for a very transparent atmosphere and 10.2 km for an atmosphere where $\tau^* = 4.0$.

The qualitative picture is therefore very satisfying as far as it goes. It does not, however, do more than predict the temperature profile for a given distribution of absorbing substance. In the real atmosphere the distribution of water vapour is itself a function of temperature profile, surface temperature and convection dynamics and must in the end be predictable *a priori*. Further, the grey-body assumption becomes obviously untenable when the profiles are examined in more detail. First, the model predicts a stratospheric temperature close to that given by the Planck function at the top of the atmosphere. This in turn is directly proportional to the net upward flux and therefore to surface temperature. Fig. 10.1 shows that the temperature T_{st} of the lower stratosphere is lowest in the tropical regions where surface temperature T_g is highest. Second, the model predicts only negative lapse rates in the stratosphere when it is known that positive lapse rates exist at low latitudes.

Qualitatively it is easy to understand how non-grey band absorption can produce an out-of-phase relation between T_g and T_{st}. Suppose for the purposes of illustration that the infrared absorption spectrum of the atmosphere is divided into only two regions — the completely clear atmospheric window between 8 and 14 μm, and an opaque or semi-opaque region outside these wavelength limits. At terrestrial temperatures the peak of the black-body spectrum occurs within the window and a rise of surface temperature yields an increase in total infrared emission from the surface and an increase in window region radiation out the top of the atmosphere. This last has no interaction with the stratosphere and is irrelevant to its radiative equilibrium. On the other hand a rise in surface temperature is (usually) accompanied by an increase in surface evaporation and/or an increase in the total water vapour in the troposphere. The optical depth of the troposphere is increased. The centre of gravity of the radiation emitted by water vapour in the troposphere is shifted upwards in height to a new level of lower temperature

T_e. Stratospheric temperature is correlated with T_e rather than T_g since radiative equilibrium is determined by the absorbing wavelengths. In effect the surface and the stratosphere are radiatively decoupled by the presence of the atmospheric window.

Goody (1949) introduced simple band absorption into his development of the basic Emden model and was able to reproduce quite well the observed latitude variation of stratospheric temperature and tropopause height (see Table 10.1). However, the agreement is again somewhat fortuitous. Goody (1964) himself points out that the agreement of his model with observation is far better than it deserves to be. More detailed computations by Manabe and Strickler (1964) point to the fact that the observed tropopause heights and lower stratosphere temperatures cannot be explained purely in terms of such a simple radiative-convective equilibrium model. It is necessary to include various aspects of the large-scale dynamics of the system in order to achieve a realistic picture.

TABLE 10.1

The temperature and height of the tropopause in a cloud-free atmosphere, without direct solar absorption (after Goody, 1949)

Ground temperature (K)	Tropopause temperature (K)		Tropopause height (km)	
	calculated	observed	calculated	observed
300	205	193	14.6	16.5
290	208	208	12.6	12.6
280	210	214	10.8	10.2
270	213	217	8.8	8.2
260	214	219	7.1	6.3

Solar absorption in the stratosphere is a small fraction of total planetary absorption but its effect on the temperature profile there can be large. It is quite possible to obtain positive stratospheric lapse rates — particularly towards the tropics where solar absorption in the ozone layer is most significant. In order to obtain quantitative figures it is necessary to employ detailed numerical calculations such as those of Manabe and Strickler. Figs. 10.3 and 10.4 are from their paper. Fig. 10.3 shows the pure radiative equilibrium profiles to be expected from an atmosphere containing the observed amounts of water vapour only, carbon dioxide only, water vapour and carbon dioxide only and finally water vapour, carbon dioxide and ozone. It is primarily the effect of ozone which produces the positive stratospheric lapse rates. Fig. 10.4 shows the profiles of temperature obtained for radiative equilibrium together with a convective adjustment to 6.5 K km^{-1}. The profiles with cloud at various levels are compared with that of a clear sky. An interesting feature here is the increase of surface temperature caused

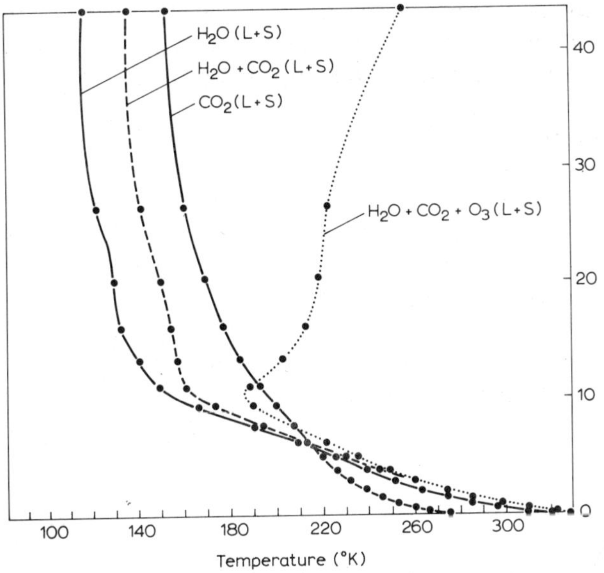

Fig. 10.3. Pure radiative equilibrium for various atmospheric constituents. $(L + S)$ indicates that both long- and short-wave radiation is considered. (From Manabe and Strickler, 1964.)

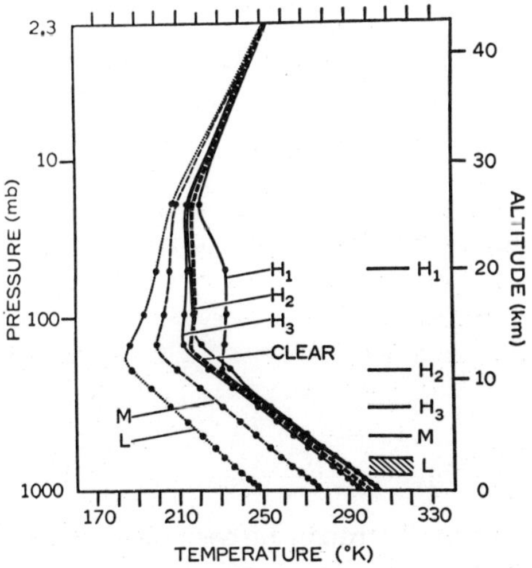

Fig. 10.4. Radiative-convective equilibrium of various atmospheres with clouds (the critical lapse rate for convective adjustment is 6.5° km^{-1}). On the right-hand side of the figure the height of overcast clouds used for each computation is shown, H_1, H_2, and H_3 denoting high clouds, M and L denoting middle and low clouds. (After Manabe and Strickler, 1964.)

by the insertion of high cirrus cloud. This possibility has already been discussed in Chapter 1.

2. RADIATION IN THE TROPOSPHERE

2.1. *Radiative cooling and profile curvature*

As mentioned in Chapter 1, calculations of the large-scale energy budget of the stratosphere show that it is not in fact in overall radiative equilibrium and that there must be some other energy inputs to ensure total energy balance. The net radiative cooling at higher latitudes in the stratosphere is not balanced by the amount of radiant heating (mostly solar heating) which exists in tropical latitudes. Nevertheless, despite these heatings and coolings, compared to the troposphere it closely approximates the definition of radiative equilibrium in a stratified medium wherein:

$$\mathrm{d}F_{\mathrm{N}}/\mathrm{d}\tau = 0 \qquad [10.16]$$

This implies that F_{N} which in practice includes the net flux density of solar radiation as well as the net infrared flux density, is constant with altitude. For simplicity we will continue with the incorrect assumption that solar absorption in the atmosphere itself is negligible so that the divergence of short-wave flux is assumed zero whether or not the atmosphere is in fact in radiative equilibrium. Thus in the following F_{N} refers only to the net flux of long-wave radiation.

Referring back to equation [10.10] of the previous section, zero divergence of net flux density at a particular wavenumber ν appears to imply that:

$$\mathrm{d}B_\nu/\mathrm{d}\tau_\nu = \text{const.} = C_1 \text{ (say)} \qquad [10.17]$$

and, in terms of the grey-body assumption where the black-body radiation is proportional to the amount of absorber $\mathrm{d}\tau$ and to the fourth power of temperature T, that:

$$\mathrm{d}(T^4)/\mathrm{d}\tau = \text{const.} = C_2 \text{ (say)} \qquad [10.18]$$

Now:

$$\mathrm{d}(T^4)/\mathrm{d}\tau = [\mathrm{d}(T^4)/\mathrm{d}T](\mathrm{d}T/\mathrm{d}\tau) = 4T^3\mathrm{d}T/\mathrm{d}\tau \qquad [10.19]$$

so that over small ranges of temperature where T^3 might be assumed roughly constant:

$$\mathrm{d}T/\mathrm{d}\tau = \text{const.} = C_3 \text{ (say)} \qquad [10.20]$$

and:

$$\mathrm{d}^2T/\mathrm{d}\tau^2 = 0 \qquad [10.21]$$

which appears then as another approximate condition for radiative equilibrium.

However, it must be emphasized at once that equation [10.17] is not in fact a definition of radiative equilibrium for a given layer of the atmosphere. The distribution of net flux density across a layer is not solely determined by the radiative temperature distribution across the layer but is determined also by the emission from all the layers above and below the level in question. The point becomes more evident in the following discussion of inversions in the troposphere. Further, it should be appreciated that the condition of equation [10.21] does not necessarily imply that d^2T/dz^2 is zero (z is the actual vertical distance in the atmosphere and in the following will be assumed positive upwards) even if the approximation of constant temperature is accepted. For instance the predicted temperature profiles in the stratosphere (see Fig. 10.2) show d^2T/dz^2 to be positive as a result of the upward decrease in mixing ratio of the absorbing substance.

If pure radiative equilibrium does not exist, the sign of the non-zero $d^2F_N/d\tau^2$ indicates whether a particular layer is being radiatively warmed or radiatively cooled. A positive sign indicates warming and vice versa. However, for equilibrium to exist there must still be total energy balance. That is, $d^2F_T/d\tau^2$ must be zero where F_T is the net of all types of vertical energy flow into and out of the layer and may include both latent and sensible heat flows. This then is the situation of "radiative-convective" equilibrium which pertains in the *mean* troposphere where dT/dt is zero and the radiative losses or gains are balanced by "convective" vertical fluxes of energy. Throughout most of the troposphere, despite the fact that d^2T/dz^2 may be slightly positive (e.g. there is a gradually increasing slope of the potential temperature profile as a function of height as a result of the release of latent heat in upper tropospheric levels) the curvature of temperature with respect to optical depth, $d^2T/d\tau^2$ is generally negative. Therefore, accepting the approximations involved in the development of equation [10.21], the troposphere is radiatively cooled almost everywhere (cf. Fig. 1.13 of Chapter 1).

There are special circumstances apart from those involving clouds when specific layers of the troposphere are radiatively warmed. The most obvious example can occur at the base of inversions where the curvature of temperature is large and positive. On the other hand the top of inversions can be more strongly cooled than the average because the curvature of temperature is large and negative. It might therefore be expected that radiation acts so as to destroy inversions. However, this simple picture is more often wrong than right because there can be very sharp gradients of absorbing substance (water vapour) within and across the inversion itself. The correspondence between $d^2T/d\tau^2$ and cooling rate (to say nothing of the correspondence between d^2T/dz^2 and cooling rate) breaks down. Detailed calculations of actual flux density divergence over very narrow increments of height are required. Since this problem occurs elsewhere also it is appropriate to elaborate on the mechanics of such calculations before proceeding.

2.2. *Calculation of flux divergence of thin layers*

In the following, let the path length be positive in the downward direction and let the subscripts g, r and 0 refer to the value of a variable at ground level, reference level and top of the absorbing atmosphere respectively. The slab or flux emissivity ϵ of an isothermal layer of optical thickness $\tau_r - \tau$ is the ratio of the total flux density emitted through one of its boundaries to that emitted if it were a black body. Therefore:

$$\epsilon(\tau_r - \tau) = \frac{1}{\sigma T^4} \int_0^\infty \tilde{B}_\nu(T)[1 - \tilde{Tr}_\nu(\tau_r - \tau)]\,d\nu \qquad [10.22]$$

where it should be remembered that $\tilde{B}_\nu(T)$ is the Planck black-body function in *flux density* units and $\tilde{Tr}_\nu$ is the flux transmittance of the layer. By ignoring the temperature dependence of $\tilde{Tr}_\nu$ and therefore of ϵ, this equation can be used to transform the flux density versions of the transfer equation [7.5] in Chapter 7 to:

$$F\!\downarrow_r = -\int_0^{\tau_r} \sigma T^4 \frac{\partial \epsilon(\tau_r - \tau)}{\partial \tau}\,d\tau \qquad [10.23]$$

and:

$$F\!\uparrow_r = \int_{\tau_r}^{\tau_g} \sigma T^4 \frac{\partial \epsilon(\tau - \tau_r)}{\partial \tau}\,d\tau + \sigma T_G^4[1 - \epsilon(\tau_g - \tau_r)] \qquad [10.24]$$

where T_G is the radiative ground temperature not necessarily the same as the actual air temperature at the ground T_g. (The flux density versions of equation [7.3] are obtained in the same manner as the flux density equations [7.11] and [7.12] were obtained from equation 7.5.) In terms of height z and net flux F_N:

$$\frac{\partial F_N}{\partial z} = \frac{\partial F_N}{\partial \tau}\left(\frac{\partial \tau}{\partial z}\right)_r = \left(\frac{\partial F\!\uparrow_r}{\partial \tau_r} - \frac{\partial F\!\downarrow_r}{\partial \tau_r}\right) \cdot \left(\frac{\partial \tau}{\partial z}\right)_r \qquad [10.25]$$

so that by differentiating equations [10.23] and [10.24] with respect to τ_r and remembering that $\partial\epsilon/\partial\tau_r = -\partial\epsilon/\partial\tau$:

$$\frac{\partial F_N}{\partial z} = -\sigma\left(\frac{\partial \tau}{\partial z}\right)_r \left[\int_0^{\tau_r} T^4 \frac{\partial^2 \epsilon(\tau_r - \tau)}{\partial \tau^2}\,d\tau + \int_{\tau_r}^{\tau_g} T^4 \frac{\partial^2 \epsilon(\tau - \tau_r)}{\partial \tau^2}\,d\tau + T_G^4 \frac{\partial \epsilon(\tau_g - \tau_r)}{\partial \tau_r}\right] \qquad [10.26]$$

or by integrating this equation by parts:

$$\frac{\partial F_N}{\partial z} = \sigma\left(\frac{\partial\tau}{\partial z}\right)_r \left[(T_g^4 - T_G^4)\frac{\partial\epsilon(\tau_g-\tau_r)}{\partial\tau_r} - T_0^4\frac{\partial\epsilon(\tau_r)}{\partial\tau_r} + \int_0^{\tau_r}\frac{\partial T^4}{\partial\tau}\frac{\partial\epsilon(\tau_r-\tau)}{\partial\tau}\mathrm{d}\tau + \int_{\tau_r}^{\tau_g}\frac{\partial T^4}{\partial\tau}\frac{\partial\epsilon(\tau-\tau_r)}{\partial\tau}\mathrm{d}\tau\right] \quad [10.27]$$

or by partial integration of this equation in turn:

$$\frac{\partial F_N}{\partial z} = \sigma\left(\frac{\partial\tau}{\partial z}\right)_r \left[(T_g^4 - T_G^4)\frac{\partial\epsilon(\tau_g-\tau_r)}{\partial\tau_r} - T_0^4\frac{\partial\epsilon(\tau_r)}{\partial\tau_r} + \left(\frac{\partial T^4}{\partial\tau}\right)_g\epsilon(\tau_g-\tau_r) - \left(\frac{\partial T^4}{\partial\tau}\right)_0\epsilon(\tau_r) - \int_0^{\tau_r}\frac{\partial^2T^4}{\partial\tau^2}\epsilon(\tau_r-\tau)\mathrm{d}\tau - \int_{\tau_r}^{\tau_g}\frac{\partial^2T^4}{\partial\tau^2}\epsilon(\tau-\tau_r)\mathrm{d}\tau\right] \quad [10.28]$$

The three equations [10.26], [10.27] and [10.28] are all equally valid and in their finite difference form (integrations replaced with sums) can in principle be used in all circumstances. However, the choice in practice depends on the accuracy with which the temperature profile is known. The form of ϵ (see, for example, Fig. 7.20 of Chapter 7) as a function of optical thickness is not known precisely. If the temperature profile is well known — as in many micrometeorological applications near the ground — equation [10.28] would be preferable since, although the integrals contain double differentials of T^4, they contain no differentials of ϵ. At the other extreme where the temperature profile is not known precisely and the differentials of T^4 would be therefore highly inaccurate, it might be preferable to use equation [10.26] which contains instead the double differentials of ϵ. In general the compromise of equation [10.27] which contains the single differentials of both T and ϵ is used. This form is the basis of the Brooks (1950) early tabular method of calculation and is the basis of Staley's (1965) calculations on the subject of inversions which will be discussed shortly.

Equation [10.28] shows clearly the reason why the curvature of temperature at the level in question is not always a good indication of radiative cooling or heating at that level — quite apart from the approximation that $\partial^2T/\partial\tau^2$ is proportional to $\partial^2T^4/\partial z^2$. The integrals in the expression for divergence contain the curvature of temperature at all levels of the atmosphere. The contribution to the cooling at one level i_0 by the curvature at other levels i_n decreases as the physical distance between i_n and i_0 increases because of the decreasing transmission between them. Thus the cooling at i_0 is dominated by the emission from the closest layers. Nevertheless it is easy to envisage situations where the local curvature over a thin layer might be positive but negative curvature elsewhere ensures that the layer cools rather than warms. The occasions on which local curvature can be

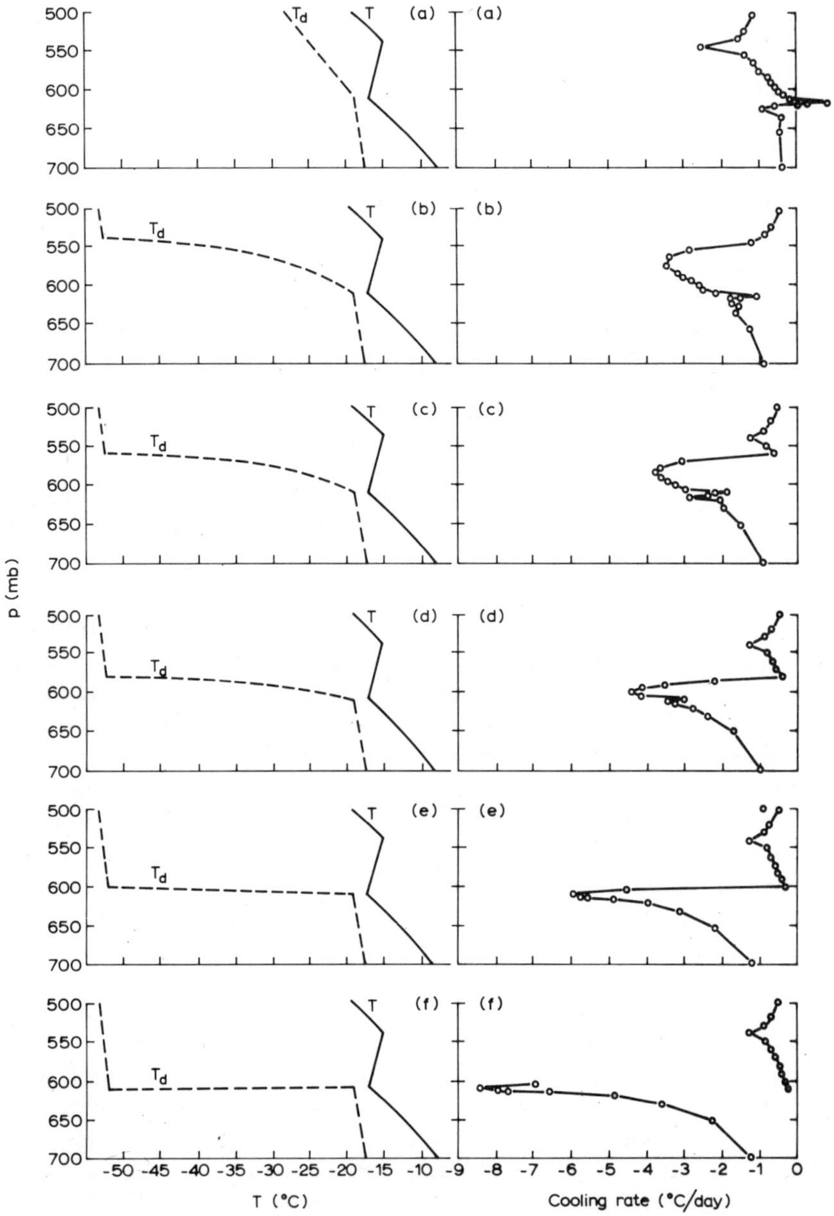

Fig. 10.5. Radiative cooling profiles associated with a temperature inversion and different degrees of "sharpness" in the water vapour profile. T and T_d are the dry and dewpoint temperatures respectively. (After Staley, 1965.)

a reliable indicator of radiative cooling or warming are restricted to cases where the water vapour mixing ratio and therefore absorption is high at the point in question and in the adjacent layers. The local curvature terms are then more likely to be dominant. Such conditions are usually present within the boundary layer close to the ground.

2.3. *Radiation and atmospheric inversions*

Fig. 10.5 is from Staley (1965) who made detailed calculations of the radiative cooling rates associated with typical inversions at about the 550-mb level in the atmosphere. The figures show various case studies of temperature and humidity profiles (dry-bulb and dewpoint temperatures are given as the left-hand graphs in each case) and display in sequence the effect of increasing the sharpness of the "drying out" within the inversion.

Fig. 10.5a shows the warming at inversion base and cooling at the top which is to be expected on the grounds of local curvature in a situation of uniformly high water vapour mixing ratios. In this specific case the cooling profile must tend to erode the inversion. The situation changes drastically as one progresses through the sequence. In Fig. 10.5b, maximum cooling is found near the middle of the inversion and the radiative warming at the base has given way to no more than a diminished cooling. The shift of maximum cooling downwards from the top is towards the region of high curvature *and* high specific humidity. At the inversion top the curvature term may be high but the humidity is low. In Fig. 10.5c the maximum cooling is forced even further towards the bottom of the inversion and a level of diminished cooling has appeared about 20 mb below the top in addition to that at the base. Fig. 10.5d, e and f show the progression as the inversion becomes drier to a state where the maximum cooling occurs at the base itself, a condition which would tend to maintain rather than dissipate the inversion.

It would not be uncommon for cooling at an inversion base to reduce the dry-bulb temperature below wet-bulb temperature with the consequent formation of cloud. Once such cloud had formed, and provided that convective activity within it did not somehow erode the inversion such that cloud top penetrated towards the inversion top, the infrared cooling to space of the cloud would greatly enhance the radiative maintenance of the inversion. Qualitatively this effect may be dominant in the almost permanent feature of the trade wind inversion of the subtropics. Quantitatively the phenomenon has yet to be examined in detail.

Staley (1965) points out that it is unlikely that radiation plays a significant role in the initial formation of an inversion since the cooling becomes significant only when the large curvatures of temperature and mixing ratio are already established.

The extremes of cooling rate indicated by Fig. 10.5 are not extraordinarily large but are nevertheless 3 or 4 times the tropospheric average of 1 or 2°C per day. Whether such rates can be expected to modify a particular atmospheric profile significantly depends on whether the detail of the initial cooling profile is maintained. In order to examine this question, Staley and Jurica (1968) studied the progressive radiative modification of several initial temperature distributions of particular interest and carried the integration forward 24 hours in time. Their initial distributions corresponded (a) to a relatively moist inversion in the middle troposphere; (b) to relatively dry inversions in the middle and lower troposphere, within which the moisture decreased rapidly upwards; (c) to a relatively dry inversion in the upper troposphere; and (d) to a smooth temperature profile having a large moisture inversion. In all cases the initial cooling tendency was maintained. In case (a) the strength of the inversion was reduced considerably after 24 hours. In case (b) the cooling rates remained more or less unchanged as the inversions intensified, with a rounding of the temperature profile at the inversion top and a sharpening of the temperature gradient discontinuity at its base. The resultant profile was very similar to those typically observed. After a few hours the lapse rate of the layers just below the inversion base reached the

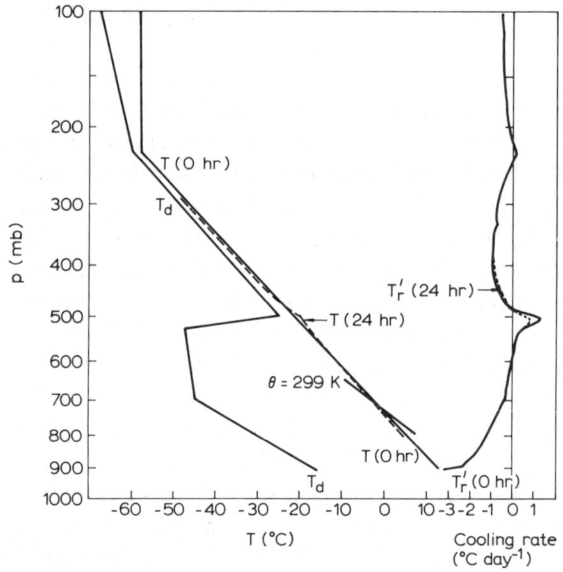

Fig. 10.6. Modification in 24 hours of an initially linear temperature profile T by radiative cooling (at rate T_r') when dewpoint temperature profile T_d is as shown. (From Staley and Jurica, 1968.)

adiabatic value (from an initially stable profile) and subsequently the region of instability progressed downwards. In case (c) the effect of radiation was negligible but was interesting in that the tendency was towards erosion of the inversion in a similar manner to case (a). The reason is that in the upper troposphere the absolute values of moisture content are small and although the change in moisture lapse may be very sharp the absolute curvature of moisture content is not sufficient to modify the cooling dictated by the curvature of temperature. The results of case (d) are reproduced here in Fig. 10.6 primarily to show how a complicated profile of cooling rate can exist even with a purely linear temperature profile throughout the entire troposphere. The 24-hour forward projection in this case shows also an example of how this not uncommon moisture profile can produce an intensely de-stabilized upper troposphere in a time significantly shorter than the period of passage of successive weather disturbances.

3. RADIATION WITHIN THE BOUNDARY LAYER

The previous section discussed situations in the troposphere where typical discontinuities in the gradients of moisture or temperature introduce layers of excess radiative cooling. The rates are not large in absolute magnitude (several degrees per day) but are presumably significant in determining the character of inversions and the overall distribution of stability within the troposphere on time scales of more than a few hours.

The discontinuity at the interface between atmosphere and ground is such that very large radiative cooling rates (many degrees per hour) can exist there. It would not therefore be surprising if the characteristics of the boundary layer were largely controlled by radiative processes. Whether this is so has been the subject of considerable discussion for many years and the problem is not yet resolved. The difficulty is that the turbulence coefficients which control the vertical fluxes of latent and sensible heat are themselves a strong function of such things as temperature gradient (i.e. of stability) and physical scale. Thus there can be tremendously complicated feedback between radiative cooling and, say, the sensible heat flux divergence within a layer. It can be quite incorrect to assert that radiative processes have no effect on a temperature profile simply on the basis that models containing a description of turbulent processes yield the observed profile and models based purely on radiative considerations do not. Almost invariably the "turbulent" models to date contain the equivalent of a pre-set profile of exchange coefficient.

Möller's (1954) treatment of the subject gives a good physical picture of radiative-convective processes near the ground despite the fact that his final conclusion is questionable.

For simplicity he invokes the assumption (tenable near the ground) that local curvature of the temperature profile is a measure of the local radiative cooling rate. Fig. 10.7 is a schematic diagram of typical temperature profiles near the ground as they might exist (a) during the day and (b) during the night. The ground can be considered as a uniformly absorbing medium extending infinitely far downwards with the constant temperature T_g of the radiating surface. (This consideration is valid only from the point of view of the atmosphere itself.) On the basis of temperature profile curvature there must therefore be regions of radiative cooling and heating close to the surface as indicated on the figure, which in both day and night profiles give way to the normal tropospheric cooling at some fairly large height above ground. Following Möller (1954), three effects of radiation can be distinguished.

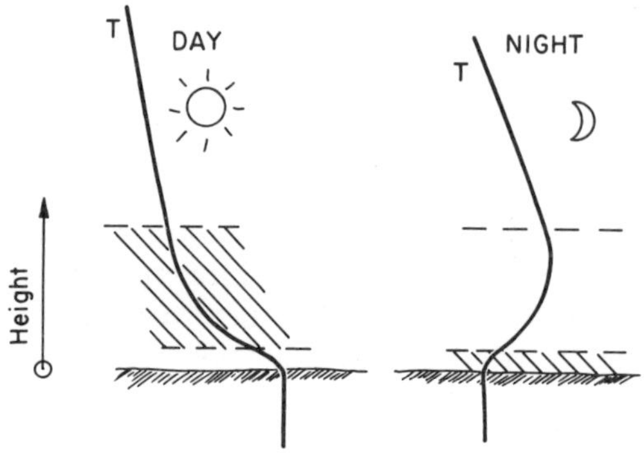

Fig. 10.7. Regions of radiative heating above the ground for typical day and night temperature profiles. Non-hatched regions are of radiative cooling, hatched regions are of radiative heating.

First, the earth's surface itself has a radiation balance different from that of the air. During the day its balance is positive because of the large solar input. During the night its balance is negative since the outgoing long-wave radiation is larger than the return of atmospheric long-wave flux. Normally therefore the temperature of the earth's surface is extreme compared with its immediate environment and there are transports of heat in one form or another between atmosphere and ground. These heat flows may balance each other in the mean over 24 hours, but would not normally do so on lesser time scales. Second, there is a very thin layer of the atmosphere immediately above the ground which is subject to radiative cooling during the day and heating during the night. Third, there is a somewhat thicker layer above this in turn, which is subject to the reverse situation of radiative heating during the day and cooling during the night.

On the scale of normal meteorology (as opposed to micro- and agricultural meteorology) the concern is primarily with the prediction of surface temperature T_g and the vertical flux densities of latent heat $\tilde{L}\tilde{E}$ and sensible heat $\tilde{H}$ into the lower atmosphere. These fluxes maintain the surface energy balance against net radiative input, and their absolute values are associated with turbulent exchange coefficients in the atmosphere itself. (By analogy with molecular conduction, this turbulent exchange coefficient A_e for sensible heat flow, say, in the atmosphere is the constant of proportionality between the heat flux density and the gradient of temperature. It is the turbulent medium equivalent of conductivity in the heat flow through metals.) Further, the divergence of these fluxes maintains the energy balance of a given layer of the mean atmosphere against the radiative energy losses or gains associated with radiative divergence. At any given instant and at any level z above the ground the rate of temperature change of the air (the net cooling rate) might be expressed:

$$\mathrm{d}T/\mathrm{d}t = \frac{1}{\rho}\frac{\partial}{\partial z}\left[A_e(z)\frac{\partial T}{\partial z}\right] + \partial T/\partial t|_{\mathrm{rad}} \qquad [10.29]$$

where $\partial T/\partial t|_{\mathrm{rad}}$ is the cooling rate attributable to the radiative flux divergence at z. The equation assumes zero horizontal divergence and zero advection.

The question of interest to the meteorologist is whether the exchange coefficient and in particular the profile of exchange coefficient $A_e(z)$ is affected by radiative processes. A number of authors have attempted to answer the question by examining various experimental case studies of the surface layers. Still following Möller (1954), equation [10.29] can be rearranged to yield:

$$A_e(z) = \rho(\partial T/\partial z)^{-1}\left[\int_0^z\left(\frac{\mathrm{d}T}{\mathrm{d}t} - \frac{\partial T}{\partial t}\bigg|_{\mathrm{rad}}\right)\mathrm{d}z + \left(\frac{A_e(z)}{\rho}\frac{\partial T}{\partial z}\right)_{z=0}\right] \qquad [10.30]$$

where the only unmeasurable quantity on the right-hand side is $A_e(0)$ — the exchange coefficient at the surface. However, this can be eliminated by making use of the surface energy balance equation:

$$F_{\mathrm{NG}} = \tilde{L}\tilde{E} + \tilde{H} + \tilde{G} \qquad [10.31]$$

which equates the net radiative input into the ground F_{NG} to the upward fluxes of latent and sensible heat from the ground plus the heat flux $\tilde{G}$ conducted downwards into the ground. The sensible heat flux is given by:

$$\tilde{H} = c_p A_e(0)(\partial T/\partial z)_{z=0} \qquad [10.32]$$

so that by substitution of equations [10.31] and [10.32] into equation [10.30]:

$$A_e(z) = \rho\left(\frac{\partial T}{\partial z}\right)^{-1}\left[\frac{1}{\rho c_p}(F_{\mathrm{NG}} - \tilde{L}\tilde{E} - \tilde{G}) + \int_0^z\left(\frac{\mathrm{d}T}{\mathrm{d}t} - \frac{\partial T}{\partial t}\bigg|_{\mathrm{rad}}\right)\mathrm{d}z\right] \qquad [10.33]$$

where all the terms on the right-hand side can in principle be measured or calculated.

Möller goes on to show from experimental data that, except in conditions of very calm winds, the integral term containing the net and radiative cooling rates is several orders of magnitude less than the first term within the square brackets. On this basis he argues that the exchange coefficient is virtually unaffected by radiative processes.

As pointed out by Kondratyev (1972) the final argument as it stands is quite fallacious. The smallness of the integral term says nothing more than that, in typical experimental situations, the radiative cooling is more or less balanced by the cooling (warming?) which results from the vertical divergence of sensible heat. Presumably the exchange coefficient profile adjusts itself so that this is so but it is still entirely possible that *a priori* prediction of $A_e(z)$ would require detailed description of the radiative processes.

Nevertheless, on the fairly broad scale of general meteorology, and if it is generally true that such an adjustment of exchange coefficient does take place, it is sensible to take advantage of the fact and parameterize the situation in such a manner as to ignore the detail of radiation interaction. Typical models of boundary layer interaction with the larger-scale weather processes do just this. They are successful enough but it should be realized that they are not completely mechanistic in the sense that the distribution of turbulent exchange coefficients is not predictable purely from independent variables.

The necessity to examine the detail of radiative processes reappears in calm conditions where, among other things, the concern may be with the formation of fog.

3.1. *Nocturnal cooling and ground fog*

The superficial explanation for the appearance of surface fog on clear nights is simply that the ground is radiatively cooled by net loss of long-wave energy to space through the atmospheric window. The low surface temperature is propagated to the atmosphere by sensible heat exchange (a downward flux which cools the atmospheric layers near the ground) so that atmospheric temperature is reduced below wet-bulb temperature with the consequent formation of liquid water drops and fog.

The occurrence of fog is not nearly as frequent as this simple picture might suggest bearing in mind that atmospheric layers near wet surfaces can be expected to be close to saturation. Two processes act to prevent supersaturation and fog formation. On the one hand much of the earth's surface is hygroscopic and is therefore a sink which tends to reduce the moisture content of the very lowest layers of the atmosphere. Even the ocean is "hygroscopic" in the sense that its associated saturated vapour pressure is lower by about 2% than that of pure water and thus the gradient of moisture content (measured upwards) can be positive and the surface becomes a sink for water vapour. On the other hand the downward flux of sensible heat itself has a divergence which tends to heat the lowest layers and maintain dry-bulb temperature above wet-bulb temperature.

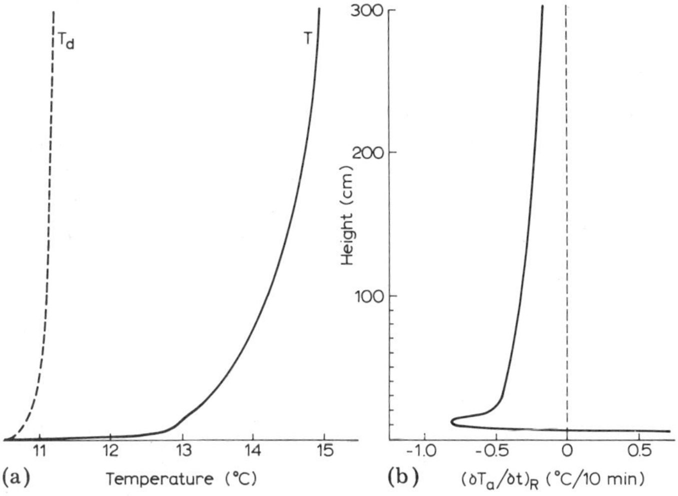

Fig. 10.8. (a) Vertical distribution of temperature (T) and dewpoint (T_d) observed on 26 August, 1954, above a water surface at Friday Harbor, Washington. (b) Vertical distribution of radiative temperature change calculated for the curves shown on the left. (From Fleagle and Businger, 1963.)

Fog formation is in fact an exceedingly complicated phenomenon and is one in which radiative cooling in the atmosphere itself can play a dominant role.

Typical profiles of dewpoint and dry-bulb temperature near the ground at night might be as shown in Fig. 10.8 (from Fleagle and Businger, 1963). Computations of the associated radiative cooling (by application of equation [10.27] for instance) yield the profile at the right of the diagram which shows a thin but powerful maximum of cooling at about the 10–20-cm level. Closer to the ground the air is subject to strong radiative warming. The radiative processes therefore are such as to drive the atmospheric temperature towards saturation at the 10–20-cm level and away from saturation at the surface itself. Under such conditions, if fog is to form it will do so not right at the surface as would be assumed on a simple model involving only sensible heat transfer but at a significant height above ground. Occurrence of the raised temperature minimum and the initial formation of fog away from the surface is quite common and is essentially a radiative phenomenon.

Of course the radiative modification of the temperature profile results in a compensating sensible heat flux by turbulence so that the net cooling may be fairly small. Experimentally it has been found (e.g. Funk, 1960) that the net cooling dT/dt is generally less than $\partial T/\partial t|_{rad}$, and that this implies a warming by sensible heat flux divergence. In any given situation the occurrence of fog will depend on both radiative and turbulent considerations. As mentioned earlier, the turbulent exchange coefficients are themselves a strong function of temperature gradient and there must be a complicated feedback between radiative cooling and the sensible heat fluxes.

Funk (1960, 1962) made an experimental investigation of the conditions existing prior to the formation of several fogs where among other things he made direct measurements of the radiative flux divergence. His composite picture drawn from several case studies is more or less consistent with the above and revealed three further significant factors. First, the radiative cooling rates actually measured were generally larger by a factor of up to 2.8 than those calculated from the temperature and humidity profiles. He attributed this phenomenon to the presence of aerosol particles. This explanation may be correct but the recent discovery that window region emissivity is a strong and increasing function of vapour pressure (see Chapter 7) provides another possibility which is more likely to be correct if this phenomenon of excess cooling rate is fairly universal. Second, long-period fluctuations of about 40-minute period in the temperature of the nocturnal inversion were apparently well correlated with the local radiation cooling rate. This tends to support the thesis of dominant radiative control of the situation. Third, the general patchiness of the measured cooling rates and of the fog itself is indicative of larger-scale advective processes which should be accounted for in any detailed examination.

Zdunkowski and Nielson (1969) developed a simplified numerical model which was applied to several case studies where the distribution of exchange coefficient with height was pre-set and fixed. The model was used to trace the development of ground fog and its transition and uplift to form stratus cloud. The radiative calculations were fairly crude but could be made more precise in a manner described in a subsequent paper (Korb and Zdunkowski, 1970). However, the limiting factor on accuracy is the inability to *calculate* the turbulent exchange coefficients as the fog develops.

According to their model the general pattern of fog development is as follows.

After the air has cooled by turbulent diffusion and radiative flux divergence to the saturation point, fog begins to form. Water droplets form first in the air overlying the ground, the region of strongest initial cooling. Radiative temperature changes are appreciable in the ground fog since the droplet density amounts to a liquid water content of about 0.1–0.2 g m^{-3}.

As the fog increases in vertical thickness it approaches the characteristics of a black body in its lower parts. Radiative temperature changes tend to be small. Moreover radiative energy losses of the soil are strongly decreased, which results in a temperature increase of the earth's surface and the air layer near the ground. Droplets now evaporate which results in a decrease in liquid content in spite of transport of moisture from regions of larger water content.

After the fog has grown to a height ranging from 5 to 15 m within a period of 30–40 minutes after initial fog formation, it slows down in its growth with height. The fog now becomes fairly dense. Moisture transport through the upper fog boundary increases the relative humidity of the air layer. Droplet transport into the same region results in evaporation which causes an increase in dewpoint temperature and a lowering in air temperature. A layer of air approximately 50 m thick now becomes saturated practically at the same time. This causes an explosive type of growth to a new and much larger thickness.

Within one hour after initial formation the fog near the ground has thinned out appreciably and the fog has virtually lifted from the ground. The maximum amount of droplet water is then found in the air layer ranging from 50 m to the upper boundary of the fog at about 250 m. It is now considered a stratus cloud. The predicted liquid water contents ranging from 0.1 to 0.2 g m^{-3} are in good agreement with observation.

The difficulty with this picture is to accept the "explosive growth" phase — or at least the physical reason for it. It is possible in the model because the preset exchange coefficient profile allows significant transport of water vapour and water drops into the layers above the fog. Intuitively one would expect the strong cooling at the top of the fog layer to create a temperature inversion with strongly stable lapse rate just above the visible boundary. This would tend to suppress any vertical transport of water (see the later discussion on stratiform clouds). It is, however, possible that the fog top acts simply as a raised ground surface so that the whole formation process can be repeated at a higher level. Multiple-level fogs are often observed.

4. RADIATION AND THE METEOROLOGY OF CLOUD

4.1. *Introduction*

Research to date has concentrated on the question — "given a cloud of specified microphysics, geometry and position in the atmosphere, how does the cloud affect the radiative streams within the atmosphere; in particular, how does the cloud affect the short- and long-wave absorption at the ground and within the atmosphere as a whole?" This is a static problem which in principle can be solved to any degree of precision by application of numerical techniques to solve the relevant transfer equation. The inverse problem has scarcely been touched — "given the radiative streams associated with cloud and its surrounding atmosphere, how does this radiation affect the character of cloud?" The question is basic to cloud prediction and therefore is bound intimately to the ultimate difficulty of *a priori* prediction of climate. On the very largest scale of general circulation models (e.g. Kasahara and Washington, 1971) some work is being done in the sense that various empirical parameterization schemes are being tried which allow very broad examination of cloud-radiation feedback. However, since "cloud" in such schemes must be defined in terms of one or two variables at a single point every few hundred kilometres, these models are scarcely a good medium for examining the basic physics of radiation interaction. In such models practicality demands that the prediction of cloud involves little more than a simple on-off process. Predefined average "cloud" is or is not present depending on, for instance, the relative humidity. More detailed work has been done on the restricted problem of fog formation (see the previous section). Feigelson (1964) has used numerical techniques to examine the development of the temperature profile within clouds and in particular the development of a strong temperature inversion at the top of stratus clouds as a result of radiative cooling. Apart from this the subject has progressed little further than the recognition (e.g. Knollenberg, 1972) that cooling and heating rates associated with radiative divergence within a cloud must be modified to take account of the release or absorption of latent heat.

The following attempts no more than to give some physical appreciation of how radiation may play a role in determining the formation and character of cloud itself. Calculations are done only at the order of magnitude level. At such a level of approximation the assumption of grey-body absorption over the long-wave spectrum is quite sufficient for the purpose.

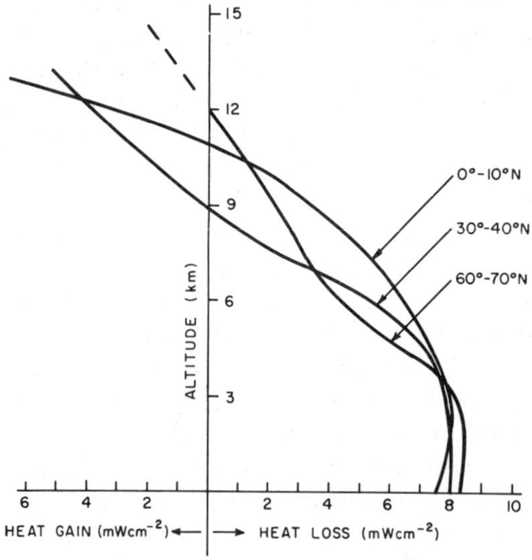

Fig. 10.9. Net radiative loss from a thin "black" cloud inserted at various heights in the mean atmospheres of three latitude zones. (After Paltridge, 1974d.)

Imagine a thin horizontal layer of temperature T at some height in the field of long-wave radiant energy in the clear atmosphere. This field has upward and downward components of flux density $F\uparrow$ and $F\downarrow$ respectively and the net radiative energy F_N absorbed by the layer is given by:

$$F_N = \epsilon F\uparrow + \epsilon F\downarrow - 2\epsilon\sigma T^4 \qquad [10.34]$$

$$= \epsilon F_{N0} \qquad [10.35]$$

where ϵ is the flux emissivity (absorptance) of the layer and $F_{N0}(=F\uparrow + F\downarrow - 2\sigma T^4)$ is the net energy which would be absorbed if the layer were optically black to infrared radiation.

Values of F_{N0} are given as a function of height for three latitude zones in Fig. 10.9. They were calculated using Elsasser charts to give $F\uparrow$ and $F\downarrow$ from the zonal mean profiles of temperature, pressure and humidity given by London (1952). F_{N0} is the maximum possible radiant energy gain or loss by a thin opaque layer or a cloud inserted at a particular level in an otherwise clear atmosphere. According to equation [10.35] the actual absolute value of the net gain or loss would be reduced in proportion to the emissivity. Note first that F_{N0} refers to energy gain or loss by the cloud itself and says nothing about the effect on absorption by the atmosphere above and below or by the planetary system as a whole. Second, a layer cloud in the low atmosphere is radiatively cooled because the upward flux from its top (mainly window radiation direct to space) greatly exceeds the net input to its base as a result of exchange with the ground. However, cloud temperature and radiative loss ($2\epsilon\sigma T^4$) decreases rapidly with height. Its radiative gain decreases far less rapidly because the upwelling flux is dominated by ground emission within the atmospheric window. Thus, depending a little on latitude, a cloud above about 10 km is radiatively warmed. Third, the calculations refer to the imaginary situation of an infinitely thin cloud whose top and bottom temperatures are the same. In practice a cloud must have a finite and significant thickness for its opacity to approach "blackness".

Ignoring this last approximation for the moment, it is necessary next to obtain a relation between cloud emissivity and some basic characteristic such as its content of liquid water or of ice. An exact relation between the two is very difficult to obtain since consideration must be given (see Chapter 8) to the variation of absorption coefficient with wavelength and particle size, to the effect of scatter, and to the absorption by water vapour. However, a relation derived from simple obstruction theory on the basis of 100% absorbing spheres is sufficient in the present context. Thus suppose there are n drops of radius r in a vertical column of air of unit cross-sectional area. The cross-sectional area of the drops in the column is $n\pi r^2$ and their mass is $\frac{4}{3}n\pi r^3\rho$ where ρ is the density of water. The numerical value of ρ is unity in c.g.s. and is ignored in the following. Therefore the obstruction (or absorption) coefficient per unit mass of water is $3/4r$ and the radiance N of a vertical beam of radiation emerging from the column would be given by $N = N_0 \exp(-3W/4r)$ where N_0 is the initial beam radiance and W is the total liquid water or ice content of the column (g cm^{-2}). Now the emissivity of the column corresponds to the quantity $(N_0 - N)/N_0$, and therefore the total flux emissivity of a cloud to hemispheric radiation would be:

$$\epsilon = 1 - \exp[-1.6(3W/4r)] \qquad [10.36]$$

where the 1.6 accounts roughly for the fact that the average path of isotropic radiation through a cloud layer would be greater than the vertical path by this amount. The mean effective droplet radius in average stratocumulus is of the order of 5–10 μm (see the Appendix) and for the ice particles in higher clouds the radius of the equivalent sphere is at least 50 μm. For the present purposes therefore, equation [10.36] becomes:

$$F_N = F_{N0}[1 - \exp(-a_1 W)] \qquad [10.37]$$

where a_1 is of order 1.2×10^3 cm^2 g^{-1} for water clouds of 10-μm drops and of order 2.4×10^2 cm^2 g^{-1} for upper level clouds of ice particles.

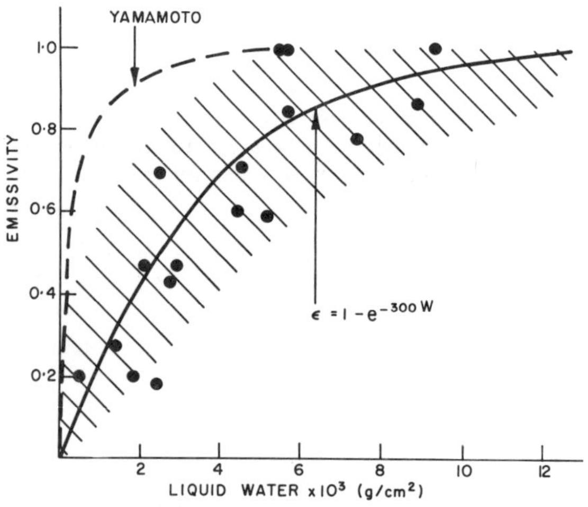

Fig. 10.10. Experimental values of total flux emissivity as a function of liquid water content in low-level stratiform cloud. The hatched area defines the possible inaccuracy, and the dashed curve is the ε versus W relation derived from the work of Yamamoto et al. (1970). (After Paltridge, 1974c.)

As far as ice clouds are concerned, the use of the pure geometrical cross-section of the particles to calculate emissivity is a fairly reasonable approximation since the absorption cross-section of large particles is very close to the geometric cross-section (see Chapter 8). The major error derives from the neglect of absorption by the water vapour within the cloud, but for high clouds at least the amount of this water vapour is relatively small since the saturated vapour pressures are low.

The same cannot be said for water clouds of the low atmosphere where the droplet sizes are much smaller and the saturated vapour pressures are much larger. Nevertheless, the above value of the flux mass absorption coefficient a_1 turns out to be a very reasonable figure in the present context. Referring to Fig. 10.10, it is bracketed by the higher theoretical figure of Yamamoto et al. (1970) and the lower measured values of Paltridge (1974c). It was calculated on the assumption of 10-μm drops on the basis that the choice of an overlarge radius tends to balance the error caused by assuming an absorption efficiency of 1.0.

Equation [10.37] relates the net radiant energy loss or gain from a cloud to the total amount of water W in the vertical column of the cloud. It is possible to relate F_N instead to the actual thickness of the cloud layer provided some assumption is made with regard to the distribution with height in the cloud of mixing ratio or liquid water content per unit mass w. In the present context the adiabatic assumption[1] is sufficient and leads to an

---

[1] The simplest picture of a stratified cloud is of a well mixed layer whose total water content (liquid plus vapour) is roughly constant with height because of the mixing. The temperature decreases with height along the saturation adiabat corresponding to the mixing ratio w_B at the cloud base and the liquid water increases with height as the saturation mixing ratio w_s decreases. The liquid water content w_{LT} per unit mass of air at the top of the cloud is given by:

$$w_{LT} = -\int_{p_B}^{p_T} \frac{dw_s}{dT}\frac{dT}{dp}\,dp = -\frac{dw_s}{dT}\frac{dT}{dp}p_h$$

where p_T and p_B are the pressures at cloud top and bottom respectively and $p_h = p_T - p_B$.

Using Clapeyron's equation to expand dw_s/dT:

$$w_{LT} = -\frac{0.62\tilde{L}}{p\alpha_v T}\frac{dT}{dp}p_h$$

where $\tilde{L}$ is the appropriate latent heat of vaporization of water or ice and α_v is the specific volume of water vapour. Then assuming the increase in liquid water with height to be linear and letting z_h be the cloud thickness, the total liquid water in the cloud is given by:

$$W = \frac{w_{LT}/\alpha}{2} z_h = \frac{w_{LT}/\alpha}{2}\frac{dz}{dp}p_h$$

where α here is the specific volume of air at cloud top and w_{LT}/α is therefore the liquid water content per unit volume there. This expression for W is in terms of quantities whose average values are well known as a function of height and these are given in Table 10.2 for three typical cloud levels in the latitude band 30–40°N (see again London, 1952). For easy reference the expressions for w_{LT} and W reduce to:

$$w_{LT} = -a_0 p_h \qquad \{\mathrm{g\,g^{-1}}\}$$

and $$W = b_0 p_h^2 \qquad \{\mathrm{g\,cm^{-2}}\}$$

TABLE 10.2

Average characteristics for three typical cloud levels in the latitude band 30–40°N

	Stratocumulus	Altocumulus	Cirrus
Altitude (km)	2.0	5.0	10.4
Pressure (mb)	792	549	261
Temperature (K)	284	267	230
$\alpha(\mathrm{cm^3\,g^{-1}})$	1.03×10^3	1.39×10^3	2.50×10^3
$\mathrm{d}T/\mathrm{d}p\,(\mathrm{K\,mb^{-1}})$	0.058	0.089	0.164
$c_p\,[\mathrm{cal\,g^{-1}\,(K)^{-1}}]$	0.240	0.240	0.240
$\mathrm{d}p/\mathrm{d}z\,(\mathrm{mb\,cm^{-1}})$	-9.0×10^{-4}	-6.3×10^{-4}	-3.9×10^{-4}
$\alpha_v(\mathrm{cm^3\,g^{-1}})$	9.98×10^4	3.15×10^5	7.69×10^6
L(liquid-vapor; cal $\mathrm{g^{-1}}$)	592	600	
L(ice-vapor; cal $\mathrm{g^{-1}}$)		677	678
a_0(water; $\mathrm{g\,g^{-1}\,mb^{-1}}$)	3.96×10^{-5}	2.99×10^{-5}	
a_0(ice; $\mathrm{g\,g^{-1}\,mb^{-1}}$)		3.39×10^{-5}	6.25×10^{-6}
b_0(water; $\mathrm{g\,cm^{-2}\,mb^{-2}}$)	2.13×10^{-5}	1.70×10^{-5}	
b_0(ice; $\mathrm{g\,cm^{-2}\,mb^{-2}}$)		1.93×10^{-5}	3.20×10^{-6}
a_1(water; $\mathrm{cm^2\,g^{-1}}$)	1.20×10^3	1.20×10^3	
a_1(ice; $\mathrm{cm^2\,g^{-1}}$)		2.40×10^2	2.40×10^2

approximately linear increase in mixing ratio with height and therefore a relation between W and cloud pressure thickness p_h of the form:

$$W = b_0 p_h^2 \qquad [10.38]$$

where b_0 is a constant for a given altitude in a mean atmosphere. In terms of pressure thickness therefore

$$F_{\mathrm{N}} = F_{\mathrm{N0}}\,[1 - \exp(a_1 b_0 p_h^2)] \qquad [10.39]$$

where typical values of a_1 and b_0 are given in Table 10.2.

Note that aside from the various approximations already mentioned, the simple model envisages a mass absorption coefficient a_1 which is constant with height in the cloud. This has been shown experimentally to be roughly correct for certain ideal stratocumulus cloud despite the typical change in drop-size distribution with height (see Chapter 8). The mean drop size increases with increasing w and therefore with height but the associated increase in absorption cross-section of the mean drop is roughly balanced by a decrease in the number of drops.

4.2. *Static energy loss, latent heat and the rate of displacement*

It is instructive to consider radiational losses in terms of their effect on the rate of vertical displacement of the individual "parcels" within a cloud when the release of latent heat is taken into account. This will be done here by envisaging a pseudo-adiabatic situation with all its attendant assumptions, by considering only the steady state where acceleration is zero and by ignoring the differential heating and cooling which takes place within the cloud itself.

The starting point is the first law of thermodynamics in the form:

$$\mathrm{d}Q = c_p \mathrm{d}T - \alpha \mathrm{d}p \qquad [10.40]$$

where $\mathrm{d}Q$ is the heat input to unit mass of the system and c_p is the specific heat at constant pressure. If in a time $\mathrm{d}t$ a cloud suffers a loss of radiant energy $-R'\mathrm{d}t$ *per unit mass* it will tend to cool and release an amount of latent heat $\tilde{L}\mathrm{d}w_s$. As before $\mathrm{d}w_s$ is the change in saturation mixing ratio. Thus $\mathrm{d}Q$ in the above is $\tilde{L}\mathrm{d}w_s - R'\mathrm{d}t$, and on rearrangement equation [10.40] becomes:

$$\mathrm{d}p = (\tilde{L}\mathrm{d}w_s - R'\mathrm{d}t)/(c_p \mathrm{d}T/\mathrm{d}p - \alpha) \qquad [10.41]$$

The factor $\mathrm{d}w_s$ can be expanded using Clapeyron's equation and we have by transposing terms an expression for the rate of change of pressure of the individual parcels of the cloud — and hence of the rate of vertical displacement of those parcels and of the cloud as a whole. That is:

$$\mathrm{d}p/\mathrm{d}t = -R'/(B_0 - A_0) \qquad [10.42]$$

where $A_0 = 0.62\tilde{L}^2(\mathrm{d}T/\mathrm{d}p)/pT\alpha_v$ and $B_0 = c_p(\mathrm{d}T/\mathrm{d}p) - \alpha$.

Now $-R'$ is the radiant energy loss per unit mass. It can be expressed in terms of total energy loss $-F_N$ as $-(F_N \alpha \mathrm{d}p/\mathrm{d}z)/p_h$, and F_N in turn can be expressed in terms of cloud pressure thickness using equation [10.39]. That is:

$$\frac{\mathrm{d}p}{\mathrm{d}t} = \frac{-F_{N0}}{p_h}[1 - \exp(-a_1 b_0 p_h^2)] \frac{\mathrm{d}p/\mathrm{d}z}{B_0 - A_0} \qquad [10.43]$$

This expression is plotted in Fig. 10.11 for the three typical cloud levels of Table 10.2. In the case of altocumulus two curves are given — one for a cloud composed of ice particles ($r = 50\ \mu$m) and the other for a cloud of water drops ($r = 10\ \mu$m).

The first point to note is that the displacement is downwards in the case of radiatively cooled clouds and is upwards for radiatively warmed clouds.

Second, the displacement rates are meteorologically significant. Suppose we envisage a deck of stratocumulus which is not being supported by an upward flux of water vapour. The individual "parcels" within the cloud would move downwards and the deck as a whole would become thinner and thinner as liquid water evaporated in the higher temperatures. In fact the cloud base would remain fixed and the top would come down to meet it at the instantaneous rate $\mathrm{d}p/\mathrm{d}t$. Theoretically, and in the absence of other factors, the cloud would take an infinite time to disappear; but we can certainly assign some sort of time constant to the process which must be calculated for each individual case since it will depend on original thickness.

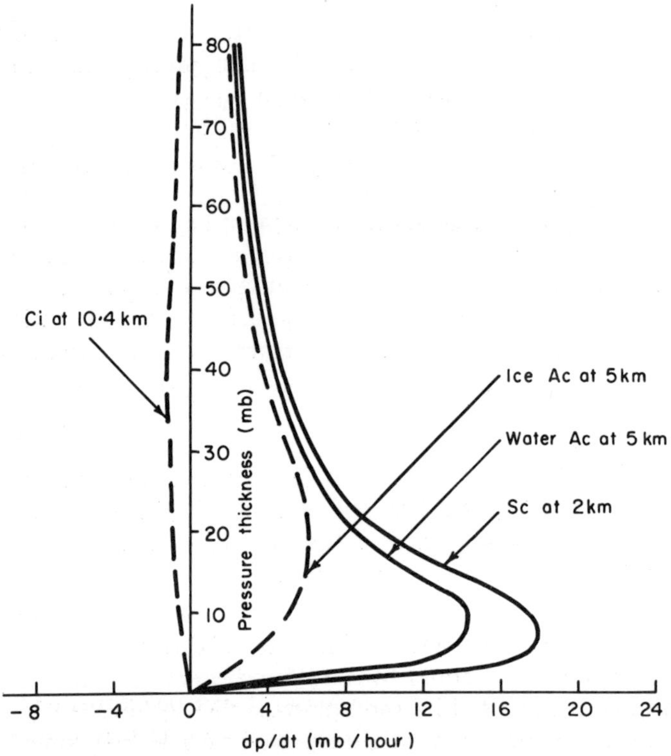

Fig. 10.11. Rate of vertical displacement of cloud as a function of cloud thickness under the influence of long-wave radiative cooling or heating.

A quick estimate can be made by selecting a rough mean of dp/dt between zero and that value corresponding to the original pressure thickness. Broadly, in the case of stratocumulus originally 10 mb thick the dissipation time would be of the order of an hour; for (water) altocumulus the time would be about 2 hours; and for (ice) altocumulus the time would be about 4 or 5 hours.

Third, the release or absorption of latent heat (which governs the value of A_0 in equation [10.43]) acts to slow down the rate of displacement. The reduction is greatest for the lowest clouds (about a factor of 3 for stratocumulus) and is least significant for high cirrus where the reduction factor is only about 1.2.

Fourth, the position of the maximum in each of the dp/dt curves is a rough measure of that thickness of cloud required for it to be optically black. Again the thicknesses involved are quite large, and in a more quantitative analysis it would obviously be necessary to alter the initial assumption here of a thin cloud. Apart from this effect, once a cloud is sufficiently thick to be optically black, further increase in thickness simply reduces the energy loss *per unit mass* and dp/dt decreases.

Finally, because of the assumptions it entails, the pseudo-adiabatic concept is probably not directly relevant to layer cloud as a whole — except in that it helps to picture the behaviour of individual parcels within such cloud.

The biggest stumbling block in modelling realistic clouds is the dynamical problem of specifying the entrainment of external air through the boundaries of the cloud. Much of the work in the field has concentrated on the entrainment associated with rising plumes and isolated convective cumulus towers (e.g. Turner, 1969). At this time, it is possible to model radiation interaction with *stratified* cloud only at the level of broad generalization. The following two sections deal with stratocumulus and with cirrus and provide qualitative examples to show how radiation may influence the short-term character of cloud.

4.3. *Stratocumulus*

A broad picture of the temperature and humidity profiles through a typical deck of stratocumulus might be as shown in Fig. 10.12. Beneath the cloud the lapse rate is approximately dry adiabatic as is characteristic of the mixing layer and for the moment we will assume the mixing ratio to be roughly constant there at a value w_B equal to the saturation mixing ratio at cloud base. The cloud top is surmounted by (and perhaps slightly penetrates) a stable temperature inversion which may originally have defined the top of the mixing layer but which in any event would be created by radiative cooling at the cloud top. The air is much drier above the inversion with a mixing ratio w_0 approximately constant with height. Within the cloud the total water content is roughly constant with height ($= w_B$) but the saturated mixing ratio w_s decreases as the liquid water increases. Broadly, the cloud would cool at the top and warm at the bottom due to radiation and for this reason alone can be expected to be internally unstable with a temperature lapse somewhat greater than saturated adiabatic. In general there will be an upward flux $\tilde{E}$ of water vapour into the cloud.

Paltridge (1974d) set up a simple model of the above situation which can be viewed qualitatively as follows. Loss of radiant energy cools the air at the top of the cloud and intensifies the inversion. Parcels of the cooled and negatively buoyant air sink toward cloud base and create turbulent eddies. These eddies erode the inversion such that as it moves upward parcels of dry air are brought down into the cloud volume. Liquid water evaporates into these parcels thereby amplifying the cooling and instability within the cloud. The entrained dry air also acts as a sink for the upward flux $\tilde{E}$ of water vapour into the base of the cloud. Since the radiant energy loss and therefore the turbulent entrainment is an increasing function of the total liquid water W within the cloud it is possible to achieve an equilibrium value of W such that $\tilde{E}$ is just balanced by the loss of liquid water into the dry air.

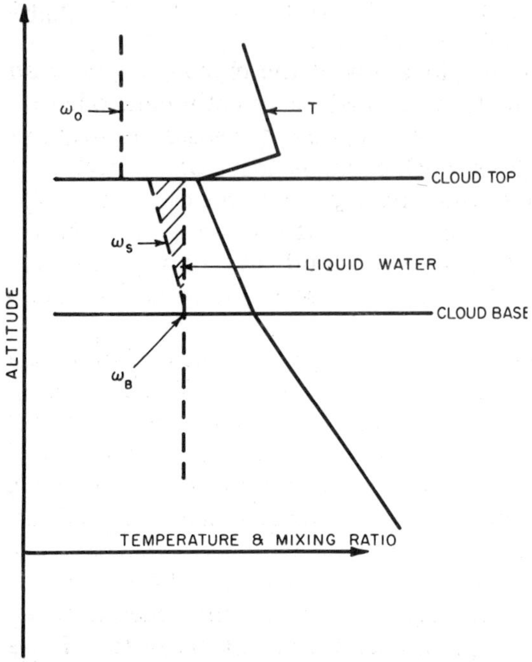

Fig. 10.12. Schematic diagram of distribution of temperature T and mixing ratio w in a stratocumulus cloud.

In order to put the model on a semi-quantitative basis it was necessary to devise some relation between radiant energy loss and the rate of entrainment of dry air through the inversion. Paltridge adopted a modification of the hypothesis of Ball (1960) that the turbulent energy created by the cooling associated with the loss of radiant energy F_N is dissipated not by viscous dissipation but by an upward erosion of the inversion equivalent to a heat flux of magnitude $F_N/2$. With this assumption he obtained an expression relating the equilibrium amount of liquid water W to the supporting vapour flux $\tilde{E}$ as follows:

$$\tilde{E} = \frac{k_1 F_{N0}(w_B - w_0)}{1 - k_1 \tilde{L}(w_B - w_0)/2} [1 - \exp(-a_1 W)] \qquad [10.44]$$

where $k_1 = 1/(2c_p \Delta T)$ and ΔT is the strength of the inversion. At this value of W the cloud will be rising overall at a rate dp/dt equal to $gk_1 F_{N0}/[1 - k_1 \tilde{L}(w_B - w_0)/2]$ where g is the acceleration due to gravity. However, the cloud will maintain a constant equilibrium value of W. Furthermore the equilibrium is stable. That is, given a particular value of $\tilde{E}$ and $w_B - w_0$, cloud thickness will tend to increase to, and remain at, the equilibrium value. If the mixing ratio difference across the inversion is too small to sustain an equilibrium, presumably the cloud will continue to build downwards from the inversion till it rains — or until w_B itself (and therefore $w_B - w_0$) becomes sufficiently large to preserve equilibrium.

According to equation [10.44] the thickness of the cloud (presumed an increasing function of W — e.g. equation [10.38]) is directly controlled by the net radiation loss such that any phenomena which increases the radiative cooling would tend to make it thinner. One might therefore expect such cloud over the sea to be thinner during the night on the basis that solar heating is removed and the effective net cooling of the upper levels of the cloud is increased. Conversely, extensive high cloud may reduce the long-wave cooling of lower atmosphere clouds, and these would respond with an increase in thickness.

There is a semi-quantitative aspect of the model which may relate to the characteristics of trade-wind stratocumulus discussed by Malkus (1962). Such cloud forms in a well-defined region below the trade-wind inversion. However, small turrets occasionally penetrate upwards and form typical cumulus towers. Now the present model of entrainment of dry air has a feedback characteristic. Evaporation of liquid water into the entrained dry air contributes to further turbulence and further entrainment and indeed the rate of entrainment dp/dt tends to infinity as the factor $k_1 \tilde{L}(w_B - w_0)/2$ tends to 1. Therefore, if over a small region the local difference in water contents across the inversion $(w_B - w_0)$ exceeds a certain amount (if ΔT was 5 K the figure would be about 8 g kg^{-1}) it is to be expected that the cloud would penetrate the inversion above that region.

4.4. *Cirrus*

Some of the ideas mentioned so far presumably apply to most forms of stratified cloud. Perhaps the most interesting case is that of cirrus which is sufficiently high in some cases to be radiatively warmed (Fig. 10.9) rather than cooled by long-wave exchange. This warming is probably augmented considerably during the day because of short-wave absorption and perhaps accounts for the distinctly convective character of some cirrus. It appears to form in clearly defined patches rather than in horizontally uniform decks. The irregularities which form the initial patches of cloud during a general uplift would tend to rise even faster than their immediate environment and the cloud would tend to thicken preferentially about the initial patches.

Thus the theoretical treatment of cirrus is much more difficult than that of stratocumulus since the amount of entrainment would so obviously be a matter of scale. One can for instance envisage cirrus as a stratified deck, but it is usually made up of patches or cells of relatively high particle density. The difficulty is to determine whether to treat the individual cells as isolated entities or to regard the deck as a whole. Furthermore, there is no evidence that cirrus is associated with such a regular feature as an inversion and concepts such as a uniform vertical flux of water vapour into the cloud base are not very meaningful.

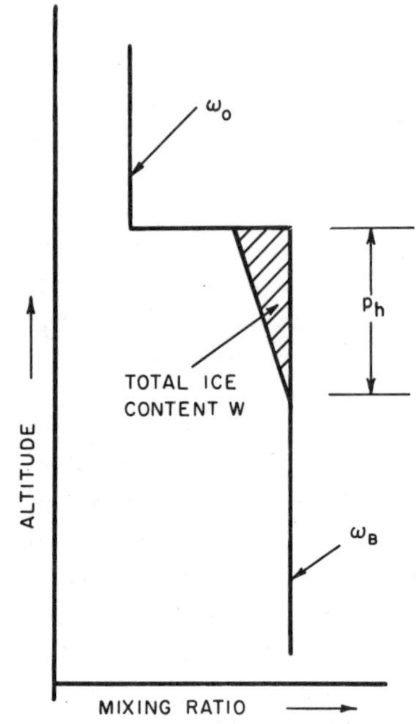

Fig. 10.13. Schematic diagram of ice content in cirrus (the hatched area) for the model discussed in the text.

The following very simple model is intended as a qualitative example of a possible situation where radiation exchange might determine cirrus character.

Imagine a situation as in Fig. 10.13 where the mixing ratio is a step function with height. Suppose that due to general uplift the pressure P at all levels is dropping at the rate $\mathrm{d}P/\mathrm{d}t$ and that the temperature is dropping adiabatically with it. Eventually cloud will form just below the step and the ice particle density would be as represented by the hatched area in the figure. In the absence of entrainment the rate of increase of ice in the cloud would be given by $\mathrm{d}W/\mathrm{d}t = (\mathrm{d}W/\mathrm{d}p_h)\cdot(\mathrm{d}p_h/\mathrm{d}t)$. On this simple model the cloud would be building downwards from the step and $\mathrm{d}p_h/\mathrm{d}t$ would equal $\mathrm{d}P/\mathrm{d}t$. Therefore, using equation [10.38]:

$$\frac{\mathrm{d}W}{\mathrm{d}t} = 2b_0 p_h \cdot \frac{\mathrm{d}P}{\mathrm{d}t} \qquad [10.45]$$

This would be the rate of ice input to the system.

Due to radiative warming the individual cloud cells would tend to rise into the drier region at a rate roughly as given by the pseudo-adiabatic assumption. That is at a rate $\mathrm{d}p/\mathrm{d}t$ in mb/hour where, from equation [10.43] and assuming F_{N0} to be a value from Fig. 10.9 appropriate to cirrus at a height of 10.4 km in the 30–40°N latitude zone:

$$\frac{dp}{dt} = \frac{68.0}{p_h} [1 - \exp(7.68 \times 10^{-4} \cdot p_h^2)] \qquad [10.46]$$

(p_h is in units of mb and is negative).

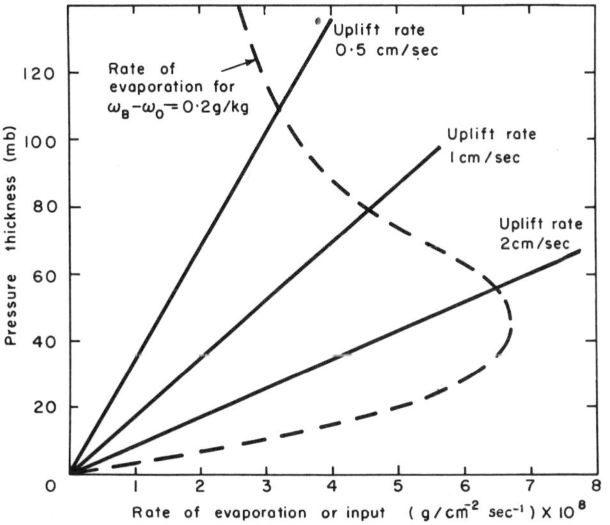

Fig. 10.14. Ice production rate for several uplift velocities (solid curves) and ice evaporation rate (dashed curve) for conditions of cirrus discussed in text.

This concept of rising air parcels suggests that air must be entrained down and around the individual cells (and we assume ultimately into the cells) in order to fill the "vacuum" created by the effective increase in volume of the cloud deck. The mass entrained would be at a rate g^{-1} dp/dt and the rate at which ice could evaporate into this entrained air would be g^{-1}(dp/dt) ($w_B - w_0$). This rate of ice removal is plotted in Fig. 10.14 as a function of pressure thickness for $w_B - w_0 = 0.2\,g\,kg^{-1}$. Also on the figure are several curves of input rates (equation [10.45]) as a function of p_h for values of dP/dt corresponding to the uplift velocities quoted. Accepting all the assumptions involved, the intersection points give the mean thickness of cloud which might exist under these conditions. The figure suggests that, if equilibria of the sort envisaged exist, then the cloud thickness would be negatively correlated with uplift velocity. The limiting upward velocity above which no equilibrium of this sort *could* exist is about 3 cm/second for the quoted conditions.

Appendix

1. AEROSOL SIZE DISTRIBUTIONS

Atmospheric aerosols are produced both in nature and by man. Naturally occurring materials include volcanic dust, meteoric dust, spores and seeds, soil particles, sea salt, ash from burning vegetation, and sulphates formed from photochemical and catalytic oxidation. Man's activities contribute materials such as fly ash from smoke stacks, mineral particles from industrial sources, soil particles due to agricultural activity and particles produced photochemically from gaseous pollutants.

Aerosols can be classified conveniently into three size ranges — Aitken particles with radii below 0.1 μm, large particles with radii between 0.1 and 1 μm, and giant particles with radii of 1 μm and upwards. The smallest particles (less than 0.005 μm radius) can be produced by gas reactions and may involve the small ions produced by cosmic radiation or the decay of radon. Atmospheric pollution can result in the production of a large number of these small particles. They have short lifetimes as they rapidly become attached to the larger aerosol particles. They can exist in large quantities only if there is a constant production rate. Although these small particles are extremely abundant, their mass is only about 10–20% of the total aerosol substance. Maximum total concentrations of over 300,000 cm^{-3} have been measured in city areas. In remote sites the figure drops to less than 500 cm^{-3}.

Junge and his colleagues (Junge et al., 1969; Junge and Jaenicke, 1971) distinguish a background aerosol, supposedly filling the upper troposphere throughout the globe. Below about 5 km altitude over continents this background aerosol is supplemented by local sources. Below about 2 km over the ocean there is a component originating from sea spray.

Fig. A.1 shows a size distribution over the ocean from measurements in the Atlantic, and Fig. A.2 shows similar curves over southern hemisphere oceans. These distributions represent our present knowledge of the so-called aerosol "background" distribution at sites remote from continents and man's activities. The peak in Fig. A.1 in the size concentration of Aitken nuclei indicates some constant source of aerosol production and a dynamic equilibrium between production and loss through coagulation. The secondary peak at about 0.2 μm is thought to be due either to a balance

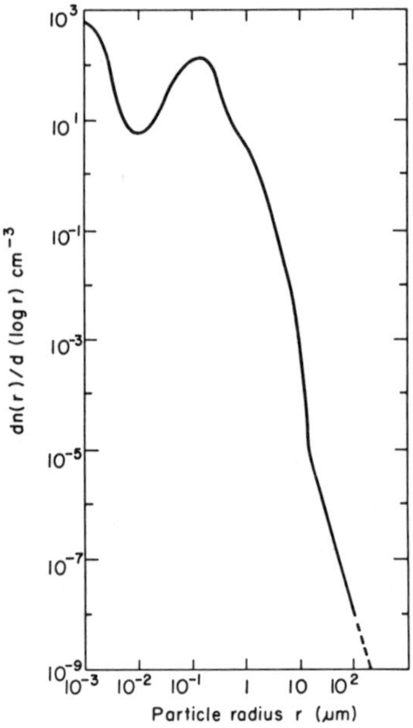

Fig. A.1. Size distribution of atmospheric aerosols in a maritime region over the Atlantic Ocean. (After Junge and Jaenicke, 1971.)

between coagulation and sedimentation processes or to condensation and evaporation of cloud and rain drops containing impurities (Junge, 1963). The upper radius limit of a size distribution is set by the progressively lower residence times of the larger aerosol particles. (Very large particles must derive from very local sources.)

Fig. A.3 shows size distributions in the large and giant particle ranges for aerosols at an ocean site and two continental sites remote from urban influences. The shapes of the distributions are rather similar to those of the background aerosol over oceans.

Fig. A.4 shows distributions (in terms of $dn(r)/dr$) representing European sites in both country and polluted urban areas. Similar distributions have been observed in U.S. areas (Clark and Whitby, 1967). In urban and moderately populated country areas both the $dn(r)/d(\log r)$ and the $dn(r)/dr$ curves appear to be fairly straight from the giant particle region to well below 0.1 μm radius. Junge (1963) therefore suggested a simple formula to fit this shape:

$$dn(r)/d(\log r) = c_1 r^{-V} \quad [A.1]$$

where V and c_1 are constants and V typically has a value of about 3. The corresponding number distribution in terms of radius is:

$$dn(r)/dr = (c_1/2.3) r^{-(V+1)} \quad [A.2]$$

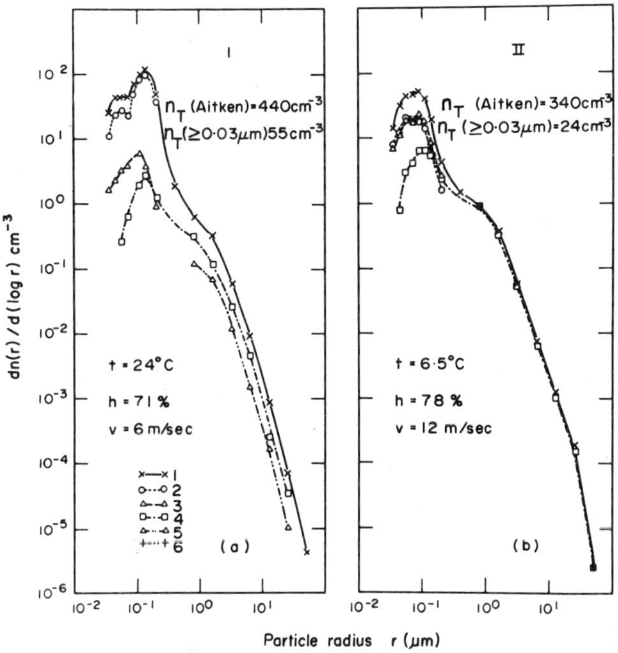

Fig. A.2. Size distributions of atmospheric aerosols in a maritime region over the southern hemisphere oceans. *1* = all particles; *2* = $(NH_4)_2SO_4$; *3* = "long-shaped particles" — supposedly caused by coagulation of $(NH_4)_2SO_4$ and NaCl particles; *4* = NaCl; *5* = non-cubic crystalline particles; *6* = H_2SO_4. (After Mészáros and Vissy, 1974.)

The Junge distribution is particularly useful in radiation problems as it is often valid, at least in moderately polluted continental situations, over that region of the radius distribution from 0.1 to 1 μm which is most effective optically.

For a typical background aerosol, the Junge distribution does not hold even in the 0.1–1-μm range. Various lognormal and gamma function distributions have been used as analytical approximations which give a single peak in the $dn(r)/d(\log r)$ distribution. These distributions are summarized in Table A.1. By changing the constants, the distributions are quite versatile. The distribution of particles originating from a single source by a single mechanism is often lognormal. Davies (1974) has fitted a polluted Los Angeles smog distribution with a combination of several lognormal distributions as shown in Fig. A.5.

The variation of size distribution with height has not been studied in great detail. With some assumptions, optical sounding methods can give the height distribution of the total number density of particles n_T but not the size distributions. These methods are well reviewed by Kondratyev (1972). Direct sampling of particles in the upper troposphere by Junge showed that the total number density of aerosols in this region was about 250 cm^{-3} and

TABLE A.1

Forms of analytical radius number distributions commonly used in radiation calculations

Modified gamma[1]	$n(r) = a_1 r^{a_2} \exp(-b_1 r^{b_2})$
(Diermendjian, 1969)	a_1, a_2, b_1, b_2 are positive and real and a_2 is an integer
Standard Gamma	$n(r) = d_1 r^{(1-3d_2)/d_2} \exp(-r/d_2 d_3)$
(Hansen and Travis, 1974)	d_1, d_2, d_3 are constants
Lognormal	$n(r) = [n_T/(2\pi)^{1/2} \ln\sigma_g] \exp - [\ln^2(r/r_g)/2\ln^2\sigma_g]$
(Porch et al., 1973)	n_T = total number of particles, r_g = geometric number mean radius, σ_g = geometric standard deviation

[1] The following values of the constants were used by Diermendjian (1969) as "typical" haze and cloud distributions:

	a_1	a_2	b_1	b_2
Haze M	5.333×10^4	1	8.9443	1/2
Haze L	4.97×10^6	2	15.1186	1/2
Cumulus C	2.383	6	3/2	1
Corona cloud C	1.085×10^{-2}	8	1/24	3

that the particles were mostly Aitken nuclei. Above the tropopause the Aitken nuclei concentration falls to a very low level but there is a maximum in the number of large particles near 20 km altitude — the well-known Junge layer. Recent *in situ* measurements by Bigg and Ono (1974) confirm the presence of this maximum. The size distribution is fairly narrow in the stratosphere and upper troposphere, consistent with the notion of an ageing, long-resident aerosol. Size distributions at various heights measured by Blifford and Ringer (1969) are shown in Fig. A.6.

The chemical composition of aerosols is highly variable and is important optically as it determines the refractive index of the particles. The background tropospheric aerosols consist mainly of sulphates and sulphuric acid. These are augmented by sea salt in maritime regions and by nitrates, silicates, carbonates and many other trace minerals in continental areas. Bigg and Ono (1974) found that stratospheric aerosols often consist mainly of ammonium sulphate, with a proportion of free sulphuric acid which increases with height.

2. PARTICLE GROWTH WITH HUMIDITY

When the relative humidity of the air rises above some critical point (the "growth point") particles of soluble substance which were initially "dry" become "wet" — that is, the particles go into solution. As the humidity

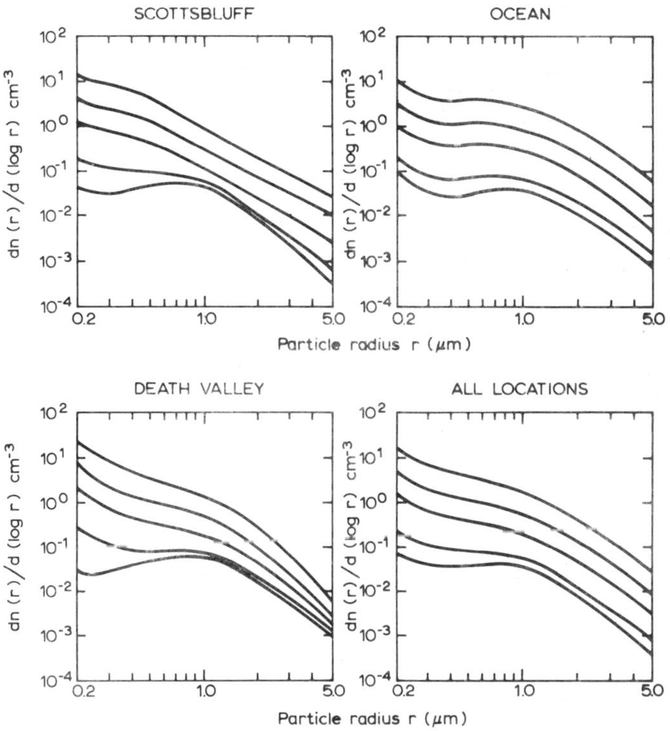

Fig. A.3. Size distribution curves of atmospheric aerosols based on measurements at remote continental and maritime sites for particle concentrations of 0.1 (lowest curve), 0.2, 1.0, 3.0 and 10.0 cm^{-3}. (After De Luisi et al., 1972.)

increases the particles take up water and increase in size. In Junge's classification maritime particles are composed entirely of hygroscopic material (sodium chloride, ammonium sulphate, etc.). On the other hand, continental particle distributions possess only a soluble fraction. They may consist of dry, insoluble dust particles, mixed coagulated particles and even maritime particles in regions not too far removed from coastal areas. Generally it is assumed that the large majority of continental particles contain several substances. One picture is of a hard insoluble core and a coating of a hygroscopic substance (Mackinnon, 1969).

The growth of particles with humidity has been thoroughly treated by Junge (1963) and Mason (1957). The measured growth of pure soluble particles beyond the growth point is found to be a function of the solute concentration in the drop and of the relative humidity. Various forms of the equation relating equilibrium drop size to relative humidity for soluble salts have been given by Junge and Mason and also by Garland (1969). Garland's formula is:

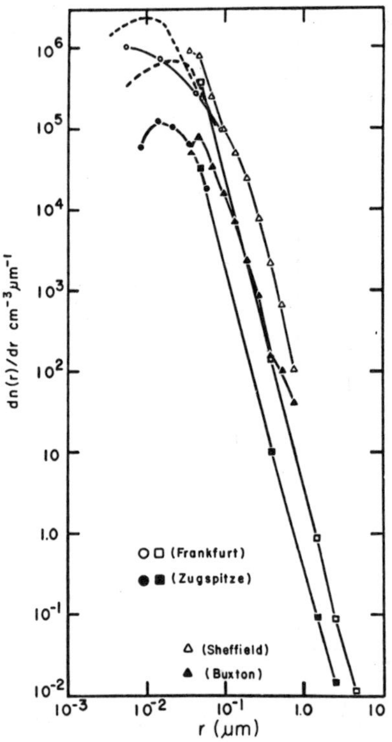

Fig. A.4. Size distributions of atmospheric aerosols measured at various sites in Europe. (After Clarke and Whitby, 1967.) Frankfurt (urban) and Zugspitze (country) measurements (Germany) were made by Junge (1963). Sheffield (urban) and Buxton (country) measurements were made by Cartwright et al. (1956).

$$RH = RH' \exp\left(\frac{2\gamma M}{\rho R T r}\right) \qquad [A.4]$$

where RH' is the relative humidity in equilibrium with a flat surface of solution of the same concentration and temperature as the droplet, RH is the atmospheric relative humidity, γ is the surface tension of the droplet solution, M is the molecular weight of water, ρ is the droplet density, R is the gas constant, T is the absolute temperature and r is the particle radius. Fig. A.7 shows the size of droplets of ammonium sulphate at different relative humidities calculated by Garland (1969) according to the above formula. Similar curves were obtained by Mason (1957) for sodium chloride. On *decreasing* the relative humidity, pure soluble particles do not return to the crystalline phase until the humidity is below 30%. This "hysteresis" effect is not evident in continental particles where the soluble fraction is found to crystallize out at the growth point of the substance.

In view of the variety of aerosol substance and the hysteresis effect for

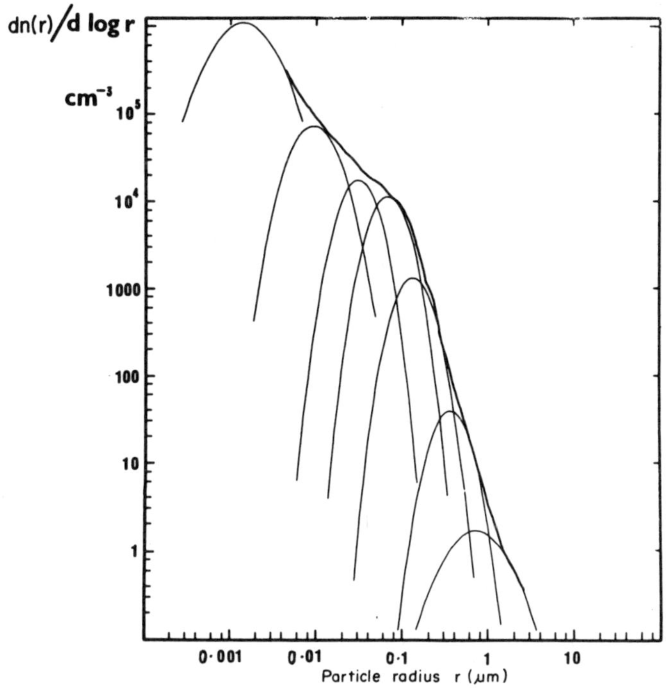

Fig. A.5. Size distribution of atmospheric aerosol in a Los Angeles smog fitted by several lognormal curves. (Davies, 1974.)

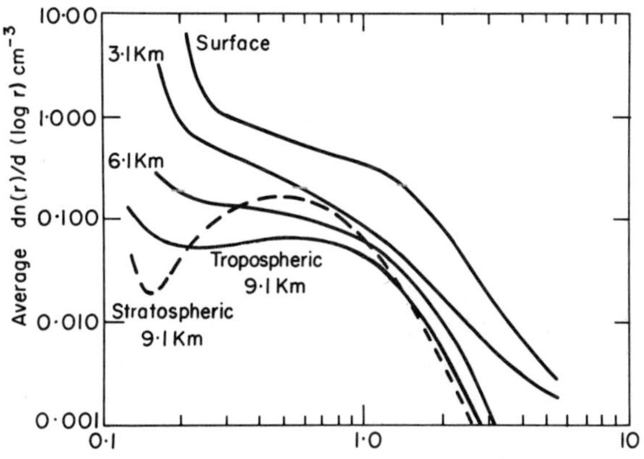

Fig. A.6. Height dependence of the size distribution of atmospheric aerosols representing average values for 22 aircraft sampling flights — radii between 0.13 and 5.5 μm. (After Blifford and Ringer, 1969.)

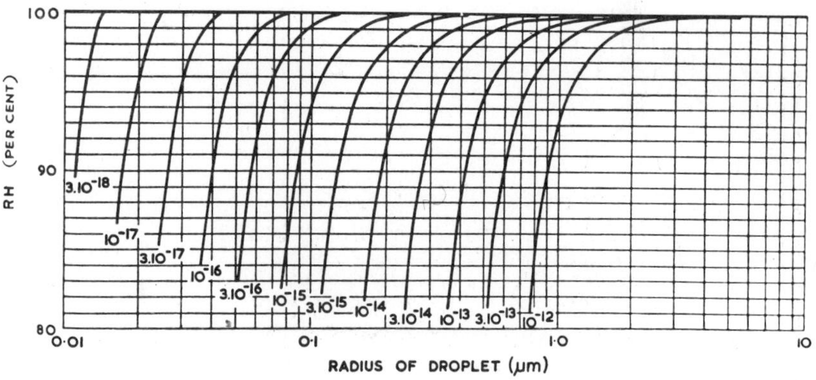

Fig. A.7. Change in radius of droplets of ammonium sulphate at different relative humidities. Each curve is labelled with the mass of ammonium sulphate in grammes. (After Garland, 1969.)

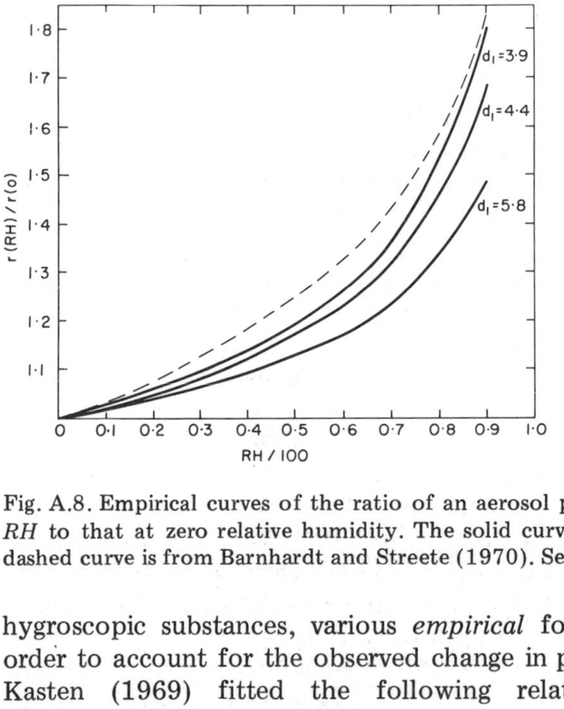

Fig. A.8. Empirical curves of the ratio of an aerosol particle radius at a relative humidity *RH* to that at zero relative humidity. The solid curves are from Kasten (1969) and the dashed curve is from Barnhardt and Streete (1970). See text for details.

hygroscopic substances, various *empirical* formulae have been derived in order to account for the observed change in particle size in natural aerosols. Kasten (1969) fitted the following relation to some experimental measurements:

$$r(RH)/r(0) = (1 - RH)^{-(1/d_1)} \qquad [A.5]$$

where $r(RH)$ and $r(0)$ are radii at relative humidities RH and 0 respectively and d_1 has values of 5.8, 4.4, 3.9 for continental, "normal" (defined by

Kasten as "mixed polluted") and maritime aerosols respectively. This formula was capable of being tuned to fit experimental measurements to within 1%. Barnhardt and Streete (1970) used the formula described in the work of Wright (1939):

$$r(RH) = G_1 r(0)$$

$$\text{where } G_1 = [1-(1/b_2 d_2)] \ln [1-(RH/100)] \quad \text{[A.6]}$$

and where for the particular particles studied, $b_2 = 2.06 \times 10^5$ cm^{-1} and $d_2 = 1.35 \times 10^{-5}$ cm. This formula assumes that particles of all sizes grow at the same rate. Garland's equation shows that this is not strictly true. Results from Kasten's and Barnhardt and Streete's formulae are plotted in Fig. A.8.

Except for the case of pure soluble particles, it is clear that the ratio $r(RH)/r(0)$ can be predicted with only a low accuracy for a situation where the aerosol composition is not well known.

3. AEROSOL REFRACTIVE INDICES

3.1. *Short-wave (solar) region*

The refractive index n_c of a substance can be written:

$$n_c = n_r - in_i \quad \text{[A.7]}$$

where n_r is the real part and n_i is the imaginary part of the refractive index. The value of n_r for dry atmospheric particles is in the region of 1.5–1.6. For water the value is about 1.33. The imaginary part of the refractive index is responsible for aerosol absorption and is not well known except for some specific substances. It is variously considered to range between 0.005 and 0.02. Sodium chloride, ammonium sulphate and water in their pure state are quite transparent; the only water absorption of any consequence occurs at wavelengths longer than 3 μm (Irvine and Pollack, 1968). Absorption in the visible region may be due to organic aerosols but is mainly due to the presence of carbon and graphite particles emitted by combustion. Twitty and Weinman (1971) have reviewed the real and imaginary refractive indices of graphites, soots, coals, etc. The results from different sources are variable but Twitty and Weinman give an average figure of $n_c = 1.8–0.5i$.

Grams et al. (1974), from optical scattering measurements, determined an upper limit of 0.005 for the n_i of typical atmospheric aerosols in the visible region. Collection and examination of their aerosol particles indicated opaque particles, calcites, spherical particles and many irregular-shaped particles in varying percentages from day to day. Grams et al. (1972) estimated the refractive index of fly ash particles as $n = 1.55–0.044i$, using direct sampling and laser radar measurements. The value of n_i was obtained

TABLE A.2

The refractive index of atmospheric aerosol materials which have a crystalline structure (Bullrich, 1964)

Particle	Refractive index	Particle	Refractive index
NH_4Cl	1.64	$CaSO_4$	1.57
NH_4NO_3	1.60	KCl	1.49
$(NH_4)_2SO_4$	1.52	Na_2SO_4	1.48
$MgCl_2$	1.54	SiO_2	1.49
$NaNO_4$	1.59	K_2SO_4	1.49

TABLE A.3

Values of the real (n_r) and imaginary (n_i) parts of the refractive index of atmospheric particles for the short-wave region observed from various experimental measurements

	n_i	n_r	Type
Twitty and Weinman (1971)	1.8	0.5	soot and coal dust
Grams et al. (1972)	1.55	0.044	fly ash
Fischer (1970)	—	0.01	atmospheric aerosol
Ivlev and Popova (1973)	1.65	0.005	atmospheric aerosol
Lin et al. (1973)	—	0.01	atmospheric aerosol
Lindberg and Laude (1974)	—	0.007	dry land dust

on the assumption that the particles were spherical — and that n_r had a value of 1.55. Fischer (1970) collected thin layers of aerosol particles with a jet impactor and estimated n_i to be about 0.01 from scattering and transmission measurements. Table A.2 gives a list of the values of n_r of some probable atmospheric particles of crystalline structure (Bullrich, 1964). From an analysis of atmospheric materials and their corresponding refractive indices, Ivlev and Popova (1973) estimated a mean n_r of 1.65 and a mean n_i of 0.005. Lin et al. (1973) used an optical transmission and scattering method together with aerosol filters and estimated a mean n_i of 0.01. A further study by Lindberg and Laude of atmospheric dust gave figures varying between 0.005 and 0.008, and quite spectrally independent. The above results are summarized in Table A.3.

The refractive indices of dry aerosols are modified when the aerosols become wet. For a hygroscopic particle, the refractive index will approach that of water as the particle grows and becomes nearer to a pure water substance. We have seen already that the growth of particles with humidity is not well known except for pure soluble particles. The same is true for the change in refractive index.

Barnhardt and Streete (1970) quote the empirical formula:

$$n_r = 1.54 + 0.030 \ln [1 - (RH/100)] \qquad [A.8]$$

Hänel (1971) gives a straight dependence on particle volume for both n_r and n_i:

$$n_r = n_{rW} + (n_{r0} - n_{rW})[r(RH)/r(0)]^{-3} \quad [A.9]$$

$$n_i = n_{iW} + (n_{i0} - n_{iW})[r(RH)/r(0)]^{-3} \quad [A.10]$$

where W and 0 refer to water and the dry particle respectively.

3.2. *Infrared region*

The imaginary part of the refractive index of aerosol substances can be quite high and variable in the infrared. Many of these substances in the solid state have absorption bands, particularly in the 7–10-μm region. The infrared refractive indices of dry aerosols and water soluble aerosols found in the atmosphere have been examined by Volz (1972). His results are shown in Figs. A.9, A.10 and A.11. The change of refractive index when aerosols take up water can be given at present only by a formula analogous to equation [A.8]. The refractive index will change from its "dry" value to the value for water as the humidity approaches 100%. This change is shown diagrammatically in Fig. A.12 for several wavelengths in the window region.

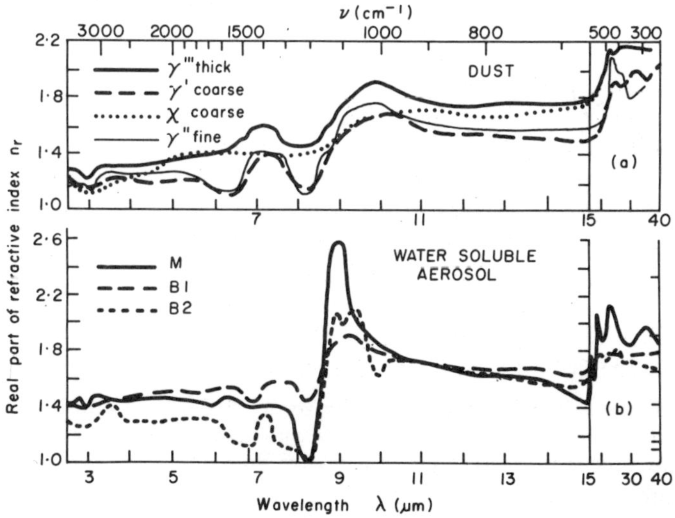

Fig. A.9. Real part of the refractive index (n_r) of materials which were obtained from rainwater by sedimentation and evaporation. n_r was calculated from measurements of the reflectance of pressed discs of the aerosol material. (a) Dust: samples γ' and χ represent unground, coarse material; γ'' and γ''' finely ground material. (b) Water soluble aerosols: sample M was obtained from continental rain; samples B_1 and B_2 were obtained from snow. (After Volz, 1972.)

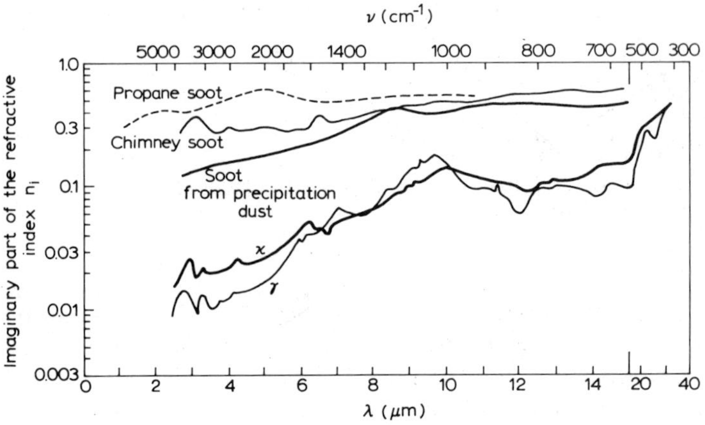

Fig. A.10. Imaginary part of the refractive index (n_i) of dust and soots which were obtained from rainwater by sedimentation. n_i was calculated from measurements of transmittance of the material pressed with Potassium bromide into a disc. Sample χ: coarse dust, sample γ: fine dust. (After Volz, 1972.)

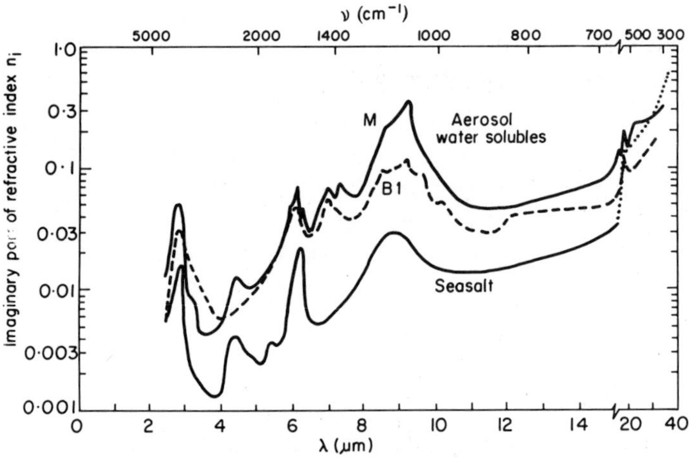

Fig. A.11. Imaginary part of the refractive index (n_i) of water-soluble aerosol substances which were obtained from rainwater by evaporation. Measurement of n_i same as in Fig. A.10. M and B_1 refer to same samples as in Fig. A.9. (After Volz, 1972.)

The review of Twitty and Weinman (1971) of the optical properties of carbonaceous aerosols indicates that at wavelengths between 1 and 20 μm, n_i and n_r remain approximately constant with the same values as in the visible. However, there is about 100% variation between reported results. Conel (1969) reviews the optical properties of silicates. Fig. A.13 gives the values of n_i and n_r for quartz (silica) from the measurements of Spitzer and

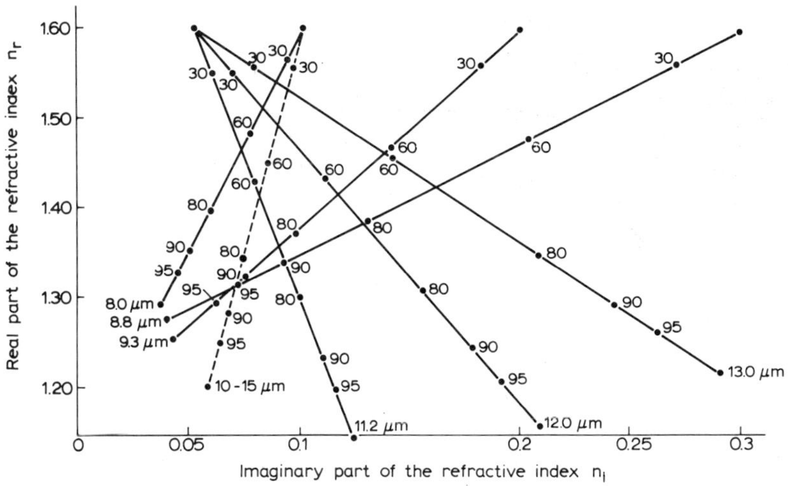

Fig. A.12. Real (n_r) and imaginary (n_i) parts of the refractive index of "dry" aerosols for different wavelengths and relative humidities. Refractive indices based on Fig. A.11. (After Grassl, 1973b.)

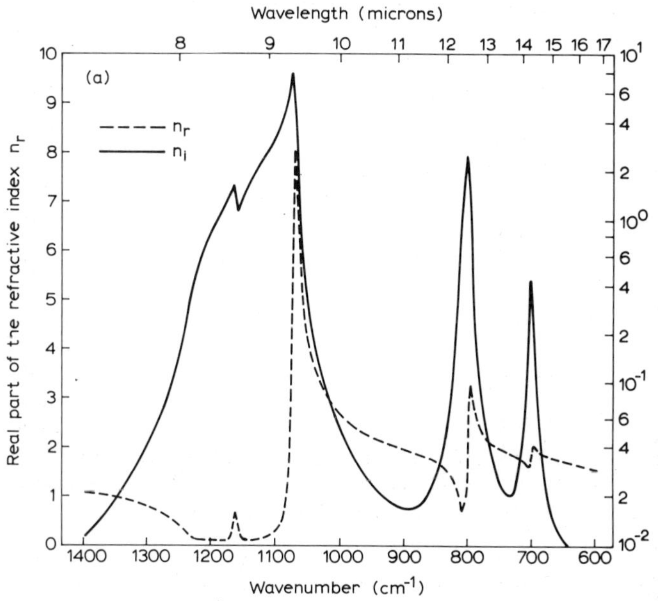

Fig. A.13. Real (n_r) and imaginary (n_i) parts of the refractive index of quartz (silica). (After Conel, 1969.)

TABLE A.4

Real (n_r) and imaginary (n_i) parts of the refractive index of water for wavelengths from 0.2 to 200 μm (after Hale and Query, 1973)

λ(μm)	n_i	n_r	λ(μm)	n_i	n_r
0.200	1.1×10^{-7}	1.396	2.90	0.268	1.201
0.225	4.9×10^{-8}	1.373	2.95	0.298	1.292
0.250	3.35×10^{-8}	1.362	3.00	0.272	1.371
0.275	2.35×10^{-8}	1.354	3.05	0.240	1.426
0.300	1.6×10^{-8}	1.349	3.10	0.192	1.467
0.325	1.08×10^{-8}	1.346	3.15	0.135	1.483
0.350	6.5×10^{-9}	1.343	3.20	0.0924	1.478
0.375	3.5×10^{-9}	1.341	3.25	0.0610	1.467
0.400	1.86×10^{-9}	1.339	3.30	0.0368	1.450
0.425	1.3×10^{-9}	1.338	3.35	0.0261	1.432
0.450	1.02×10^{-9}	1.337	3.40	0.0195	1.420
0.475	9.35×10^{-10}	1.336	3.45	0.0132	1.410
0.500	1.00×10^{-9}	1.335	3.50	0.0094	1.400
0.525	1.32×10^{-9}	1.334	3.6	0.00515	1.385
0.550	1.96×10^{-9}	1.333	3.7	0.00360	1.374
0.575	3.60×10^{-9}	1.333	3.8	0.00340	1.364
0.600	1.09×10^{-8}	1.332	3.9	0.00380	1.357
0.625	1.39×10^{-8}	1.332	4.0	0.00460	1.351
0.650	1.64×10^{-8}	1.331	4.1	0.00562	1.346
0.675	2.23×10^{-8}	1.331	4.2	0.00688	1.342
0.700	3.35×10^{-8}	1.331	4.3	0.00845	1.338
0.725	9.15×10^{-8}	1.330	4.4	0.0103	1.334
0.750	1.56×10^{-7}	1.330	4.5	0.0134	1.332
0.775	1.48×10^{-7}	1.330	4.6	0.0147	1.330
0.800	1.25×10^{-7}	1.329	4.7	0.0157	1.330
0.825	1.82×10^{-7}	1.329	4.8	0.0150	1.330
0.850	2.93×10^{-7}	1.329	4.9	0.0137	1.328
0.875	3.91×10^{-7}	1.328	5.0	0.0124	1.325
0.900	4.86×10^{-7}	1.328	5.1	0.0111	1.322
0.925	1.06×10^{-6}	1.328	5.2	0.0101	1.317
0.950	2.93×10^{-6}	1.327	5.3	0.0098	1.312
0.975	3.48×10^{-6}	1.327	5.4	0.0103	1.305
1.0	2.89×10^{-6}	1.327	5.5	0.0116	1.298
1.2	9.89×10^{-6}	1.324	5.6	0.0142	1.289
1.4	1.38×10^{-4}	1.321	5.7	0.0203	1.277
1.6	8.55×10^{-5}	1.317	5.8	0.0330	1.262
1.8	1.15×10^{-4}	1.312	5.9	0.0622	1.248
2.0	1.1×10^{-3}	1.306	6.0	0.107	1.265
2.2	2.89×10^{-4}	1.296	6.1	0.131	1.319
2.4	9.56×10^{-4}	1.279	6.2	0.0880	1.363
2.6	3.17×10^{-3}	1.242	6.3	0.0570	1.357
2.65	6.7×10^{-3}	1.219	6.4	0.0449	1.347
2.70	0.019	1.188	6.5	0.0392	1.339
2.75	0.059	1.157	6.6	0.0356	1.334
2.80	0.115	1.142	6.7	0.0337	1.329
2.85	0.185	1.149	6.8	0.0327	1.324

TABLE A.4 (continued)

$\lambda(\mu m)$	n_i	n_r	$\lambda(\mu m)$	n_i	n_r
6.9	0.0322	1.321	19.0	0.414	1.461
7.0	0.0320	1.317	19.5	0.404	1.476
7.1	0.0320	1.314	20.0	0.393	1.480
7.2	0.0321	1.312	21.0	0.382	1.487
7.3	0.0322	1.309	22	0.373	1.500
7.4	0.0324	1.307	23	0.367	1.511
7.5	0.0326	1.304	24	0.361	1.521
7.6	0.0328	1.302	25	0.356	1.531
7.7	0.0331	1.299	26	0.350	1.539
7.8	0.0335	1.297	27	0.344	1.545
7.9	0.0339	1.294	28	0.338	1.549
8.0	0.0343	1.291	29	0.333	1.551
8.2	0.0351	1.286	30	0.328	1.551
8.4	0.0361	1.281	32	0.324	1.546
8.6	0.0372	1.275	34	0.329	1.536
8.8	0.0385	1.269	36	0.343	1.527
9.0	0.0399	1.262	38	0.361	1.522
9.2	0.0415	1.255	40	0.385	1.519
9.4	0.0433	1.247	42	0.409	1.522
9.6	0.0454	1.239	44	0.436	1.530
9.8	0.0479	1.229	46	0.462	1.541
10.0	0.0508	1.218	48	0.488	1.555
10.5	0.0662	1.185	50	0.514	1.587
11.0	0.0968	1.153	60	0.587	1.703
11.5	0.142	1.126	70	0.576	1.821
12.0	0.199	1.111	80	0.547	1.886
12.5	0.259	1.123	90	0.536	1.924
13.0	0.305	1.146	100	0.532	1.957
13.5	0.343	1.177	110	0.531	1.966
14.0	0.370	1.210	120	0.526	2.004
14.5	0.388	1.241	130	0.514	2.036
15.0	0.402	1.270	140	0.500	2.056
15.5	0.414	1.297	150	0.495	2.069
16.0	0.422	1.325	160	0.496	2.081
16.5	0.428	1.351	170	0.497	2.094
17.0	0.429	1.376	180	0.499	2.107
17.5	0.429	1.401	190	0.501	2.119
18.0	0.426	1.423	200	0.504	2.130
18.5	0.421	1.443			

Kleinman (1961) and reported by Conel. The rapid spectral variations of refractive indices in silicate materials such as quartz show up as corresponding variations in surface emissivities of desert sands. For instance the radiance curve of Fig. 7.1 indicates a desert surface emissivity of less than unity in the 8–9-μm spectral region.

In "wet" aerosols, values of n_i and n_r tend to those of water (or ice, for

TABLE A.5

Volume absorption coefficient (β_a) and real (n_r) and imaginary (n_i) parts of the refractive index of ice for wavelengths from 0.95 to 152 μm (after Irvine and Pollack, 1968)

λ (μm)	n_r	β_a (cm^{-1})	n_i	λ (μm)	n_r	β_a (cm^{-1})	n_i	λ (μm)	n_r	β_a (cm^{-1})	n_i
0.95	1.302	0.11	8.3×10^{-7}	3.0	1.130	9522	0.2273	6.4	1.226	1000	0.0555
1.00	1.302	0.25	1.99×10^{-6}	3.05	1.192	13093	0.3178	6.5	1.225	1065	0.0551
1.03	1.301	0.34	2.8×10^{-6}	3.075	1.225	14008	0.3428	6.6	1.223	1065	0.0559
1.05	1.301	0.30	2.5×10^{-6}	3.1	1.280	13184	0.3252	6.65	1.222	1065	0.0564
1.10	1.300	0.21	1.84×10^{-6}	3.15	1.547	9979	0.2502	6.7	1.222	1039	0.0554
1.15	1.299	0.32	2.93×10^{-6}	3.2	1.557	6135	0.1562	6.8	1.221	988	0.0535
1.20	1.298	0.96	9.18×10^{-6}	3.25	1.550	3479	0.0900	6.9	1.221	925	0.0508
1.25	1.297	1.42	1.41×10^{-5}	3.3	1.530	2381	0.0625	7.0	1.221	881	0.0491
1.30	1.296	1.20	1.24×10^{-5}	3.35	1.515	1649	0.0440	7.1	1.221	837	0.0473
1.40	1.295	1.84	2.04×10^{-5}	3.4	1.490	1135	0.0307	7.2	1.221	790	0.0453
1.45	1.294	10.2	1.18×10^{-4}	3.45	1.445	824	0.0226	7.5	1.220	669	0.0399
1.50	1.294	46.8	5.58×10^{-4}	3.5	1.422	586	0.0163	8.0	1.219	580	0.0369
1.52	1.294	53.0	6.41×10^{-4}	3.55	1.408	458	0.0129	8.5	1.217	520	0.0352
1.55	1.294	47.4	5.83×10^{-4}	3.6	1.395	367	0.0105	9.0	1.210	510	0.0365
1.60	1.293	29.9	3.80×10^{-4}	3.8	1.356	270	0.0082	9.5	1.192	410	0.0310
1.65	1.293	23.1	3.03×10^{-4}	3.9	1.340	335	0.0104	10	1.152	520	0.0413
1.70	1.2925	14.2	1.92×10^{-4}	4.0	1.327	389	0.0124	10.5	1.195	720	0.0602
1.75	1.292	10.2	1.42×10^{-4}	4.1	1.316	459	0.0150	11	1.290	1090	0.0954
1.80	1.292	7.9	1.13×10^{-4}	4.2	1.307	525	0.0175	11.5	1.393	1240	0.0114
1.85	1.292	4.3	6.33×10^{-5}	4.3	1.299	636	0.0218	12	1.480	1260	0.0120
1.90	1.292	25.9	3.91×10^{-4}	4.4	1.288	806	0.0282	12.5	1.565	1190	0.0119
1.95	1.291	71.7	1.11×10^{-3}	4.5	1.280	921	0.0330	13	1.612	1040	0.0108
2.00	1.291	101.0	1.61×10^{-3}	4.6	1.273	783	0.0287	13.5	1.613	870	0.0935
2.05	1.289	85.8	1.40×10^{-3}	4.7	1.266	576	0.0215	15	1.550	640	0.0762
2.10	1.288	50.1	8.35×10^{-4}	4.8	1.258	454	0.0173	17.5	1.486	250	0.0347
2.15	1.286	39.5	6.76×10^{-4}	4.9	1.252	377	0.0147	20	1.455	160	0.0255
2.20	1.282	17.6	3.08×10^{-4}	5.0	1.247	334	0.0133	25	1.425	150	0.0298
2.25	1.278	11.9	2.13×10^{-4}	5.1	1.241	321	0.0130	30	1.427	220	0.0525

2.30	1.275	10.8	1.98×10^{-4}	5.2	1.236	322	0.0133	35	1.440	400	0.111
2.35	1.270	18.7	3.50×10^{-4}	5.3	1.231	360	0.0152	40	1.460	560	0.178
2.40	1.258	29.9	5.71×10^{-4}	5.4	1.227	408	0.0175	43.4	1.49	930	0.321
2.45	1.247	36.2	7.07×10^{-4}	5.5	1.226	479	0.0210	44.8	1.49	1630	0.581
2.50	1.235	40.1	7.95×10^{-4}	5.6	1.226	548	0.0244	46.5	1.50	1170	0.433
2.55	1.220	41.8	8.50×10^{-4}	5.7	1.226	690	0.0313	52	1.530	650	0.269
2.60	1.206	39.0	8.02×10^{-4}	5.8	1.227	918	0.0424	56	1.545	340	0.152
2.625	1.200	39.8	8.35×10^{-4}	5.9	1.232	1120	0.0526	62	1.57	580	0.286
2.65	1.193	67.8	1.43×10^{-3}	6.0	1.235	1293	0.0617	83	1.620	330	0.22
2.8	1.152	550	0.0123	6.05	1.235	1341	0.0646	100	1.650	100	0.08
2.85	1.140	1557	0.0353	6.1	1.234	1325	0.0643	117	1.690	32	0.03
2.9	1.132	4394	0.1014	6.2	1.232	1248	0.0616	150	1.722	25	0.03
2.95	1.125	7690	0.1805	6.3	1.228	1164	0.0584	152	1.730	25	0.03

frozen aerosols). These values are shown in Table A.4 and A.5 for water and ice respectively. Remsberg (1973) has reported optical constants for sulphuric acid in different parts by weight. In the 8–14-μm region, n_i and n_r are about 0.5 and 1.8 respectively — very similar to Twitty and Weinman's figures for carbonaceous aerosol.

4. CLOUD DROPLET SIZE DISTRIBUTIONS

One of the most awkward aspects of atmospheric radiation is that clouds dominate the consideration of radiant energy flow. On the one hand it is difficult to predict their macroscopic character — their mean height, thickness, geographical position or even whether they should exist at all. On the other, the effect of cloud on both short- and long-wave radiation is extremely dependent on the microphysics of the cloud — that is, on the liquid water contents and the drop size distributions which vary from cloud to cloud and from one position in a cloud to another. Both these microphysical characteristics are related but in a fairly non-predictable manner. Most work to date has assumed homogeneous clouds of fixed liquid water content and drop size distribution — an unlikely state of affairs but one which is at least simple in concept.

As with aerosols, it is unlikely in the near future that anything more than the liquid water content w per unit volume will be predictable *a priori* in general meteorological models, so that one will be forced to use pre-set size distributions appropriate to the average cloud of a particular type. It may prove possible to include also some relation between w and the mean effective radius of the associated size distribution.

Carrier et al. (1967) surveyed the literature and produced the assembly of model drop size distributions of Fig. A.14 and Table A.6. One can argue about how representative these distributions may be, but to no great effect until a specific program of drop size sampling is mounted and tuned to the need of radiation modelling. Most sampling to date has concentrated on clouds which do, or are likely to, precipitate. Raining clouds are a small percentage of the total.

The stratus I model is based on observations by Neiburger and Chien (1960) off the coast of California. The altostratus and stratocumulus models are taken from a series of measurements made by Diem (1948); altitudes of the cloud bases varied from 1520 to 2750 m, tops ranged from 1825 to 3970 m in altitude and typical cloud thicknesses were 1825 m. The nimbostratus model is that reported by Weickmann and aufm Kampe (1953). The fair weather cumulus cloud model is similar to the Standard Weather Code "Low Cloud, Type 2", defined as a cumulus cloud having little vertical development and flattened in appearance; over the oceans these clouds are often called trade-wind cumulus. The stratus II model, after Diem (1948), is

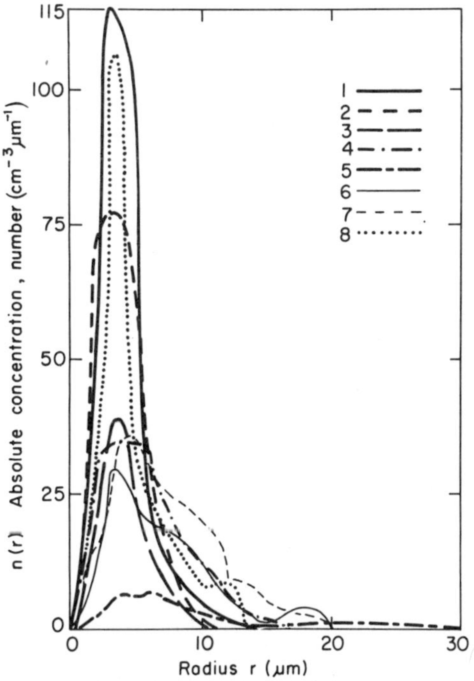

Fig. A.14. Model cloud drop spectra. *1* = stratus I, *2* = stratocumulus, *3* = fair-weather cumulus, *4* = stratus II, *5* = cumulonimbus, *6* = cumulus congestus, *7* = nimbostratus, and *8* = altostratus. (After Carrier et al., 1967.)

TABLE A.6

Model cloud drop size and concentration[1] (Carrier et al., 1967)

Cloud Type	N^a (No. cm^{-3})	r_{mode} (μm)	r_{min} (μm)	r_{max} (μm)	Δr (μm)
Stratus I	464	3.5	0	16.0	3.0
Altostratus	450	4.5	0	13.0	4.5
Stratocumulus	350	3.5	0	11.2	4.4
Nimbostratus	330	3.5	0	19.8	9.5
Fair-weather cumulus	300	3.5	0.5	10.0	3.0
Stratus II	260	4.5	0	20.0	5.7
Cumulus congestus	207	3.5	0	16.2	6.7
Cumulonimbus	72	5.0	0	30.0	7.0

[1] N^a = total concentration, r_{mode} = mode radius = radius corresponding to the maximum number of droplets, r_{min} = minimum radius, r_{max} = maximum radius, Δr = bandwidth of the drop-size distribution at half-value points.

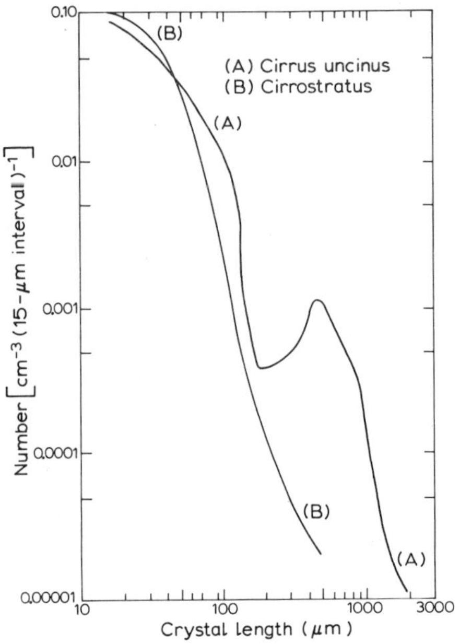

Fig. A.15. Particle spectra taken in the downshear region near the base of the head of a cirrus "uncinus" cloud (A) and near the top of a cirrostratus deck (B). (After Heymsfield, 1975.)

representative of a stratus deck over land. Considerable variation exists among stratus clouds at different locations, and the stratus I and II models represent extremes. The distribution for the cumulus congestus, a very common stormy weather cloud, was taken from a series of 150 samples of cumulus clouds which were reported by Durbin (1959); in severe storms, clouds have been found with a total concentration of only 12 drops cm^{-3}, but with an average radius as large as 25 μm. The cumulonimbus, which is associated with thunderstorms, has a double mode at 4 μm and 6 μm with a significant number of large drops of radii up to 30 μm; some researchers report a significant number of drops in cumulonimbus with radii as large as 100 μm, but these larger droplets would be encountered less frequently than the model with radii up to 30 μm.

To these distributions can be added the few ice cloud distributions obtained recently by Heymsfield (1975) and his colleagues (e.g. Heymsfield and Knollenberg, 1972). Fig. A.15 shows typical spectra obtained in two cirrus cloud types, where it should be remembered that the distributions are in terms of crystal length. Bearing in mind that the equivalent spherical radius of a particle is approximately $(A/4\pi)^{1/2}$ where A is the surface area of the (convex) particle, this radius of cirrus ice crystals is probably about a factor of 5 down on the actual crystal length. The average ice water contents

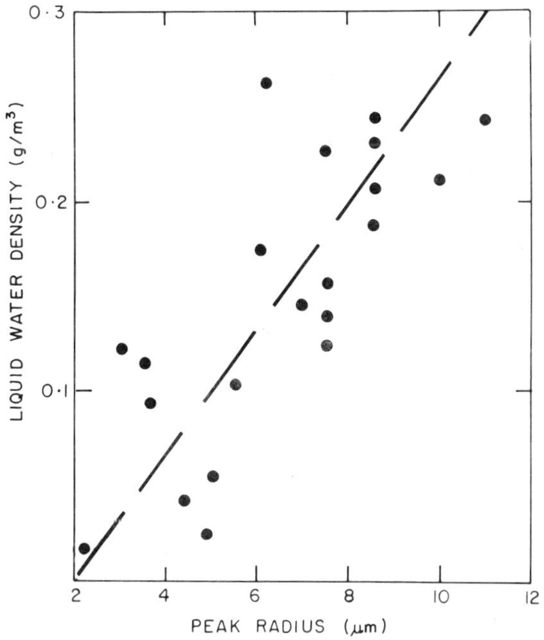

Fig. A.16. Experimentally measured peak radius (the radius of the peak in number density of a distribution) of water droplet size distributions plotted as a function of the associated liquid water content in a number of "ideal" stratocumulus clouds off the east coast of Australia. (From Paltridge, 1974c.)

in the eight clouds investigated by Heymsfield ranged from 0.025 to 0.24 $g\,m^{-3}$.

In general one might expect that the mean radius of a drop distribution should increase with w. This is borne out to some extent by the experimental measurements of Fig. A.16 which were obtained at different levels in a number of "ideal" stratiform clouds over the sea. "Ideal" in this context refers to the fact that they were isolated decks of non-raining clouds below 3 km with uniform tops and bases. They were surmounted by strong inversions. The mode radius increases markedly with w. It is also significant that in these ideal clouds the liquid water content had the characteristic one might expect on pure "adiabatic lifting" theory — namely that w increased almost linearly with height. One can find numerous specific examples where such a relation is not valid. However, if one is to assign a w versus height character for operational and test-bed models, such a variation of w (and of effective radius) would be preferable to the simple assumption of a homogeneous cloud. The increasing w with height appears also in the work of Weickmann and aufm Kampe (1953) and of Nieburger (1949). Such a characteristic leads to certain simplifying assumptions in long-wave transfer, in particular a possible constancy of *mass* absorption coefficient within a cloud (Chapter 8).

Zdunkowski and Crandall (1971) have discussed the often-used empirical formulae for various cloud drop distributions produced by Best (1951) and Diermendjian (1964) in terms of the differences they produce on long-wave transfer. These differences are not very critical as far as the effect of the overall cloud is concerned.

References

Adem, J., 1962. On the theory of the general circulation of the atmosphere. *Tellus*, 14: 102–115.

AFCRL, 1967. Appendix B of Revision of Chapter 22 of Handbook of Geophysics and Space Environments. *Air Force Surv. Geophys.*, No. 199. Office of Aerospace Research, U.S. Air Force, Bedford, Mass.

Albrecht, T., 1949. Über die Wärme und Wasserblanz der Erde. *Ann. Meteorol.*, 2: 129–143.

Allen, J.R., 1971. Measurements of cloud emissivity in the 8–13 μ waveband. *J. Appl. Meteorol.*, 10: 260–265.

Ångström, A., 1958. On pyrheliometric measurements. *Tellus*, 10: 342–354.

Ångström, A., 1964. The parameters of atmospheric turbidity. *Tellus*, 16: 64–75.

Ångström, A., 1974. Circumsolar radiation as a measure of the turbidity of the atmosphere. *Appl. Opt.*, 13: 1477–1480.

Ångström, A.K. and Ångström, K.H., 1971. The distribution of radiation over the sun's disk. *Sol. Energy*, 13: 243–250.

Ångström, K., 1899. The absolute determination of the radiation of heat with the electric compensation pyrheliometer. *Astrophys. J.*, 9: 332–346.

Ball, F.K., 1960. Control of inversion height by surface heating. *Q. J. R. Meteorol. Soc.*, 86: 483–494.

Barnhardt, E.A. and Streete, J.L., 1970. A method for predicting atmospheric aerosol scattering coefficients in the infrared. *Appl. Opt.*, 9: 1337–1344.

Baur, F. and Phillipps, H., 1935. Der Wärmehaushalt der Luftülle der Nordhalbkugel im Januar und Juli, 2. *Gerlands Beitr. Geophys.*, 45: 82–132.

Benedict, W.S. and Kaplan, L.D., 1959. Calculation of linewidths in H_2O—N_2 collisions. *J. Chem. Phys.*, 30: 388–399.

Bergström, R.W., 1973. Extinction and absorption coefficients of the atmospheric aerosol as a function of particle size. *Contrib. Atmos. Phys.*, 46: 223–234.

Berson, F.A. and Kulkarni, R.N., 1968. Sunspot cycle and the quasi-biennial stratospheric oscillation. *Nature (London)*, 217: 1133–1134.

Best, A.C., 1951. Drop-size distribution in cloud and fog. *Q. J. R. Meteorol. Soc.*, 77: 418–426.

Beuttner, K.J.K. and Kern, C.D., 1965. The determination of infrared emissivities of terrestrial surfaces. *J. Geophys. Res.*, 70(6): 1329–1337.

Bigg, E.K. and Ono, A., 1974. Size distribution and nature of stratospheric aerosols. In: *Proc. Int. Conf. on Structure and Composition and General Circulation of the Upper and Lower Atmospheres and Possible Anthropogenic Perturbations*. IAMAP, Atmospheric Environment Service, Downsview, Ont., 1: 144–157.

Bignell, K.J., 1970. The water vapour infra-red continuum. *Q. J. R. Meteorol. Soc.*, 96: 390–403.

Bignell, K.J., Saiedy, F. and Sheppard, P.A., 1963. On the atmospheric infrared continuum. *J. Opt. Soc. Am.*, 53: 466–479.

Blevin, W.R. and Brown, W.J., 1971. A precise measurement of the Stefan–Boltzmann constant. *Metrologia*, 7(1): 15–29.

Blifford, I.H. and Ringer, L.D., 1969. The size and number distribution of aerosols in the continental troposphere. *J. Atmos. Sci.*, 26: 716–726.

Bolle, H.-J., 1965. Investigation of the infrared emission spectrum of the atmosphere and earth. *Final Rep. Contract AF61(052)-481*. AFCRL, U.S. Air Force, Bedford, Mass., 173 pp.

Bolle, H.-J., Möller, F. and Zdunkowski, W., 1963. Investigation of the infrared emission spectrum of the atmosphere and earth. *Techn. Rep. No. 2. Contract AF61(052)-488*. AFCRL, U.S. Air Force, Bedford, Mass.

Braslau, N., 1972. Overlap emissivity of CO_2 and H_2O in the 15 μm spectral region. *IBM J. Res. Dev.*, 17: 180–183.

Braslau, N. and Dave, J.V., 1973. Effect of aerosols on the transfer of solar energy through realistic model atmospheres, 1. Non-absorbing aerosols. *J. Appl. Meteorol.*, 12: 601–615.

Bramson, M.A., 1968. *Infrared Radiation–A Handbook for Applications*. Plenum Press, New York, N.Y., 622 pp. (translated from Russian).

Brooks, D.L., 1950. A tabular method for the computation of temperature change by infrared radiation in the free atmosphere. *J. Meteorol.*, 7: 313–321.

Brunt, D., 1932. Notes on radiation in the atmosphere, 1. *Q. J. R. Meteorol. Soc.*, 58: 389–420.

Budyko, M.I., 1969. The effect of solar radiation variations on the climate of the earth. *Tellus*, 21: 611–619.

Bullrich, K., 1964. Scattered radiation in the atmosphere and the natural aerosol. *Adv. Geophys.*, 10: 101–257.

Burch, D.E., 1970. Investigation of the absorption of infrared radiation by atmospheric gases. *Semi-annual Rep. Publ. U-4784*. Aeronutronic Div., Philco-Ford Corporation, Newport Beach, Calif.

Burch, D.E., Gryvnak, D. and Williams, D., 1961. Infrared absorption by carbon dioxide. *Contract AF19(604)-2633*, AFCRL-255, Ohio State University, Colombus, Ohio.

Burch, D.E., Gryvnak, D., Singleton, E.H., France, W.L. and Williams, D., 1962. Infrared absorption by carbon dioxide, water vapour, and minor atmospheric constituents. *Res. Rep. Geophys. Res. Dir., AFCRL 62-698.*

Burroughs, W.J., Jones, R.G. and Gebbie, H.A., 1969. A study of submillimetre atmospheric absorption using the HCN Maser. *J. Quant. Spectrosc. Radiat. Transfer*, 9: 809–824.

Carrier, L.W., Cato, G.A. and Von Essen, K.J., 1967. The back-scattering and extinction of visible and infrared radiation by selected major cloud models. *Appl. Opt.*, 6: 1209–1216.

Cartwright, J., Nagelschmidt, G. and Skidmore, J., 1956. The study of air pollution with the electron microscope. *Q. J. R. Meteorol. Soc.*, 82: 82–86.

Cashwell, E.D. and Everett, C.J., 1959. *A Practical Manual on the Monte Carlo Method for Random Walk Problems*. Pergamon Press, New York, N.Y., 153 pp.

Chandrasekhar, S., 1950. *Radiative Transfer*. Clarendon Press, Oxford, 393 pp.

Chandrasekhar, S. and Elbert, D.D., 1954. The illumination and polarization of sunlit sky in Rayleigh scattering. *Trans Am. Philos. Soc.*, 44(6): 643–728.

Charlson, R.J. and Pilat, M.J., 1969. Climate: the influence of aerosols. *J. Appl. Meteorol.*, 8: 1001–1002.

Chýleck, P. and Coackley, Jr., J.A., 1974. Aerosols and climate. *Science*, 183: 75–77.

Clapp, P.F., 1964. Global cloud cover for seasons using TIROS nephanalyses. *Mon. Weather Rev.*, 92: 495–507.

Clark, W.E. and Whitby, K.T., 1967. Concentration and size distribution measurements of atmospheric aerosols and a test of the theory of self-preserving size distributions. *J. Atmos. Sci.*, 24: 677–687.

Cole, A.E., Court, A. and Kantor, A.J., 1965. Model atmospheres. In: S.L. Valley (Editor), *Handbook of Geophysics and Space Environments*. AFCRL, U.S. Air Force, Bedford, Mass., chapter 2.

Collins, D.G., Blattner, W.G., Wells, M.B. and Borak, H.G., 1972. Backward Monte Carlo calculations of the polarization characteristics of the radiation emerging from spherical-shell atmospheres. *Appl. Opt.*, 11: 2684–2695.

Collins, D.G. and Wells, M.B., 1965. *Monte Carlo Codes for Study of Light Transport in the Atmosphere*. U.S. Department of Commerce, Institute for Applied Technology, Washington, D.C., 100 pp.

Conel, J.E., 1969. Infrared emissivities of silicates: experimental results and a cloudy atmosphere model of spectral emission from condensed particulate mediums. *J. Geophys. Res.*, 74: 1614–1634.

Conover, J.H., 1965. Cloud and terrestrial albedo determinations from TIROS satellite pictures. *J. Appl. Meteorol.*, 4: 378–386.

Coulson, K.L., 1959. Characteristics of the radiation emerging from the top of a Rayleigh atmosphere, 1 and 2. *Planet. Space Sci.*, 1: 256–284.

Cowling, T.G., 1950. Atmospheric absorption of heat radiation by water vapour. *Philos. Mag.*, 41: 109–123.

Cox, C. and Munk, W., 1954. Measurement of the roughness of the sea surface from photographs of the sun's glitter. *J. Opt. Soc. Am.*, 44: 838–850.

Cox, S.K., 1969. Radiation models of mid-latitude synoptic features. *Mon. Weather Rev.*, 97: 637–651.

Cox, S.K., 1971. Cirrus clouds and the climate. *J. Atmos. Sci.*, 28: 1513–1515.

Cox, S.K., 1973. Infrared heating calculations with a water vapour pressure broadened continuum. *Q. J. R. Meteorol. Soc.*, 99: 669–679.

Crutcher, H.L. and Meserve, J.M., 1970. Selected level heights, temperatures and dew-points for the northern hemisphere. *NAVAIR Publ. No. 50-IC-52*. U.S. Government Printing Office, Washington, D.C.

Curtis, A.R., 1956. The computation of radiative heating rates in the atmosphere. *Proc. R. Soc. London, Ser. A*, 236: 156–159.

Curtis, A.R. and Goody, R.M., 1956. Thermal radiation in the upper atmosphere. *Proc. R. Soc. London, Ser. A*, 236: 193–206.

Davies, C.N., 1974. Size distribution of atmospheric particles. *J. Aerosol Sci.*, 5: 293–300.

Davis, P.A., 1971. Applications of an airborne ruby lidar during a BOMEX program of cirrus observations. *J. Appl. Meteorol.*, 10: 1314–1323.

Deacon, E.L., 1969. Physical processes near the surface of the earth. In: H. Flohn (Editor), *World Survey of Climatology, 2. General Climatology*. Elsevier, Amsterdam, 2: 39–104.

Deacon, E.L., 1970. The derivation of Swinbank's long-wave radiation formula. *Q. J. R. Meteorol. Soc.*, 96: 313–319.

De Luisi, J.J., Blifford, Jr., I.H. and Takamine, J.A., 1972. Models of tropospheric aerosol size distributions derived from measurements at three locations. *J. Geophys. Res.*, 77: 4529–4538.

Diem, M., 1948. The measurement of cloud elements, 2. *Meteorol. Rundsch.*, 9/10: 261–273.

Diermendjian, D., 1964. Scattering and polarization properties of water clouds and hazes in the visible and infrared. *Appl. Opt.*, 3: 187–196.

Diermendjian, D., 1969. *Electromagnetic Scattering on Spherical Polydispersions*. American Elsevier, New York, N.Y., 290 pp.

Diermendjian, D. and Sekera, Z., 1954. Global radiation resulting from multiple scattering in a Rayleigh atmosphere. *Tellus*, 6: 382–398.

Dilke, F.W.W. and Gough, D.O., 1972. The solar spoon. *Nature (London)*, 240: 262–264; 293–294.

Dines, W.H., 1917. The heat balance of the atmosphere. *Q. J. R. Meteorol. Soc.*, 43: 151–158.

Dopplick, T.G., 1972. Radiative heating of the global atmosphere. *J. Atmos. Sci.*, 29: 1278–1294.

Drayson, S.R., 1966. Atmospheric transmission in the CO_2 bands between 12μ and 18μ. *Appl. Opt.*, 5: 385–391.

Drayson, S.R., 1967. *The Calculation of Long-wave Radiative Transfer in Planetary Atmosphere.* Ph.D. Thesis, University of Michigan, Ann Arbor, Mich.

Durbin, W., 1959. Droplet sampling in cumulus clouds. *Tellus*, 7: 202–215.

Edlén, B., 1953. The dispersion of standard air. *J. Opt. Soc. Am.*, 43: 339–344.

Ellingson, R.G. and Gille, J.C., 1972. A new long-wave radiative transfer model: calibration and application to the tropical atmosphere. In: *Proc. Int. Radiation Symposium, Sendai, May–June, 1972.* WMO/JMA, Japanese Meteorological Agency, Tokyo, pp. 31–32.

Elsasser, W.M., 1942. *Heat Transfer by Infrared Radiation in the Atmosphere. Harvard Meteorological Studies, 6.* Harvard University Press, Cambridge, Mass., 107 pp.

Elsasser, W.M. and Culbertson, M.F., 1960. *Atmospheric Radiation Tables. Meteorological Monographs, 4.* American Meteorological Society, Boston, Mass., 43 pp.

Elterman, L. and Toolin, R.B., 1965. Atmospheric optics. In: S.L. Valley (Editor), *Handbook of Geophysics and Space Environments.* AFCRL, U.S. Air Force, Bedford, Mass., chapter 7.

Emden, R., 1913. Über Strahlungsgleichgewicht und atmosphärische Strahlung. *Sitzungsber. Bayer. Akad., Wiss. Math.-Phys. Kl.*, 1: 55.

Ensor, D.S., Porch, W.M., Pilat, K.J. and Charlson, R.J., 1971. Influence of the atmospheric aerosol on albedo. *J. Appl. Meteorol.*, 10: 1303–1306.

Feigelson, E.M., 1964. *Light and Heat Radiation in Stratus Clouds.* English version by Israel Program for Scientific Translations, Jerusalem, 1966, 245 pp.

Fillipov, V.L. and Mirumyants, S.O., 1972. Aerosol extinction of visible and infrared radiation as a function of air humidity. *Izv. Acad. Sci. U.S.S.R., Atmos. Oceanic Phys.*, 9: 988–993.

Fischer, K., 1970. Bestimmung der absorption von sichtbarer strahlung durch aerosolpartikeln. *Contrib. Atmos. Phys.*, 43: 244–254.

Fleagle, R.G. and Businger, J.A., 1963. *An Introduction to Atmospheric Physics.* Academic Press, New York, N.Y. and London, 346 pp.

Flohn, H., 1959. Bemerkungen zur Klimatologie von Hochasien. *Akad. Wiss. Lit., Mainz, Abh. Math.-Naturwiss. Kl.*, 14: 309–331.

Flowers, E.C., McCormick, R.A. and Kurfis, K.R., 1969. Atmospheric turbidity over the United States. *J. Appl. Meteorol.*, 8: 955–962.

Fowlé, F.E., 1912. The spectroscopic determination of aqueous vapour. *Astrophys. J.*, 35: 149–162.

Fowle, F.E., 1915. The transparency of aqueous vapor. *Astrophys. J.*, 42: 394–411.

Fritz, S., 1954. Scattering of solar energy by clouds of "large drops". *J. Meteorol.*, 11: 291–300.

Fritz, S., 1958. Absorption and scattering of solar energy in clouds of "large water drops", 2. *J. Meteorol.*, 15: 51–58.

Funk, J.P., 1960. Measured radiative flux divergence near the ground at night. *Q. J. R. Meteorol. Soc.*, 86: 382–389.

Funk, J.P., 1962. Radiative flux divergence in radiation fog. *Q. J. R. Meteorol. Soc.*, 88: 233–249.

Gambling, D.J., 1972. An aerosol attenuation model for the visible and infrared regions. *WRE-Tech. Note*, No. 593(AP). Weapons Research Establishment, Salisbury, S.A., 11 p.

Garland, J.A., 1969. Condensation on ammonium sulphate particles and its effect on visibility. *Atmos. Environ.*, 3: 347–354.

Gebbie, H.A., Burroughs, W.J., Chamberlain, J., Harries, J.E. and Jones, R.G., 1969. Dimers of the water vapour molecule in the earth's atmosphere. *Nature (London)*, 221: 143–145.

Georgievskiy, Yu.S. and Rozenberg, G.V., 1973. Humidity as a factor in aerosol variability. *Izv. Acad. Sci. U.S.S.R., Atmos. Oceanic Phys.*, 9: 126–137.

Gold, E., 1909. The isothermal layer of the atmosphere and atmospheric radiation. *Proc. R. Soc. London, Ser. A*, 82: 70.

Goldberg, L. and Pierce, A.K., 1959. The photosphere of the sun. In: S. Flügge (Editor), *Encyclopedia of Physics*. Springer-Verlag, Berlin, Vol. 52, 601 pp.

Goldman, A., Murcray, D.G., Murcray, G.H., Williams, W.J. and Brooks, J.N., 1973. Distribution of water vapour in the stratosphere as determined from balloon measurements of atmospheric emission spectra in the 24–29 μm region. *Appl. Opt.*, 12: 1045–1053.

Goody, R.M., 1949. The thermal equilibrium at the tropopause and the temperature of the lower stratosphere. *Proc. R. Soc. London, Ser. A*, 197: 505.

Goody, R.M., 1952. A statistical model for water vapour absorption. *Q. J. R. Meteorol. Soc.*, 78: 165–169.

Goody, R.M., 1964. *Atmospheric Radiation, 1. Theoretical Basis*. Clarendon Press, Oxford, 436 pp.

Grams, G.W., 1974. Complex index of refraction of airborne soot particles. *J. Appl. Meteorol.*, 13: 459–471.

Grams, G.W., Blifford, Jr., I.H., Schuster, B.G. and De Luisi, J.J., 1972. Complex index of refraction of airborne fly ash determined by laser radar and collection of particles at 13km. *J. Atmos. Sci.*, 29: 900–905.

Grams, G.W., Blifford, Jr., I.H., Gillette, D.A. and Russell, P.B., 1974. Complex index of refraction of Airborne Soil particles. *J. Appl. Meteorol.* 13: 459–471.

Grassl, H., 1973a. Separation of atmospheric absorbers in the 8–13 micrometer region. *Contrib. Atmos. Phys.*, 46: 75–88.

Grassl, H., 1973b. Aerosol influence on radiative cooling. *Tellus*, 25: 386–395.

Grassl, H., 1974. Influence of different absorbers in the window region on radiative cooling (and on surface temperature determination). *Contrib. Atmos. Phys.*, 47: 1–13.

Green, A.E.S., Deepak, A. and Lipofsky, B.J., 1971. Interpretation of the sun's aureole based on atmospheric aerosol models. *Appl. Opt.*, 10: 1263–1279.

Green, J.S.A., 1967. Division of radiative streams into internal transfer and cooling to space. *Q. J. R. Meteorol. Soc.*, 93: 371–372.

Griggs, M. 1968. Aircraft measurements of albedo and absorption of stratus clouds, and surface albedos. *J. Appl. Meteorol.*, 7: 1012–1017.

Hale, G.M. and Query, M.R., 1973. Optical constants of water in the 200 nm to 200 μm wavelength region. *Appl. Opt.*, 12: 555–563.

Hammad, A., 1948. The primary and secondary scattering of sunlight in a plane-stratified atmosphere of uniform composition. *Astrophys. J.*, 108: 338–346.

Hammad, A. and Chapman, S., 1939. The primary and secondary scattering of sunlight in a plane-stratified atmosphere of uniform composition. *Philos. Mag. Ser. 7*, 28: 99–110.

Hamon, R.W., Weiss, L.W. and Wilson, W.T., 1954. Insolation as an empirical function of daily sunshine duration. *Mon. Weather Rev.*, 82: 141–146.

Hänel, G., 1971. New results concerning the dependence of visibility on relative humidity and their significance in a model for visibility forecast. *Contrib. Atmos. Phys.*, 44: 137–167.

Hänel, G. and Bullrich, K., 1970. Berechnungen der spektralen Strahlungsextinktion an atmosphärischen Aerosolpartikeln mit verschiedenen komplexen Berechnungsindices. *Contrib. Atmos. Phys.*, 43: 202–207.

Hanel, R.A., Conrath, B.J., Kunde, V.G., Prabhakara, C., Revah, I., Salomonson, V.V. and Wolford, G., 1972. The Nimbus 4 infrared spectroscopy experiment, 1. Calibrated thermal emission spectra. *J. Geophys. Res.*, 11: 2629–2641.

Hansen, J.E. and Travis, L.D., 1974. Light scattering in planetary atmospheres. *Space Sci. Rev.*, 16: 527–610.

Harries, J.E. and Burroughs, W.J., 1971. Measurements of sub-millimeter wavelength radiation emitted by the stratosphere. *Q. J. R. Meteorol. Soc.*, 97: 519–536.

Herman, B.M., 1962. Infrared absorption, scattering, and total attenuation cross-sections for water spheres. *Q. J. R. Meteorol. Soc.*, 88: 143–150.

Hewson, E.W., 1943. The reflection, absorption, and transmission of solar radiation by fog and cloud. *Q. J. R. Meteorol. Soc.*, 69: 47–62.

Heymsfield, A., 1975. Cirrus Unicinus generating cells and the evolution of cirriform clouds, 1. *J. Atmos. Sci.*, 32: 799–808.

Heymsfield, A.J. and Knollenberg, R.G., 1972. Properties of cirrus generating cells. *J. Atmos. Sci.*, 29: 1358–1366.

Houghton, H.J., 1954. On the annual heat balance of the northern hemisphere. *J. Meteorol.*, 11: 1–9.

Houghton, J.T. and Lee, A.C.L., 1972. Atmospheric transmission in the 10–12 μm window. *Nature (London), Phys. Sci.*, 238: 117–118.

Houghton, J.T. and Smith, S.D., 1966. *Infrared Physics*. Clarendon Press, Oxford, 319 pp.

Howard, J.N., Burch, D.E. and Williams, D., 1955. Near-infrared transmission through synthetic atmospheres. *Geophys. Res. Pap.*, No. 40, AFCRC-TR-55-213.

Howard, J.N., King, J.I.F. and Gast, P.R., 1961. Thermal radiation. In: *Handbook of Geophysics*. MacMillan, New York, N.Y., chapter 16.

Humphreys, W.J., 1909. Vertical temperature gradient of the atmosphere, especially in the region of the upper inversion. *Astrophys. J.*, 29: 14.

Hunt, B.G., 1972. Photochemical heating of the mesosphere and lower thermosphere. *Tellus*, 24: 47–54.

Hunt, G.E., 1973. Radiative properties of terrestrial clouds at visible and infrared thermal window wavelengths. *Q. J. R. Meteorol. Soc.*, 99: 346–369.

Hunt, G.E. and Grant, I.P., 1969. Discrete space theory of radiative transfer and its application to problems in planetary atmospheres. *J. Atmos. Sci.*, 26: 963–972.

Irvine, W.M. and Pollack, J.B., 1968. Infrared optical properties of water and ice spheres. *Icarus*, 8: 324–360.

Ivlev, L.S. and Popova, S.I., 1973. The complex refractive indices of substances in the atmospheric-aerosol dispersed phase. *Izv. Acad. Sci. U.S.S.R., Atmos. Oceanic Phys.*, 9: 1034–1043.

Jerlov, N.G., 1968. *Optical Oceanography*. Elsevier, Amsterdam, 194 pp.

Johnson, F.S., 1954. The solar constant. *J. Meteorol.*, 11: 431–439.

Junge, C.E., 1963. *Air Chemistry and Radioactivity*. Academic Press, New York, N.Y., 117 pp.

Junge, C.E. and Jaenicke, R., 1971. New results in background aerosols studies from the Atlantic expedition of R.V. "Meteor", Spring, 1969. *J. Aerosol Sci.*, 2: 305–314.

Junge, C.E., Robinson, E. and Ludwig, F.L., 1969. A study of aerosols in Pacific air masses. *J. Appl. Meteorol.*, 8: 340–347.

Kaplan, L.D., 1952. On the pressure dependence of radiative heat transfer in the atmosphere. *J. Meteorol.*, 9: 1–12.

Kasahara, A. and Washington, W.M., 1971. General circulation experiments with a six-layer NCAR model. *J. Atmos. Sci.*, 28: 657–701.

Kasten, F., 1966. A new table and approximate formula for the relative optical air mass. *Arch. Meteorol. Geophys. Bioklimatol. Ser. B*, 14(2): 206–223.

Kasten, F., 1969. Visibility forecast in the phase of precondensation. *Tellus*, 21: 631–635.

Kasten, F. and Raschke, E., 1974. Reflection and transmission terminology by analogy with scattering. *Appl. Opt.*, 13: 460–464.

Kattawar, G.W. and Plass, G.N., 1967. Electromagnetic scattering from absorbing spheres. *Appl. Opt.*, 6: 1377–1382.

Kelton, G. et al., 1963. Infrared target and background radiometric measurements — concepts, units, and techniques. *Infrared Phys.*, 3: 139–169.

Kendall, J.M. and Berdahl, C.M., 1970. Two blackbody radiometers of high accuracy. *Appl. Opt.*, 9: 1082–1091.

Kimball, H.H., 1927. Measurements of solar radiation intensity and determinations of its depletion by the atmosphere with bibliography of pyrheliometric measurements. *Mon. Weather Rev.*, 55: 155–169.

Kimball, H.H., 1930. Measurements of solar radiation intensity and determination of its depletion by the atmosphere. *Mon. Weather Rev.*, 58: 43–52.

Knollenberg, R.G., 1972. On radiational cooling computations in clouds. *J. Atmos. Sci.*, 29: 212–214.

Kondratyev, K.Ya., 1969. *Radiation in the Atmosphere*. Academic Press, New York N.Y., and London, 912 pp.

Kondratyev, K.Ya., 1972. *Radiation Processes in the Atmosphere*. World Meteorological Organization, Geneva, WMO No. 309, 214 pp.

Kondratyev, K.Ya., (Editor), 1973. *Radiation Characteristics of the Atmosphere and the Earth's Surface*. Amerind, New Delhi, 580 pp. (translated from Russian).

Kondratyev, K.Ya. and Nikolsky, G.A., 1970. Solar radiation and solar activity. *Q. J. R. Meteorol. Soc.*, 96: 509–522.

Kondratyev, K.Ya., Nikolsky, G.A., Murcray, D.G., Kosters, J.J. and Gast, P.R., 1971. The solar constant from data of balloon investigations in the USSR and the USA. *Space Res.*, 11: 695–703.

Korb, G. and Möller, F., 1962. Theoretical investigations on energy gain by absorption of solar radiation in clouds. *Final Rep. Contract No. DA-91-591-EUC-1612*. Meteorologisches Institüt, München, 185 pp.

Korb, G. and Zdunkowski, W., 1970. Distribution of radiative energy in ground fog. *Tellus*, 22: 298–320.

Kuhn, P.M., 1970. Airborne observations of contrail effects on the thermal radiation budget. *J. Atmos. Sci.*, 27: 337–342.

Kuhn, P.M., 1972. Radiation transfer calculations and direct observations in the presence of clouds and aerosols (abstract) *Bull. Am. Meteorol. Soc.*, 53: 500–501.

Kuhn, P.M. and Weickmann, H.K., 1969. High altitude radiometric measurements of cirrus. *J. Atmos. Sci.*, 8: 147–154.

Kuhn, P.M., Weickmann, H.K., Lojko, J. and Stearns, L.P., 1974. Transfer of infrared radiation through clouds. *Appl. Opt.*, 13: 512–517.

Kuhn, W.R. and London, J., 1969. Infrared radiative cooling in the middle atmosphere (30–110 km). *J. Atmos. Sci.*, 26: 189–204.

Kunde, V.G. and Maguire, W.C., 1974. Direct integration transmittance model. *J. Quant. Spectrosc. Radiat. Transfer*, 14: 803–814.

Kuriyan, J.G. and Sekera, Z., 1974. Scattering in liquid haze — analytic approximations. *Q. J. R. Meteorol. Soc.*, 100: 67–75.

Kyle, T.G., 1969. Calculations of atmospheric transmittance from 1.7 to 20 μ. *J. Quant. Spectrosc. Radiat. Transfer*, 9: 1477–1488.

Lacis, A.A. and Hansen, J.E., 1974. A parameterization for the absorption of solar radiation in the earth's atmosphere. *J. Atmos. Sci.*, 31: 118–133.

Latimer, J.R., 1972. On the Ångström and Smithsonian pyrheliometric scales. In: *Proc. Int. Radiation Symposium, Sendai, May–June, 1972*. WMO/JMA, Japanese Meteorological Agency, Tokyo, pp. 232–233.

Latimer, J.R., 1973. On the Ångström and Smithsonian absolute pyrheliometric scales and the International Pyrheliometric Scale, 1956. *Tellus*, 25: 586–592.

Laulainen, N.S. and Taylor, B.J., 1974. The precision and accuracy of Volz sunphotometry. *J. Appl. Meteorol.*, 13: 298–302.

Lee, A.C.L., 1973. A study of the continuum absorption within the 8–13 μm atmospheric window. *Q. J. R. Meteorol. Soc.*, 99: 490–505.

Lettau, H., 1954. A study of the mass, momentum and energy budget of the atmosphere. *Arch. Meteorol. Geophys. Bioklimatol., Ser. A*, 7: 133–157.

Lettau, H., 1965. Atmospheric temperature, density, pressure and moisture. In: S.L. Valley (Editor), *Handbook of Geophysics and Space Environments*. AFCRL, U.S. Air Force, Bedford, Mass., chapter 3.

Lin, Chin-i, Baker, M. and Charlson, R.J., 1973. Absorption coefficient of atmospheric aerosol: a method of measurement. *Appl. Opt.*, 12: 1356–1363.

Lindberg, J.D. and Laude, L.S., 1974. Measurement of the absorption coefficient of atmospheric dust. *Appl. Opt.*, 13: 1923–1927.

Lindholm, F., 1958. On the Ångström absolute pyrheliometric scale. *Tellus*, 10: 249–255.

Liou, K., 1974. Analytic two-stream and four-stream solutions for radiative transfer. *J. Atmos. Sci.*, 31: 1473–1475.

List, R.J., 1968. *Smithsonian Meteorological Tables*. Smithsonian Press, Washington, D.C., 6th ed., 527 pp.

London, J., 1952. The distribution of radiational temperature change in the northern hemisphere during March. *J. Meteorol.*, 9: 145–151.

London, J. and Sasamori, T., 1968. A brief note concerning radiative cooling as a function of temperature and water vapour. In: *Proc. WMO-IUGG Symp. on Radiation Including Satellite Techniques, Bergen, 1968*. World Meteorological Organization, Geneva, WMO Techn. Note No. 104, pp. 488–489.

London, J. and Sasamori, T., 1971. Radiative energy budget of the atmosphere. *Space Res.*, 11: 639–649.

Longhurst, R.S., 1960. *Geometrical and Physical Optics*. Jarrold and Sons, Norwich, 534 pp.

MacKinnon, P.J., 1969. The effect of hygroscopic particles on the backscattered power from a laser beam. *J. Atmos. Sci.*, 26: 500–510.

McClatchey, R.A., Fenn, R.W., Selby, J.E.A., Volz, F.E. and Garing, J.S., 1970. Optical properties of the atmosphere. *AFCRL-70-0527*, Air Force Cambridge Research Laboratories, Bedford, Mass., 85 pp.

McClatchey, R.A., Fenn, R.W., Selby, J.E.A., Volz, F.E. and Garing, J.S., 1972. Optical properties of the atmosphere. *AFCRL-72-0497*, Air Force Cambridge Research Laboratories, Bedford, Mass., 3rd ed., 108 pp.

McCoy, J.H., Rensch, D.B. and Long, R.K., 1969. Water vapour continuum absorption of carbon dioxide laser radiation near 10 μ. *Appl. Opt.*, 8: 1471–1478.

McDonald, J.E., 1960. Direct absorption of solar radiation by atmospheric water vapor. *J. Meteorol.*, 17: 319–328.

McKee, T.B. and Cox, S.K., 1974. Scattering of visible radiation by finite clouds. *J. Atmos. Sci.*, 31: 1885–1892.

Malkus, J.S., 1962. Large-scale interactions. In: M.N. Hill (Editor), *The Sea, 1. Physical Oceanography*. Interscience, New York, N.Y., chapter 4.

Manabe, S., 1970. Cloudiness and the radiative convective equilibrium. In: S.F. Singer (Editor), *Global Effects of Environmental Pollution*. D. Reidel, Dordrecht, pp. 156–157.

Manabe, S. and Hunt, B.G., 1968. Experiments with a stratospheric general circulation model, 1. *Mon. Weather Rev.*, 96: 477–502.

Manabe, S. and Möller, F., 1961. On the radiative equilibrium and heat balance of the atmosphere. *Mon Weather Rev.*, 89: 503–532.

Manabe, S. and Strickler, R.F., 1964. Thermal equilibrium of an atmosphere with a convective adjustment. *J. Atmos. Sci.*, 21: 361–385.

Manabe, S. and Wetherald, R., 1967. Thermal equilibrium of the atmosphere with a given distribution of relative humidity. *J. Atmos. Sci.*, 24: 241–259.

Marchgraber, R.M. and Frölich, C. 1972. The third international pyrheliometer comparisons. In: *Proc. Int. Radiation Symposium, Sendai, May–June, 1972*. WMO/JMA, Japanese Meteorological Agency, Tokyo, pp. 292–295.

Margenau, H. and Murphy, G.M., 1956. *The Mathematics of Physics and Chemistry*. D. van Nostrand, Princeton, N.J., 2nd ed., 604 pp.

Marggraf, W.A. and Griggs, M., 1969. Aircraft measurements and calculations of the total downward flux of solar radiation as a function of altitude. *J. Atmos. Sci.*, 26: 469–477.

Margulis, L. and Lovelock, J.E., 1974. Biological modulation of the earth's atmosphere. *Icarus*, 21: 471–489.

Mason, B.J., 1957. *The Physics of Clouds*. Clarendon Press, Oxford, 481 pp.

Mészáros, A. and Vissy, K., 1974. Concentration, size distribution and chemical nature of atmospheric aerosol particles in remote oceanic areas. *J. Aerosol Sci.*, 5: 101–109.

Mitchell, J.M., 1971. The effect of atmospheric aerosols on climate with special reference to temperature near the earth's surface. *J. Appl. Meteorol.*, 10: 703–714.

Möller, F., 1954. Radiation processes near the surface of the earth. *Sci. Rep.*, No. 2, *Contract AF19(122)-165*. College of Engineering, New York University, New York, N.Y., 19 pp.

Möller, F., 1957. Strahlung in der unteren Atmosphäre. In: *Handbuch der Physik, 48. Geophysik 2*. Springer-Verlag, Berlin, pp. 155–253.

Mügge, R. and Möller, F., 1932. Zur Berechnung von Strahlungsströmen und Temperaturänderungen in Atmosphären von beliebigem Aufbau. *Z. Geophys.*, 8: 53–64.

Namias, J., 1952. The annual course of month-to-month persistence in climatic anomalies. *Bull. Am. Meteorol. Soc.*, 33: 279–285.

Namias, J., 1954. Further aspects of month-to-month climatic anomalies. *Bull. Am. Meteorol. Soc.*, 35: 112–117.

NASA, 1971. Solar electromagnetic radiation. *NASA Publ. SP-8005*. National Aeronautics and Space Administration, Greenbelt, Md.

Neiburger, M., 1949. Reflection, absorption, and transmission of insolation by stratus cloud. *J. Meteorol.*, 6: 98–104.

Neiburger, M. and Chien, C.W., 1960. Computations of the growth of cloud drops by condensation using an electronic digital computer. In: H. Weickmann (Editor), *Physics of Precipitation. Geophysical Monograph, 5*. American Geophysical Union, Washington, D.C., pp. 191–210.

Newkirk, Jr., G.A., 1956. Photometry of the solar aureole. *J. Opt. Soc. Am.*, 46: 1028–1037.

Palmer, C.H., 1960. Experimental transmission functions for the pure rotation band of water vapour. *J. Opt. Soc. Am.*, 50: 1232–1242.

Paltridge, G.W., 1969. A note on the calculation of aerosol absorption from solar radiation measurements. *Q. J. R. Meteorol. Soc.*, 95: 650–652.

Paltridge, G.W., 1970. Daytime long-wave radiation from the sky. *Q. J. R. Meteorol. Soc.*, 96: 645–653.

Paltridge, G.W., 1973. Direct measurement of water vapour absorption of solar radiation in the free atmosphere. *J. Atmos. Sci.*, 30: 156–160.

Paltridge, G.W., 1974a. Global cloud cover and earth surface temperature. *J. Atmos. Sci.*, 31: 1571–1576.

Paltridge, G.W., 1974b. Solar radiation statistics for Australia. *CSIRO Div. Atmos. Phys. Tech. Paper* No. 23, 22 pp.

Paltridge, G.W., 1974c. Infrared emissivity, shortwave albedo, and the microphysics of stratiform water clouds. *J. Geophys. Res.*, 79: 4053–4058.

Paltridge, G.W., 1974d. Atmospheric radiation and the gross character of stratiform cloud. *J. Atmos. Sci.*, 31: 244–250.

Paltridge, G.W., 1975. Global dynamics and climate — a system of minimum entropy exchange. *Q. J. R. Meteorol. Soc.*, 101: 475–484.

Paltridge, G.W. and Platt, C.M.R., 1973. Absorption and scatter of radiation by an aerosol layer in the free atmosphere. *J. Atmos. Sci.*, 30: 734–737.

Paltridge, G.W. and Sargeant, S.L., 1971. Solar and thermal radiation measurements to 32 km at low solar elevations. *J. Atmos. Sci.*, 28: 242–253.

Pekeris, C.L., 1932. The development and present status of the theory of the heat balance of the atmosphere. *MIT Prof. Notes*, No. 5, Massachusetts Institute of Technology, Cambridge, Mass., 85 pp.

Penndorf, R.B., 1956. New tables of Mie scattering functions for spherical particles. *Geophys. Res. Pap.*, No. 45, *AFCRC-TR-56-204(6)* Air Force Cambridge Research Centre, U.S. Air Force, Bedford, Mass., 98 pp.

Penndorf, R., 1957. Tables of refractive index for standard air and the Rayleigh scattering coefficient for the spectral region between 0.2 and 20 μ and their application to atmospheric optics. *J. Opt. Soc. Am.*, 47: 176–182.

Petrova, L.V. and Feygel'son, Ye.M., 1969. Calculation of longwave influx in the case of rough vertical resolution. *Izv. Acad. Sci. U.S.S.R., Atmos. Oceanic Phys.*, 5: 585–595.

Plass, G.N. and Kattawar, G.W., 1969. Radiative transfer in an atmosphere-ocean system. *Appl. Opt.*, 8: 455–466.

Platt, C.M.R., 1972. Airborne infrared radiance measurements (10–12 micron wavelength) off tropical East Coast Australia. *J. Geophys. Res.*, 77: 1597–1609.

Platt, C.M.R., 1973. Lidar and radiometric observations of cirrus clouds. *J. Atmos. Sci.*, 30: 1191–1204.

Platt, C.M.R., 1975. Infrared emissivity of cirrus — simultaneous satellite, lidar and radiometric observations. *Q. J. R. Meteorol. Soc.*, 101: 119–126.

Platt, C.M.R. and Bartusek, K., 1974. Structure and optical properties of some middle-level clouds. *J. Atmos. Sci.*, 31: 1079–1088.

Platt, C.M.R. and Gambling, D.J., 1971. Laser radar reflections and downward infrared flux enhancement near small cumulus clouds. *Nature (London)*, 232: 182–185.

Pontier, L. and Dechambenoy, C., 1966. Détermination des constantes optiques de l'eau liquide entre 1 et 40 μ. Application au calcul de son pouvoir réflecteur et de son émissivité. *Ann. Geophys.*, 22(4): 633–641.

Porch, W.M., Ensor, D.S., Charlson, R.J. and Heintzenberg, T., 1973. Blue moon: is this a property of background aerosol? *Appl. Opt.*, 12: 34–36.

Posey, J.W. and Clapp, P.F., 1964. Global distribution of normal surface albedo. *Geofis. Int.*, 4(1): 33–48.

Quenzel, H., 1970. Determination of size distribution of atmospheric aerosol particles from spectral solar radiation measurements. *J. Geophys. Res.*, 75: 2915–2921.

Radiation Commission, 1975. *Standard Procedures to Compute Atmospheric Radiative Transfer in a Scattering Atmosphere, 1* (J. Lenoble, Editor). International Association of Meteorology and Atmospheric Physics (IUGG), WMO, Geneva, 105 pp.

Raschke, E., Vonder Haar, T.H., Baudeen, W.R. and Pasternak, M., 1973. The annual radiation balance of the earth-atmosphere system during 1969–1970 from Nimbus 3 measurements. *J. Atmos. Sci.*, 30: 341–364.

Reck, R.A., 1974. Influence of surface albedo on the change in the atmospheric radiation balance due to aerosols. *Atmos. Environ.*, 8: 823–833.

Reck, R.A., 1975. Influence of aerosol cloud height on the change in the atmospheric radiation balance due to aerosols. *Atmos. Environ.*, 9: 89–99.

Remsberg, E.E., 1973. Stratospheric aerosol properties and their effects on infrared radiation. *J. Geophys. Res.*, 78: 1401–1408.

Roach, W.T., 1961. The absorption of solar radiation by water vapour and carbon dioxide in a cloudless atmosphere. *Q. J. R. Meteorol. Soc.*, 87: 364–373.

Roach, W.T. and Goody, R.M., 1958. Absorption and emission in the atmospheric window from 770 to 1250 cm^{-1}. *Q. J. R. Meteorol. Soc.*, 84: 319–333.

Robinson, G.D., 1958. Some observations from aircraft of surface albedo and the albedo and absorption of cloud. *Arch. Meteorol. Geophys. Bioklimatol., Ser. B.*, 9: 28–41.

Robinson, G.D., 1966. Some determinations of atmospheric absorption by measurement of solar radiation from aircraft and at the surface. *Q. J. R. Meteorol. Soc.*, 92: 263–269.

Robinson, N., 1966. *Solar Radiation*. Elsevier, Amsterdam, 347 pp.

Rodgers, C.D., 1967a. The use of emissivity in atmospheric radiation calculations. *Q. J. R. Meteorol. Soc.*, 93: 43–54.

Rodgers, C.D., 1967b. The radiative heat budget of the troposphere and lower stratosphere. *Planet. Circ. Proj. Dep. Meteorol., Rep.*, No. A2, Massachusetts Institute of Technology, Cambridge, Mass., 99 pp.

Rodgers, C.D. and Walshaw, C.D., 1966. The computation of infrared cooling rate in planetary atmospheres. *Q.J. R. Meteorol. Soc.*, 92: 67–92.

Sagan, C. and Pollack, J.B., 1967. Anisotropic nonconservative scattering and the clouds of Venus. *J. Geophys. Res.*, 72: 469–477.

Saltzmann, B., 1968. Steady state solutions for axially symmetric climatic variables. *Pure Appl. Geophys.*, 69: 237–259.

Sasamori, T., 1959. The temperature effect on the absorption of the 15 microns carbon dioxide band. *Sci. Rep. Tohoku Univ., Ser. 5*, 11: 149–161.

Sasamori, T., 1968. The radiative cooling calculation for application to general circulation experiments. *J. Appl. Meteorol.*, 7: 721–729.

Sasamori, T., London, J. and Hoyt, D.V., 1972. Radiation budget of the southern hemisphere. In: C.W. Newton (Editor), *Meteorology of the Southern Hemisphere*. American Meteorological Society, Boston, Mass., chapter 2.

Schneider, S.H., 1972. Cloudiness as a global climatic feedback mechanism: the effects on the radiation balance and surface temperature of variations in cloudiness. *J. Atmos. Sci.*, 29: 1413–1422.

Sellars, W.D., 1965. *Physical Climatology*. University of Chicago Press, Chicago, Ill., and London, 272 pp.

Sellars, W.D., 1969. A global climatic model based on the energy balance of the earth-atmosphere system. *J. Appl. Meteorol.*, 8: 392–400.

Shaw, G.E., Reagan, J.A. and Herman, B.M., 1973. Investigations of atmospheric extinction using direct solar radiation measurements made with a multiple wavelength radiometer. *J. Appl. Meteorol.*, 12: 374–380.

Sherhag, R., 1948. *Neue Methoden der Wetteranalyse und Wetterprognose*. Springer-Verlag, Berlin, Göttingen, Heidelberg, 424 pp. (p. 64 in particular).

Shettle, E.P. and Weinman, J.A., 1970. The transfer of solar irradiance through inhomogeneous turbid atmospheres evaluated by Eddington's approximation. *J. Atmos. Sci.*, 27: 1048–1055.

Shuster, A., 1905. Radiation through a foggy atmosphere. *Astrophys. J.*, 21: 1–22.

Simpson, G.C., 1928. Further studies in terrestrial radiation. *Mem. R. Meteorol. Soc.*, 3(21): 1–26.

Simpson, G.C., 1940. Possible causes of change in climate and their limitations. *Proc. Linn. Soc. London*, 152: 190–219.

Sissenwine, N., 1969. Standard and supplemental atmospheres. In: D.F. Rex (Editor), *World Survey of Climatology, 4. Climate of the Free Atmosphere*. Elsevier, Amsterdam, pp. 5–44.

SMIC, 1971. *Inadvertent Climate Modification: Report of the Study of Man's Impact on Climate*. Massachusetts Institute of Technology Press, Cambridge, Mass., 308 pp.

Smith, E.V.P. and Gottlieb, D.M., 1974. Solar flux and its variations. *Space Sci. Rev.*, 16: 771–802.

Spencer, J.W., 1971. Fourier series representation of the position of the sun. *Search*, 2: 172.

Spiro, I.J., Jones, R.C. and Wark, D.Q., 1965. Atmospheric transmission: concepts, symbols, units and nomenclature. *Infrared Phys.*, 5: 11–36.

Spitzer, W.G. and Kleinman, D.A., 1961. Infrared lattice bands of quartz. *Phys. Rev.*, 121: 1324–1335.

Staley, D.O., 1965. Radiative cooling in the vicinity of inversions and the tropopause. *Q. J. R. Meteorol. Soc.*, 91: 282–301.

Staley, D.O. and Jurica, G.M., 1968. Radiative modification of selected vertical temperature distributions. *Q. J. R. Meteorol. Soc.*, 94: 310–317.

Staley, D.O. and Jurica, G.M., 1970. Flux exissivity tables for water vapour, carbon dioxide and ozone. *J. Appl. Meteorol.*, 9: 365–372.

Stensrud, G., Heir, O. and Pleym, H., 1971. An application of Rodgers' method to compute the infrared cooling rate in the atmosphere over Southern Norway. *Meteorol. Ann. Oslo*, 6: 1–29.

Stergis, C.G., Hering, W.S., Huffman, R.E., Hunt, W.W. and Paulson, J.F., 1965. Atmospheric composition. In: S.L. Valley (Editor), *Handbook of Geophysics and Space Environments*. AFCRL, U.S. Air Force, Bedford, Mass., chapter 6.

Stone, H.M. and Manabe, S., 1968. Comparison among various numerical models designed for computing infrared cooling. *Mon. Weather Rev.*, 96: 735–741.

Stringfellow, M.F., 1974. Lightning incidence in Britain and the solar cycle. *Nature (London)*, 249: 332–333.

Stull, V.R., Wyatt, P.J. and Plass, G.N., 1964. The infrared transmittance of carbon dioxide. *Appl. Opt.*, 3: 243–254.

Swinbank, W.C., 1963. Long-wave radiation from clear skies. *Q. J. R. Meteorol. Soc.*, 89: 339–348.

Taljaard, J.J., Van Loon, H., Crutcher, H.L. and Jenne, R.L., 1969. *Climate of the Upper Air, 1. Southern Hemisphere*. U.S. Department of Commerce, ESSA, Environmental Data Service, Washington, D.C.

Telegadas, K. and London, J., 1954. A physical model of the northern hemisphere troposphere for winter and summer. *Sci. Rep. No. 1, Contract AF19(122)-165*. College of Engineering, New York University, New York, N.Y., 55 pp.

Thekaekara, M.P. and Drummond, A.J., 1971. Standard values for the solar constant and its spectral components. *Nature (London), Phys. Sci.*, 229: 6–9.

Turner, J.S., 1969. Buoyant plumes and thermals. In: A.A. Sears and L. van Dyke

(Editors), *Annual Review of Fluid Mechanics, 1*. Annual Reviews Inc., Palo Alto, Calif., pp. 29–44.
Twitty, J.T. and Weinman, J.A., 1971. Radiative properties of carbonaceous aerosols. *J. Appl. Meteorol.*, 10: 725–731.
Twomey, S., 1970. On the possible absorption of visible light by clouds. *J. Atmos. Sci.*, 27: 514–515.
Twomey, S., Jacobowitz, H. and Howell, H.B., 1967. Light scattering by cloud layers. *J. Atmos. Sci.*, 24: 70–79.

Van de Hulst, H.C., 1957. *Light scattering by Small Particles*. John Wiley and Sons, New York, N.Y., Chapman and Hall, London, 470 pp.
Van Woerkom, A.J.J., 1953. The astronomical theory of climate change. In: H. Shapley (Editor), *Climate Change*. Harvard University Press, Cambridge, Mass., chapter 11.
Volz, F.E., 1970. Spectral skylight and solar radiance measurements in the Carribean: maritime aerosols and Sahara dust. *J. Atmos. Sci.*, 27: 1041–1047.
Volz, F.E., 1972. Infrared refractive index of atmospheric aerosol substances. *Appl. Opt.*, 11: 755–759.
Vonder Haar, T.H. and Oort, A.H., 1973. New estimate of annual poleward energy transport by northern hemisphere oceans. *J. Phys. Oceanogr.*, 3(2): 169–172.
Vonder Haar, T.H. and Suomi, V.E., 1971. Measurements of the earth's radiation budget from satellites during a five-year period, 1. Extended time and space means. *J. Atmos. Sci.*, 28: 305–314.

Walshaw, C.D., 1957. Integrated absorption by the 9.6μ band of ozone. *Q. J. R. Meteorol. Soc.*, 83: 315–321.
Walshaw, C.D. and Rodgers, C.D., 1963. The effect of the Curtis–Godson approximation on the accuracy of radiative heating-rate calculations. *Q. J. R. Meteorol. Soc.*, 89: 122–130.
Weickmann, H., and H.'aufm Kampe, 1953. Physical properties of cumulus clouds. *J. Meteorol.*, 10: 204–245.
Wilcox, J.M., Scherrer, P.H., Svalgaard, L., Roberts, W.O., Olson, R.H. and Jenne, R.L., 1974. Influence of solar magnetic sector structure on terrestrial atmospheric vorticity. *J. Atmos. Sci.*, 31: 581–587.
Willson, R.C., 1972. Experimental comparisons of the international pyrheliometric scale with the absolute radiation scale. *Nature (London)*, 239(5369): 208–209.
WMO, 1971. *Guide to Meteorological Instruments and Observing Practices*. World Meteorological Organization, WMO No. 8 TP3, Geneva (in particular chapter 9).
Wright, H.L., 1939. Atmospheric opacity: a study of visibility observations in the British Isles. *Q. J. R. Meteorol. Soc.*, 65: 411–442.
Wyatt, P.J., Stull, V.R. and Plass, G.N., 1962. Quasi-random model of band absorption. *J. Opt. Soc. Am.*, 52: 1209–1217.
Wyatt, P.J., Stull, V.R. and Plass, G.N., 1964. The infrared transmittance of water vapour. *Appl. Opt.*, 3: 229–241.

Yamamoto, G., 1952. On a radiation chart. *Sci. Rep. Tohoku Univ., Ser. 5*, 4: 9–23.
Yamamoto, G., 1962. Direct absorption of solar radiation by atmospheric water vapour, carbon dioxide and molecular oxygen. *J. Atmos. Sci.*, 19: 182–188.
Yamamoto, G. and Onishi, G., 1952. Absorption of solar radiation by water vapour in the atmosphere. *J. Meteorol.*, 9: 415–421.
Yamamoto, G. and Onishi, G., 1953. A chart for the calculation of radiative temperature changes. *Sci. Rep. Tohoku Univ., Ser. 5, Geophys.*, 4: 108.
Yamamoto, G. and Tanaka, M., 1972. Increase of global albedo due to air pollution. *J. Atmos. Sci.*, 29: 1405–1412.

Yamamoto, G., Tanaka, M. and Asano, S., 1970. Radiative transfer in water clouds in the infrared region. *J. Atmos. Sci.*, 27: 282–292.

Zdunkowski, W.G., Barth, R.E. and Lombardo, F.A., 1966. Discussion on the atmospheric radiation tables by Elsasser and Culbertson. *Pure Appl. Geophys.*, 63: 211–219.

Zdunkowski, W.G. and Crandall, W.K., 1971. Radiative transfer of infrared radiation in model clouds. *Tellus*, 23: 517–527.

Zdunkowski, W.G. and Davis, C.T., 1974. Radiative transfer in vertically inhomogeneous layer clouds. *Contrib. Atmos. Phys.*, 47: 187–212.

Zdunkowski, W.G. and Nielson, B.C., 1969. A preliminary prediction analysis of radiation fog. *Pure Appl. Geophys.*, 75: 278–299.

Zeuner, F.E., 1959. *The Pleistocene Period.* Hutchinson, London, 447 pp. (see chapter 5).

Index